The Extraction and Refining of Metals

Colin Bodsworth

CRC PRESS SERIES IN

Materials Science and Technology

Brian Ralph, Series Editor

Control of Microstructures and Properties in Steel Arc Welds
Lars-Erik Svensson

The Extraction and Refining of Metals
Colin Bodsworth

Grain Boundary Properties and the Evolution of Microstructures and Texture
G. Gottstein and L.S. Shvindlerman

Quantitative Description of Microstructures of Materials
K.J. Kurdzydlowski and Brian Ralph

The Extraction and Refining of Metals

Colin Bodsworth
Department of Materials Technology
Brunel University
Uxbridge, Middlesex
United Kingdom

CRC Press
Boca Raton Ann Arbor London Tokyo

Library of Congress Cataloging-in-Publication Data

Bodsworth, Colin.
The extraction and refining of metals / Colin Bodsworth.
p. cm.
Includes bibliographical references and index.
ISBN 0-8493-4433-6
1. Metals—Refining. 2. Extraction (Chemistry). I. Title.
TN665.B67 1994
669—dc20 94-8937
CIP

Direct all inquiries to CRC Press, Inc., 2000 Corporate Blvd., N.W., Boca Raton, Florida 33431.

International Standard Book Number 0-8493-4433-6
Library of Congress Card Number 94-8937
Printed in the United States of America 1 2 3 4 5 6 7 8 9 0
Printed on acid-free paper

TABLE OF CONTENTS

Preface ix
List of Symbols xi

Chapter 1
The Economics of Metal Production 1
1.1. The Cost of Metals 1
1.2. Mineral Deposits 3
1.2.1. Average Concentrations of Metals 3
1.2.2. Igneous Intrusions 3
1.2.3. Sedimentary Deposits 4
1.2.4. Workable Deposits 7
1.3. Mineral Dressing 10
1.3.1. Comminution 11
1.3.2. Separation 12
1.4. Agglomeration 21
1.4.1. Pelletizing 21
1.4.2. Sintering 22
1.4.3. Sintering versus Pelletizing 25
1.5. Reduction and Refining 25
1.5.1. Pyrometallurgy 26
1.5.2. Hydrometallurgy 30
1.5.3. Electrometallurgy 32
1.6. Choice of Process 33
1.7. Typical Production Routes 35
1.7.1. Aluminum 35
1.7.2. Iron 38
1.7.3. Copper 39
1.7.4. Lead 40
1.7.5. Tin 41
1.7.6. Zinc 42
1.7.7. Other Metals 42
1.8. Factors Affecting the Price of Metals 43
1.8.1. Cost 43
1.8.2. Price 46
Further Reading 49

Chapter 2
Theoretical Principles—1, Thermodynamics 51
2.1. Thermodynamic Relations 51
2.1.1. Thermodynamic Systems 52
2.1.2. Free Energy, Enthalpy and Entropy 52

2.2. Solutions 56
2.2.1. Partial Molar Free Energy 56
2.2.2. Fugacity 58
2.2.3. Activity 59
2.2.4. The Equilibrium Constant 60
2.2.5. Solution Models 62
2.2.6. Dilute Solutions 65
2.2.7. Alternative Standard States 67
2.2.8. Multicomponent Solutions 70
2.3. Graphical Representation of Thermodynamic Data 72
2.3.1. Phase Diagrams 72
2.3.2. Free Energy-Temperature Diagrams 78
2.3.3. Predominance Area Diagrams 84
2.3.4. Pourbaix Diagrams 88
2.3.5. Other Diagrams and Applications 93

Chapter 3
Theoretical Principles—2, Kinetics 95
3.1. Chemical Reaction Kinetics 96
3.1.1. Kinetic Theory 96
3.1.2. Reaction Rate Theories 98
3.2. Transport Kinetics 100
3.2.1. Diffusion 100
3.2.2. Heat Transfer 104
3.2.3. Mass Transport 105
3.3. Reactions Across Interfaces 106
3.3.1. Gas-Liquid Reactions 106
3.3.2. Gas-Solid Reactions 108
3.3.3. Liquid-Liquid Reactions 112
3.4. Nucleation 116
3.4.1. Nucleation of Precipitates 116
3.4.2. Nucleation of Gas Bubbles 120
Further Reading 122

Chapter 4
Pyrometallurgical Extraction 123
4.1. Reducing Agents 123
4.1.1. Carbon Reduction 126
4.1.2. Hydrogen Reduction 128
4.1.3. Metallothermic Reduction 129
4.2. Roasting 130
4.2.1. Dead Roasting 132
4.2.2. Environmental Constraints 134

4.3. Agglomeration 135
4.3.1. Sintering 137
4.3.2. Pelletizing 140
4.4. Carbothermic Reduction 141
4.4.1. The Blast Furnace Process 141
4.4.2. Carbothermic Reduction without Coke 163
4.5. Reduction of Sulfide Ores 171
4.5.1. Treatment of Copper Sulfides 172
4.5.2. Treatment of Nickel Sulfides 179
4.5.3. Treatment of Lead Sulfides 180
4.6. Reduction of the More Reactive Metals 181
4.6.1. Metallothermic Reduction of Magnesium 181
4.6.2. The Halide Extraction Route 183
Further Reading 188

Chapter 5
Pyrorefining 189
5.1. Objectives and Principles 189
5.1.1. Overview 189
5.1.2. Standard Specifications 190
5.1.3. Partition Ratios 191
5.1.4. Refining Kinetics 192
5.2. Refining Slags 193
5.2.1. Slag Structures 194
5.2.2. Thermodynamic Properties 197
5.3. Slag-Metal Reactions 201
5.3.1. Iron Refining 201
5.3.2. Steelmaking Practice 211
5.4. Metal-Metal Equilibria 218
5.4.1. Liquation and Drossing 218
5.4.2. Deoxidation 222
5.5. Metal-Gas Reactions 227
5.5.1. Volatilization 227
5.5.2. Degassing 232
5.5.3. Melt-Refractory Reactions 235
Further Reading 236

Chapter 6
Hydrometallurgy 237
6.1. Overview 237
6.2. Leaching 239
6.2.1. Leaching Practice 246
6.2.2. Leaching Kinetics 249

6.3. Recovery of the Metal from Solution 250
6.3.1. Cementation 250
6.3.2. Hydrogen Reduction 253
6.3.3. Solvent Extraction 255
6.3.4. Ion Exchange 259
Further Reading 260

Chapter 7
Electrolytic Extraction and Refining 261
7.1. Overview 261
7.2. Electrowinning 262
7.2.1. Aqueous Electrowinning 263
7.2.2. Electrowinning from Molten Halides 273
7.3. Electrorefining 278
7.4. Economic Aspects 281
Further Reading 281

Chapter 8
Environmental Issues 283
8.1. Introduction 283
8.2. Pollution Sources and Control 285
8.2.1. Solids 285
8.2.2. Gases 287
8.2.3. Liquids 293
8.3. Scrap Metal Recycling 294
8.3.1. Sources and Preparation of Scrap Metal 295
8.3.2. Recirculation Rates 298
8.3.3. Scrap Melting and Refining 300
8.4. Postscript 302
Further Reading 303

Chapter 9
Computer Control of Metal Production 305
9.1. Sensors 305
9.1.1. Traditional Measurement Techniques 306
9.1.2. Modern Techniques 307
9.2. Computer Control 310
9.2.1. Fundamental Models 311
9.2.2. Empirical Models 314
9.2.3. Dynamic Models 315
9.2.4. Examples of Control Systems 316

Appendix: Worked Examples 319
References 339
Index 343

PREFACE

The procedures used to convert minerals containing metallic elements into metals with the purity required for specific applications are referred to collectively as extractive metallurgy. A wide variety of production routes is available for this purpose. At first glance, it is not easy to see why certain routes are selected in preference to others for the preparation of a particular metal. A clue is contained in the first sentence. The purity of the metal that can be obtained is often dependent on the processes used for its production. Some processes can only produce a relatively impure metal, which is not suitable for the more demanding applications, while others can readily produce metals with close to 100% purity.

Each process can be used to produce metal with a wide range of purity. An understanding of how a process can be manipulated to obtain a metal of a required composition can be obtained from consideration of the theoretical principles that control the rate and the extent of the chemical reactions which occur. Thermodynamics defines the end point or equilibrium state for these reactions and can be used to explore how the end point can be changed by varying the imposed conditions such as temperature, pressure and the solid, liquid or gas composition. Kinetics control the rate at which equilibrium is approached and hence the time required for the reactions to go to completion.

The other important factor affecting the selection of the process route is the cost of production. Metals are traded on an open market and there is no commercial future in selecting a process route that produces metal of a given composition at a much higher cost than can be obtained by some other route.

Since cost is a major factor in process selection, the first chapter sets the scene by examining why some metals command a higher selling price than others. The major factors influencing the production costs are indicated, and an outline is provided of the most common routes normally selected for the production of the common engineering metals. Costs are referred to frequently in subsequent chapters. No attempt is made to quote actual costs, since these can vary over a wide range between different companies operating basically the same process. Companies are not usually prepared to disclose details of cost, since this information would be useful to their competitors; and, in any case, the information would rapidly become out of date. The selling prices are given here merely as an indication of the relative differences between the values of the metals, since these also change frequently.

The theoretical principles are addressed in Chapters 2 and 3, and the following four chapters show how they are applied to pyrometallurgical extraction and refining and to hydro- and electrometallurgy. A variety of techniques is considered under each of these headings to illustrate how the operating variables can be manipulated to obtain a required metal composition. This serves as a basis for an explanation of why specific processes can be applied for the production of some metals but, for fundamental or economic reasons, are not used for the production of other metals. The consideration of pyrometallurgical techniques may appear to be biased toward the production of iron and steel. This is because the output of ferrous metal greatly exceeds the cumulative output of all other metals and a vast amount of research effort has been devoted to the study of the extraction and refining of iron and steel. The principles explored can be applied to the production of a wide range of metals.

Detailed descriptions of the production of individual metals are not given. Numerous texts have been written for this purpose and a selection of books for further reading is given at the end of most chapters.

Until comparatively recent times, extractive metallurgy was a major source of environmental pollution. Mankind has been alerted to the damage being caused to the environment, and strenuous efforts are now being made to prevent ecological degradation by the industry. The remelting of scrap metal generated when structures and artifacts reach the end of their useful life is one of the ways in which the earth's natural resources of minerals and fossil fuels can be conserved. In recent years, the advent of high-powered computers at relatively low cost has led to more sophisticated control of the processes. This is resulting in closer control of the composition of the product, with lower operating costs and reduced pollution. These two issues are considered briefly in the last two chapters.

A basic knowledge of the principles of thermodynamics and kinetics is assumed. The objective here is to show how these concepts can be applied to gain a fuller understanding and control of the processes. An appendix contains worked examples to illustrate how thermodynamics can be used in practice.

ACKNOWLEDGMENTS

The author is grateful to the many colleagues who have provided information and comments during the preparation of this text. In particular, the author records his appreciation of the many helpful comments and suggestions provided by Professor Brian Ralph, Dr. David Talbot and Dr. Animesh Jha. I am indebted to Mrs. A. Ralph for typing the equations and to Mr. J. Furley for the preparation of the line drawings. Thanks are due also to the publishers who have given permission for the reproduction of illustrations as indicated in the captions to the appropriate diagrams.

LIST OF SYMBOLS

The symbols used for the various quantities are primarily those recommended by the International Union of Pure and Applied Chemistry (IUPAC). Changes have been made only where the conventional use of other symbols could cause confusion. Pressures are quoted in terms of standard atmospheres for simplicity.

Symbol	Meaning	SI units
a	Activity	
C	Concentration	
C_p	Molar heat capacity	$J\ mol^{-1}\ deg^{-1}$
D	Diffusion coefficient	$m^2\ s^{-1}$
d	Diameter	m
E	Electrode potential (emf)	V
E^θ	Standard electrode potential	V
E_A	Activation energy	$J\ mol^{-1}$
F	Faraday	$C\ mol^{-1}$
f	Fugacity	
h	Activity coefficient (Henry)	
h_y^x	Interaction coefficient: the effect of element X on the activity coefficient of Y	
G	Gibbs free energy	$J\ mol^{-1}$
G^θ	Standard free energy	$J\ mol^{-1}$
$\overline{G}$	Partial molar free energy	$J\ mol^{-1}$
G^E	Excess molar free energy	$J\ mol^{-1}$
G^M	Integral molar free energy of mixing	$J\ mol^{-1}$
$\overline{G}_A^M$	Relative partial molar free energy of component A (similar suffixes and subscripts identify the corresponding enthalpy and entropy quantities)	$J\ mol^{-1}$
H	Enthalpy, heat content	$J\ mol^{-1}$
I	Current	A
I_0	Exchange current density	A
J	Joule	
J_x	Diffusive flux in the x direction	mol
K	Equilibrium constant	
k	Rate constant	
L_e	Latent heat (enthalpy) of evaporation	$J\ mol^{-1}$
L_f	Latent heat of fusion	$J\ mol^{-1}$
ln	Natural logarithm	
log	Common logarithm: $\log_{10}$	
M	Molecular weight	g
m	Molality	$mol\ kg^{-1}$
N	Mole fraction	
n	Number of moles	
P	Pressure	$N\ m^{-2}$
p	Partial pressure	$N\ m^{-2}$
Q	Activation energy for diffusion	$J\ mol^{-1}$
R	Gas constant	$J\ deg^{-1}$
r	Rate of reaction	$mol\ m^{-3}\ s^{-1}$
r^*	Critical nucleus size	m

Symbol	Meaning	SI units
S	Entropy	J mol^{-1} deg^{-1}
S_f	Entropy of fusion	J mol^{-1} deg^{-1}
T	Temperature	K
t	Tonne	
V	Volume	mol; m^3
γ	Activity coefficient (Raoult)	
δ	Boundary layer thickness	m
ε_Y^X	Interaction parameter; $\partial \ln h_X / \partial N_Y$	
η	Overpotential	V
θ	Contact angle	
κ	Henry's Law constant	
μ	Chemical potential	J mol^{-1}
τ	Excess surface concentration	mol m^{-2}
ω	Mass of vapor molecules	kg

CONVERSION FACTORS

	Traditional unit	SI equivalent
Atmosphere, standard	1 atm	101,325 N m^{-2}
Calorie		4.1868 J
Degree Celsius (°C)		T°C + 273.13 K
Gas constant	1.987 cal deg^{-1} mol^{-1}	8.3143 J mol^{-1} K^{-1}
Ton		1.01605 Mg
Ton, metric		1 Mg

Chapter 1

THE ECONOMICS OF METAL PRODUCTION

1.1. THE COST OF METALS

Many new materials for use in manufacture have been invented, developed and produced in the 20th century. Plastics, composites and ceramics are now well established for a wide range of applications. But metals, together with wood and concrete, still retain a dominant position in terms of the tonnages which are used annually. Although the new materials may provide a better combination of physical and mechanical properties for some applications, or simply improved aesthetic appeal, they have not made significant impact on the quantities of metal used. If wood and cement are excluded, the annual production of metals exceeds the quantity of all other construction and manufacturing materials. Steel continues to dominate the market; in tonnage terms the world production of just under 800 million tonnes of steel per annum is 93% of the total output of metals and is roughly five times the total production of composites, plastics, cermets and all the other metals. In contrast, only 11 million tonnes of plastics were made in 1989.

A number of factors account for the predominance of metals including properties (strength, toughness, corrosion resistance, electrical and thermal conductivity, etc.), formability and surface appearance, but a major factor is the cost to the customer. The metal extraction industries are well established throughout the world and have attained a high degree of efficiency, which is still improving, thus holding down production costs and maintaining or reducing the real costs when corrected for the effects of inflation. But metals will only retain their preeminent position for as long as the total cost (including maintenance and replacement cost) of artifacts produced from them is lower than that for artifacts produced from alternative materials. Over the past few decades there has been extensive replacement of metals by plastics in domestic appliances. There is increasing substitution of plastics, composites and ceramics in motorcar and aerospace vehicles. The metallurgical industries are making major efforts to retain their share of the market and their success is demonstrated by the continuing rise in the total consumption of metals.

This chapter examines the major metallurgical factors which account for the cost of the production of metals and indicates how these influence the selection of the extraction and refining routes. The major manufacturing routes for the principal engineering metals are briefly described. This provides a background for exploration in subsequent chapters of the scientific principles and constraints inherent in the various processes, consideration of which can lead to further improvements in efficiency of manufacture.

TABLE 1.1
The Price of High Grade Metals and their Average Concentrations in the Earth's Crust

Metal	Average price 1992	Average content wt%	Minimum workable grade %	Concentration factor	Ref.
Al	$1254	8.23	30	4	1
		8.0	25	3.1	2
Fe		5.63	25	4	1
		5.8	25	4.3	2
Mn		0.1	15	150	2
Cr		0.096	15	150	2
Zn	$1240	0.007	4	570	1
		0.0082	2.5	300	2
Ni	$7002	0.007	1	140	2
Cu	£1297	0.0055	0.4	73	1
		0.0058	0.5	86	2
Pb	£307	0.0013	4	3000	1
		0.001	2	2000	2
Sn	$6107	0.0002	0.5	2500	1
		0.00015	0.2	1300	2
W		0.0001	1.35	13500	2
Ag	$0.12×10^6	0.0000007	0.01	1250	1
Pt	$11.1×10^6	0.0000004	0.0001	250	1
Au	$10.80×10^6	0.0000002	0.0008	4000	1

The selling price of metals (i.e., cost of production plus profit) varies over a very wide range from gold and the precious metals at one extreme to iron at the other. The price quoted for any one metal can also vary over a wide range, depending on the purity or the value of specific properties (e.g., the electrical conductivity of copper) and the cast or wrought form in which it is marketed. Many metals are traded through Metal Exchanges, where the prices are fixed daily in response to supply and demand. The average selling prices throughout 1992 for some high-grade metals are listed in Table 1.1. The prices given for the base metals are the average settlement prices on the London Metal Exchange where, by tradition, the metals are quoted in dollars (U.S.) per tonne except for Cu and Pb which are always quoted in pounds (U.K.) per tonne. The equivalent annual price in dollars is about $2270 for Cu and $537 for Pb. Iron is not normally traded through an Exchange and is not traded in bulk as a high purity material equivalent to the high-grade nonferrous metals. The nearest equivalent is very low carbon, unalloyed steel containing about 99.98% iron, which is valued at about $350 per tonne in ingot form.

Inevitably, the price of a metal is high when it first becomes available and falls as the scale of production is increased. For example, in antiquity when silver was first available in usable quantities, it was valued almost as highly as gold. Iron held a similar position with the Pharaoh's of ancient Egypt. In more recent times, aluminum was valued at about £130/kg when it was exhibited at the Paris Exhibition in 1855. The real cost of metals, after correction for inflation, has fallen steadily during the 20th century as scientific and technological knowledge has contributed to closer control of the processes and as the scale of production has increased. But the price of one metal relative to another has fluctuated widely during that time. Thus, in 1951, high-grade tin sold for over five times the price of copper. By 1990 the price differential had decreased to a factor of a little over 2 but had increased again one year later to about 2.5 and to 2.7 in 1992. So, a scale of relative prices is appropriate only to the moment in time to which it refers. It is a useful concept, however, against which the factors affecting cost can be considered.

1.2. MINERAL DEPOSITS

1.2.1. Average Concentrations of Metals

The earth is the source of almost all the metals used by mankind. Large quantities of some metals have, over time, dissolved to form aqueous solutions in surface waters and are now distributed throughout the seas but, with the exception of magnesium, metals are not normally produced from seawater. The major source is the thin layer of continental crust of the earth. The average amounts of some common metals within this surface layer are listed in Table 1.1 in order of decreasing concentration. The reference numbers identify the source of the data listed in the last three columns of the table.

There is poor correlation between price and the average concentration of the metals in the earth. From its position in the table one would expect, for example, that Pb would be more expensive than Cu, Ni or Zn, but it is the lowest cost metal in this group. In fact, metals would be too expensive for general use if they had to be produced from earth containing only the average amounts of the elements. With the exception of Al and Fe, most would have production costs similar to, or even higher than, the present price of gold. Fortunately for mankind, natural phenomena have produced local areas of enrichment of metallic minerals in the form of ore bodies and these are the source of the metals. The enrichments have been produced by the selective transport of metals from the core of the earth by igneous intrusions and by the weathering of the surface rocks to produce selective accumulation of metals in sedimentary deposits.

1.2.2. Igneous Intrusions

The starting point in the formation of an ore body is an intrusion of molten magma from the core of the earth into the surface mantle of solid rock. If the intrusion reaches the outer surface as a volcanic eruption, the volatile species escape into the atmosphere and there is little time for the remaining minerals to segregate before the lava solidifies. When the intrusion does not reach the surface, however, the volatiles cannot escape and fractional solidification occurs from both the gas and the liquids as the temperature falls. Some metals solidify from the magma as oxides combined with other inorganic compounds (e.g., as silicates) or as sulfides that are immiscible with the oxides. Differences in density may cause preferential settling and concentration of the heavier minerals such as chromite ($FeO{\cdot}Cr_2O_3$), ilmenite ($FeO{\cdot}TiO_2$) and magnetite (Fe_3O_4) as in the formation of the Swedish magnetite deposits. Metals that form oxides, sulfides, etc. with lower melting points remain in the melt at this stage.

The solubility of a gas in a liquid decreases as the temperature falls. Consequently, the volume of gas associated with the intrusion increases as the magma cools. The heat released from the magma is transferred to the surrounding bedrock and volatile species are also released from the rock as it reacts with the hydrothermal gases. The sum effect is an increase in the pressure and in the fluidity of the liquids, forcing the remaining liquid magma and the associated gases away from the intrusion into the cracks and pores of the bedrock. The minerals that crystallize from the magma and the hydrothermal solutions at this stage appear as seams, lodes or veins infilling the cracks and as fine disseminations infilling the original pores in the rock (Figure 1.1a). In general, Au, Sn and W tend to occur in the aureole nearest the intrusion, with Cu in the next zone and Ag, Hg, Pb and Zn in the outer zone. Frequently, however, an intimate mixture or a solid solution of minerals is found in the same zone. The sulfides of Cu, Pb and Zn often occur together and, similarly, two or more of Fe, Co, Cu and Ni sulfides may be found in an intimate mixture. The latter type is found in the Ni deposits at Sudbury, Ontario.

Magmatic intrusions are formed only rarely today. The vast majority of igneous mineral deposits were formed hundreds of millions of years ago when the crust of the earth was less stable. Many deposits were formed when molten rock was formed by the collision of tectonic plates, between 200 and 400 million years ago, to form the present distribution of the continents.

1.2.3. Sedimentary Deposits

Ore bodies are also formed by weathering of bedrock. The action of frost, rain, wind and, particularly, the movement of glaciers during the ice ages resulted in the erosion of the surface rocks. Fragments of rock released by this action were ground down to finer sizes by fast-flowing rivers and by tidal action. The particles suspended in the water were deposited when the velocity of the transporting water decreased, as when a river emerged from hills onto a plain, or discharged into a lake or the sea. The zone of deposition varied with the size and density of the particles. For particles of

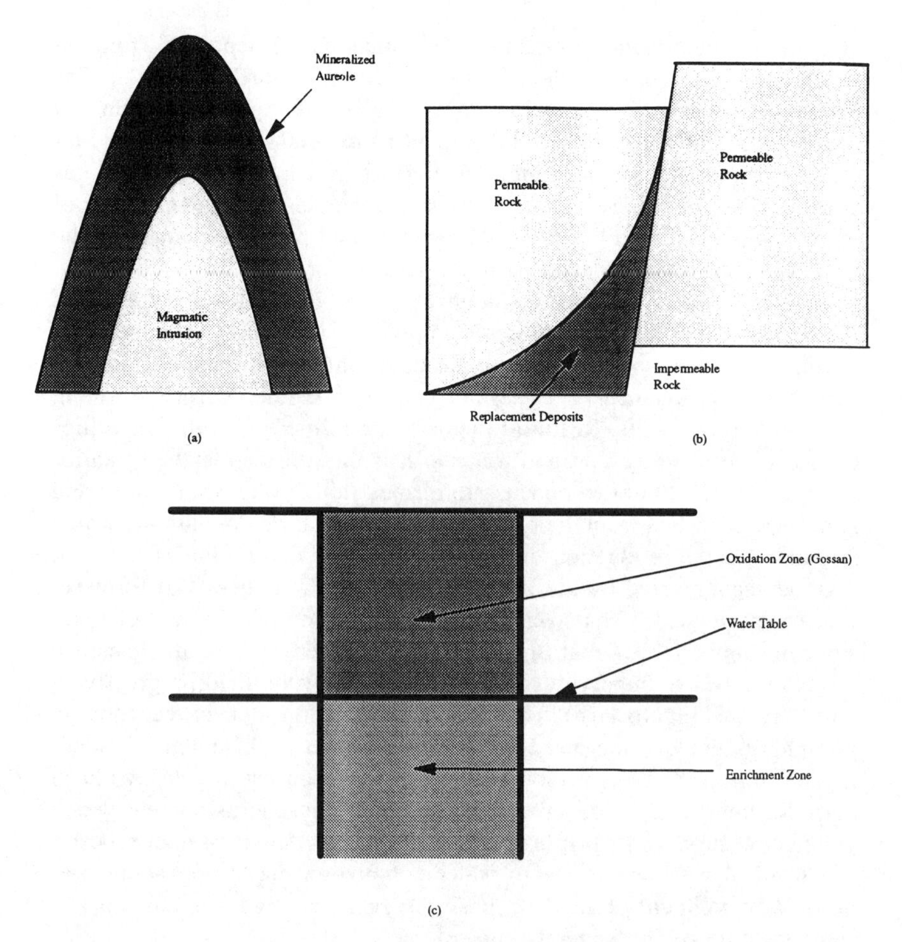

FIGURE 1.1. Formation of ore deposits. (a) Molten intrusion; (b) replacement or sedimentary deposit; (c) chemical weathering.

similar size, those of highest density were deposited nearer to the source while the lighter particles were carried farther away. This process is still continuing and can often be observed in river beds. **Placer deposits** are formed in this way. The miners, panning for gold in the days of the ''gold rush,'' were exploiting this method of concentration and panning is still practiced today in some parts of South America. It was the original source of Cornish tin, which was first recovered from river beds some 2000 years ago and about two thirds of the tin extracted around the world is recovered from placer deposits.

The concentration of minerals is further enhanced by dissolution of components of the bedrock in surface waters. Carbon dioxide dissolves in water to form a weak solution of carbonic acid which, in turn, can dissolve small amounts of some metallic minerals. The minerals are deposited from solution when the acidity of the water is decreased by further reaction. The ochre color of some rivers due to iron in suspension and in solution is a manifestation of the process. The amount of mineral transported per liter of water is minute, but the cumulative effect over hundreds of years has often resulted in massive accumulation of minerals. Subsequent burial of the deposits by silt, followed by temperature and pressure changes in the earth, has converted the deposits into a sedimentary rock.

Although this is the most common way in which sedimentary deposits are formed, other mechanisms can also produce mineral enrichments. For example, surface water may percolate down through the rock strata and carbonic acid contained in the water can dissolve the bedrock, forming soluble inorganic salts. **Residual deposits** are formed when the impurities have been leached by acidified water to leave a massive residue of an insoluble mineral. Bauxite deposits are formed in this way, iron oxides and silica being leached from the rock to leave a deposit rich in alumina which is mined for the production of aluminum. Or the dissolution of the rock may be accompanied by the deposition of metallic minerals to form **replacement deposits.** Thus, water penetrating down faults in beds of limestone in Cumberland, Great Britain, dissolved caverns where the limestone rested on a bed of impervious slate. The caverns were filled progressively with iron ore (Figure 1.1b). This ore which was, fortuitously, very low in phosphorus content was used by Bessemer when he invented his steelmaking process. The development of the steel industry might have been very different if he had used ores with a high phosphorus content which would have produced a brittle metal in his original acid Bessemer process.

Chemical weathering due to water percolation may bring about **secondary enrichment** of an ore deposit. Oxygen-bearing water can convert some metallic sulfide minerals into soluble sulfates, e.g.,

$$CuS + 2O_2 = CuSO_4 \tag{1.1}$$

while other sulfides are converted into insoluble oxides:

$$2FeS + 9O_2 + 4H_2O = 2Fe_2O_3 + 4H_2SO_4 \tag{1.2}$$

The soluble sulfates are carried downward by the water until they encounter more reducing conditions, where the sulfides are reprecipitated (Figure 1.1c). Different sulfides react at different rates. Thus, in a mixture of ZnS and PbS, the zinc mineral is less readily reduced, resulting in zinc accumulating below the lead. Other minerals, such as nickel, are transported as soluble carbonates, etc. and are deposited where the water encounters more alkaline conditions. The insoluble elements remain on top of the deposit as a cap or Gossan.

1.2.4. Workable Deposits

Whether or not an ore body can be exploited to produce metal at a sellable price depends on a number of factors. These include the metal content of the ore, the commercial value of the metal, the concentrations of undesirable impurities and of other valuable metals (such as silver), the location of the deposit in relation to supplies of coal, electricity and labor for metal production and the cost of transportation to an extraction plant where these are available. Several assessments have been made of the minimum metal content required in an ore body to give an economically workable deposit and, from this, the extent of natural concentration which must have occurred in the earth to produce an exploitable ore. Examples of two of these assessments are given in columns 5 and 6 of Table 1.1. The concentration factors show poor correlation with the selling price. On this basis, lead would be expected to cost as much as tin and gold. There is a better correlation in order (although not in magnitude) between price and the minimum workable content of an ore body. But this correlation suggests that the price for Al and Fe should be similar, while Zn should be no more expensive than Pb.

The average metal content of ores that currently are being mined is generally about twice the minimum contents listed in Table 1.1. Iron ores, for example, are being processed with iron contents ranging from about 45 to 65%.

Metals do not normally exist, however, in the metallic form in the ore. Metals such as Cu, Au and Pt (which form the less stable oxides) have been found in metallic or native form, but mineral compounds with other nonmetallic elements are the usual form. The most common compounds in which the metals are found are listed in Table 1.2. The weight of the compound is significantly greater than the weight of the metal that it contains. For example, the gram atomic weights of iron and oxygen are 55.85 and 16.00, respectively; thus, the atomic weight of hematite (Fe_2O_3) is

$$(2 \times 55.85) + (3 \times 16.00) = 159.70$$

Hence, an ore deposit that contains 65 wt% Fe as hematite actually consists of

$$(159.70/111.70) \times 65 = 92.93 \text{ wt\% of } Fe_2O_3$$

and only 7.07 wt% nonmetallic material, or gangue, which is primarily a mixture of silica, alumina, lime and magnesia. Similarly, the atomic weights of Cu and S are 63.57 and 32.06, respectively; thus, chalcopyrite, $(CuFe)S_2$, contains

$$100 \times 63.57 \,/\, [63.57 + 55.85 + (2 \times 32.06)] = 34.6 \text{ wt\% Cu}$$

TABLE 1.2
The Principal Minerals from Which Metals Are Extracted and Other Metals Which May Be Present in the Same Ore Body

Metal	Density $Mg\ m^{-3}$	Melting point °C	Principle ores	Average density $Mg\ m^{-3}$	Associated metals
Al	2.7	660	Bauxite $Al_2O_3{\cdot}2H_2O$	2.6	Ti,Fe,Ga
			Gibbsite $Al_2O_3{\cdot}3H_2O$	2.4	
Cr	7.1	1850	Chromite $FeCr_2O_4$	4.6	
Cu	8.9	1083	Chalcopyrite $(CuFe)S_2$	4.2	Ag,Au,Ni,
			Chalcocite Cu_2S	5.6	Pt,Pd,Re,
			Bornite Cu_5FeS_4	5.1	As,Bi,Se
Fe	7.9	1537	Hematite Fe_2O_3	5.3	
			Magnetite Fe_3O_4	5.8	
			Limonite $Fe_2O_3{\cdot}nH_2O$	4.0	
			Goethite $Fe_2O_3{\cdot}H_2O$	4.2	
			Siderite $FeCO_3$	3.8	
Pb	11.3	327	Galena PbS	7.5	Zn,Ag,Cu,
			Anglesite $PbSO_4$	6.3	Fe,Sn,As,
			Cerussite $PbCO_3$	6.6	Bi,In,Sb
Mg	1.7	650	Sea water		
			Dolomite		
			$MgCO_3{\cdot}CaCO_3$	2.8	
Ni	8.9	1454	Pentlandite $(NiFe)_9S_8$	5.0	Co,Cu,Fe,
			Pyrrhotite $(NiFe)_7S_8$	5.2	
Sn	7.3	232	Cassiterite SnO_2	6.9	
			Stannite Cu_2SnFeS_4	4.4	
Ti	4.5	1660	Rutile TiO_2	4.2	
			Ilmenite $FeTiO_3$	4.8	Fe
Zn	7.1	420	Zinc blende ZnS	4.1	Pb,Ag,Cu
			Calamine $ZnCO_3$	4.4	As,Cd,Ga

A typical Cu ore containing 0.8% Cu in this mineral form actually contains $0.8 \times (100/34.6) = 2.3$ wt% chalcopyrite and 97.7% gangue. These two examples illustrate the marked difference between iron ores, which often contain very little gangue, and nonferrous metal deposits where, with the exception of aluminum ores, the mineral part of the ore body is a very small part of the total mass.

In general, as the selling price of the metal increases, it becomes economical to mine ores with a lower mineral content. For any one metal, the smallest size deposit that is exploitable increases as the mineral content of the ore decreases. An ore which can be recovered by surface (open cast) working is more valuable than a deposit of similar size and mineral content that requires underground mining, with the attendant problems of ore transport to the surface, support of the roof, ventilation, drainage, etc.

The value of a deposit is dependent also on the chemical composition of the ore. The difficulty in recovering a metal from the ore often varies with the mineral form in which it is present. For example, some metals (e.g., Cu) are extracted more readily from sulfide than from oxide minerals. The exothermic oxidation of FeS in minerals such as chalcopyrite can generate heat which reduces or eliminates the need for external fuel supplies to melt the metal (see Chapter 4). Iron and aluminum ores and placer deposits of the tin mineral, cassiterite, rarely contain significant amounts of other metals. Many of the other nonferrous metal minerals are often associated with one or more other metals in sufficient concentration to justify the recovery of more than one metal. Examples of metals which are often associated in the same ore body are given in the last column of Table 1.2. A lower grade deposit of metal can often be processed economically when other metals are also recovered from the same ore.

Metal ores may contain small amounts of other metallic and nonmetallic elements (such as As, Sb, Cd, Bi, P, Se and Te) that are detrimental if they dissolve in the metal and are classed as impurities or tramp elements. The solubility of most solutes is significantly higher in a liquid than in a solid metal. On solidification, concentrations as low as a few parts per million (ppm) of an impurity may separate out as precipitates or as brittle grain boundary films which can have a serious effect on the physical properties (e.g., electrical conductivity), chemical properties (e.g., corrosion resistance) and mechanical properties (e.g., toughness) of the metal. Thus, the specification for the highest grade of copper requires a total impurity content not greater than 0.0065 wt% and the total content of Bi, Se and Te must not exceed 0.0003%. The value of an ore body is related inversely to the amount of work that has to be done in the extraction and refining processes to remove any impurities present in the ore to achieve the purity required. This, in turn, is related to the metal content of the ore that is being processed. If the whole of an impurity that occurs at a concentration of 10 ppm in the ore is extracted in the metal, its concentration will increase to only 16.7 ppm in an ore containing 60% of the metal (e.g., iron), but it will increase to 1000 ppm or 0.1% if the ore contains only 1.0% of the metal (e.g., copper).

The composition of the other gangue materials is also important. The difficulty of separating a metallic mineral from the earth is usually greater when the ore is embedded in a gangue that has crystallized from a molten magma than when it is recovered from a more friable sedimentary deposit. The melting temperature of the gangue is important when the metal is extracted in the molten state (i.e., by pyrometallurgy). Alumina, lime, magnesia and silica, which are the principal components of some gangues, melt at higher temperatures than any of the common engineering metals. So the temperature required for separating the metal and gangue by melting is often determined by the temperature at which the gangue melts and not by the melting temperature of the metal. This temperature can be lowered markedly if two or more of the pure oxides are mixed in suitable proportions (see Chapter 4). If the gangue is comprised of an appropriate mixture of these oxides, that is a bonus. Otherwise, ores from two or more different sources may have to be mixed together (blended) to obtain the required composition of the gangue. Alternatively, other minerals may be added to the furnace as fluxes to lower the melting point, but the thermal load is raised and the thermal efficiency of the extraction process is lowered if large amounts of fluxes are required.

1.3. MINERAL DRESSING

After the minerals have been extracted from the earth, they must be converted into metallic form. Until nearly the end of the first half of the 20th century, this conversion was achieved almost exclusively by pyrometallurgical means and this still remains the dominant technique. The lowest temperature at which even a blended or a fluxed ore can be melted is usually well in excess of 1000° C. If the ore contains only a low concentration of a metallic mineral, an enormous amount of energy would be consumed in melting all of the gangue and large amounts of CO_2 would be released into the atmosphere when the energy is supplied by combustion of gas, liquid or solid fuels containing carbon. Since the earliest days of bulk metal production, therefore, a wide range of mineral dressing or beneficiation techniques have been developed to separate the useful minerals from the unwanted constituents, prior to melting.

The cost of transportation of the ore from the mine to the reduction plant also decreases with increases in the degree of beneficiation accomplished at the mine. Up until about 50 years ago, the reduction plants were of small capacity and were often located in close proximity to the mine, so that transportation was not a problem. During the last 50 years there has been a rapid increase in the size of the extraction furnaces in order to reduce costs by increasing throughput, decreasing man-hours per tonne of product and increasing thermal efficiency. This trend was accelerated by the drastic increase in the price of oil in the early 1970s, which was reflected in the price of other metallurgical fuels. The modern blast furnace for the extraction of iron can produce about 10,000 tonnes (t) of molten Fe per day, while a copper smelting furnace may have an output of 300 t per day. Each of these production units consumes enormous quantities of ore. If the iron furnace is supplied with an ore containing 60% Fe, then about 17,000 t of ore per day is required. A copper smelter fed with concentrate from an ore containing 0.6% Cu would require the mining of 50,000 t of ore per day. Few ore deposits can provide ore at the appropriate rate to satisfy these demands for a sufficiently long period of time to justify the location of the smelting unit in close proximity to the mine. Thus it is normal today for the smelting plant to be remote from the ore source, often at a coastal site for ease of transportation, and for it to be fed with ore from several sources. Beneficiation is then very important to reduce transportation costs. For any given ore deposit there is an optimum degree to which the mineral can be concentrated. Transport costs increase as the extent of concentration decreases. But the loss of unseparated mineral in the rejected material (tailings) increases almost exponentially with increasing concentration, with a consequent decrease in the productivity of the mine.

Beneficiation techniques have been developed that allow recovery in separate streams of different metallic minerals and rejection of other minerals containing elements that would form undesirable impurities in the metal. But the separation is never complete and the cost tends to increase as the subdivision is increased. A further advantage of mineral dressing is that a concentrate of fairly uniform size can be fed to the smelter.

1.3.1. Comminution

The first stage in mineral dressing is a reduction in the size (comminution) of the lumps of ore recovered from the mine to particles of a similar size to the crystal or grain size of the mineral being recovered. This is achieved by first crushing and then grinding the ore to finer and finer sizes, promoting intergranular fractures and minimizing transgranular fractures in order to release the mineral grains in the largest possible size range.

Some type of sizing device, such as vibrating sieves or screens, is inserted between each stage of comminution, the finer material passing through the screen and on to the next stage, while the oversized pieces are recirculated through the previous stage. This closed-circuit operation reduces the total energy expended. Dry ore is usually preferred for crushing. Grinding can also be completed with a dry ore, but dust losses may then be severe and constitute an environmental hazard. More commonly the ore is suspended in water and surfactants or dispersants are added to separate the particles during grinding to finer sizes.

The minimum particle size required to release the mineral from the gangue depends upon the type of ore. Relatively coarse grains may be released by grinding down to particles of a few millimeters in diameter. Those minerals formed, however, as fine disseminations within the rock are not usually released until the particle size is less than 200 μm. The loss of metal values is increased if grinding is halted at too large a particle size to release all the minerals. Excessive grinding, on the other hand, is a waste of time and energy. Comminution often consumes a major part of the energy expended in beneficiation. If a significant fraction of the particles is reduced to below 10 to 20 μm in a wet grinding circuit, there is also an increased risk of loss of the mineral as a slime which is difficult to separate from the water and which is lost by adherence to the larger particles of gangue. So there is a strong economic incentive to ensure that the ore is reduced in

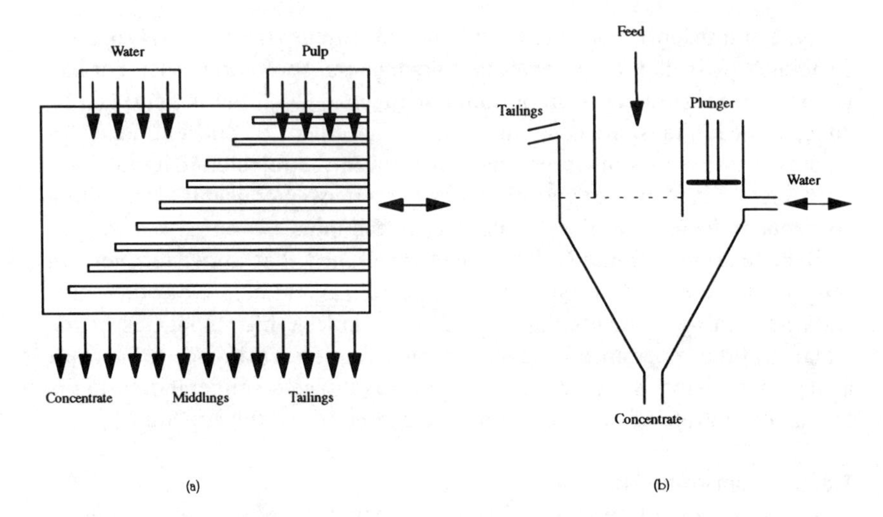

FIGURE 1.2. Examples of classification processes. (a) Wilfley table; (b) jig classifier.

size just sufficiently to release all the minerals, while avoiding overreduction of some of the ore and the generation of unrecoverable fine particles.

1.3.2. Separation

After the sizing operation has been completed the minerals are separated from the unattached gangue, and different metallic minerals are separated from each other, using a variety of techniques that depend on physical or chemical differences between the species.

1.3.2.1. Classification

Separation by exploiting the differences in the settling rates of the particles is the simplest and oldest means of concentration. The techniques in this group, which are called classification methods, utilize the same principle that occurred naturally in the formation of the placer deposits. In the simplest form, separation is achieved by pouring a waterborne suspension of material down sluices, as used by gold miners; or the suspension is poured onto the highest point of an inclined, corrugated and reciprocating table (Figure 1.2a). As the particles fall from the liquid, they become trapped between the corrugations and are carried along the table by the reciprocating motion. The heavy metallic minerals are discharged at the far end of the table while the lighter gangue materials are discharged along the side opposite to the entry point. Separation is never complete and the "middlings," which may have a high mineral content, are recycled to increase the recovery. More advanced forms use rakes to move the deposited material to the lowest point of a shallow bowl, while the lighter particles are discharged at the periphery before they have time to settle. Other devices include jigs (Figure 1.2b), in which a pulsating motion aids separation, spiral concentrators and centrifuges. The smallest particle size that can be separated by gravity is about 100 μm. As the particle size is decreased, the effect of gravity is diminished and there is more interference to the settling rates due to surface friction between the particles.

A further improvement is achieved by **heavy media separation** in which the density of the suspending liquid is intermediate between the densities of the mineral and the gangue. Thus, the density of silicate and aluminate compounds is usually in the range from 2.2 to 3.4 Mg m^{-3} whereas the density of the tin ore, cassiterite, is over 6 Mg m^{-3} (see Table 1.2). Hence, if the density of the fluid is maintained at, say 3 Mg m^{-3} the cassiterite will sink while most of the gangue floats to the top. Fine particles (<50 μm diameter) of magnetite, (Fe_3O_4), or ferrosilicon are added to the water to obtain the required density of the fluid. Since these materials are magnetic they can be recovered with a magnet from the discharge liquid and recycled. The maximum fluid density that can be achieved by these suspensions is about 3.3. Somewhat higher densities can be obtained by applying a magnetic field to a colloidal suspension of magnetite in water, the density increasing with increasing field strength. It is then possible to separate the minerals from gangue constituents that have a higher density.

The principle of this mode of separation can be understood from consideration of Stokes' Law. This states that the terminal velocity, V, of a particle which is descending through a stationary liquid under the influence of acceleration, g, due to gravity is proportional to the square of the diameter, d, of the particle and the difference in density between the particle, ρ_p, and the liquid, ρ_l, and is inversely proportional to the dynamic viscosity, η, i.e.:

$$V = \frac{g \cdot d^2 \cdot (\rho_p - \rho_l)}{9\,\eta} \qquad (1.3)$$

If the liquid is water with a density of unity, V is positive for all the ore particles and, if they are all spheres of identical diameter, the terminal velocity of each particle should be proportional to its density.

In practice, deviations from the predicted rate of settlement may arise from turbulence in the fluid, the irregular (nonspherical) shape of the particles, interparticle interference from collisions (since the pulp may contain more than 30% of solids) and from variations in the fluid viscosity due to changes in the amount of suspended solids. There is usually a difference of 20% or more in the mesh aperture between any one sieve and the next largest sieve in the series. So, even if all the particles were spherical, there would be an unavoidable spread in particle size, resulting also in either more gangue being recovered with the mineral or more mineral rejected with the gangue than would be expected from simple application of Stokes'

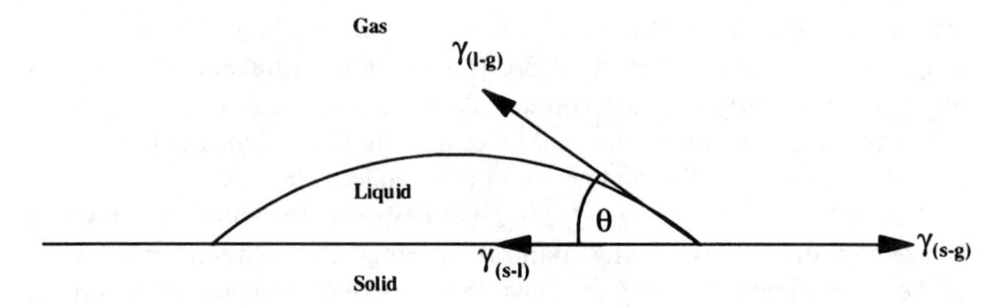

FIGURE 1.3. Surface tension forces between gas, liquid and a solid substrate.

Law. The deviations from the expected behavior become greater as the density difference between the mineral and the gangue decreases. In practice, a difference in density of at least 1 Mg m^{-3} is required to achieve adequate separation. The densities of most common metallic minerals are rather similar and, while separation by gravity from the lighter gangue materials is relatively straightforward, it is less easy to separate the metallic minerals from each other by this method.

In the heavy media sink-float process the density of most of the gangue is less than that of the fluid and hence, according to Stokes' Law, its velocity is negative irrespective of particle size (i.e., it will float upward in the medium). The efficiency of separation is then dependent both on the density difference between gangue and mineral and the constancy of the density of the fluid. Efficient mineral concentration can be achieved if the former is large and the latter is maintained constant. It is claimed that particles with a density difference of only 0.2 to 0.3 Mg m^{-3} can be separated in this way. But the cost of operation is higher due to the initial cost of the substances added to increase the density of the fluid and the cost of the recovery and recycling of those substances from the overflow.

1.3.2.2. Flotation

The process which is most widely used for mineral concentration depends on modification of the surface tension of the particles to alter their rate of settling in water and cause the required mineral to float to the surface. Consider a drop of water on the surface of a solid in contact with a gas. The spread of the water over the solid surface is determined by the balance of the three forces of surface tension, $\gamma_{(solid-liquid)}$, $\gamma_{(solid-gas)}$ and $\gamma_{(liquid-gas)}$ at the point of contact with the solid, Figure 1.3:

$$\gamma_{(s-g)} - \gamma_{(s-l)} = \gamma_{(l-g)}\cos\theta \tag{1.4}$$

If $(\gamma_{(s-g)} - \gamma_{(s-l)})$ is similar in magnitude to $\gamma_{(l-g)}$, the contact angle, θ, is approximately zero and the water spreads over the surface of the solid (i.e., the liquid wets the surface). Conversely, when $\gamma_{(s-g)}$ is greater than $\gamma_{(s-l)}$, θ is greater than 90° and the water remains as a droplet on the surface.

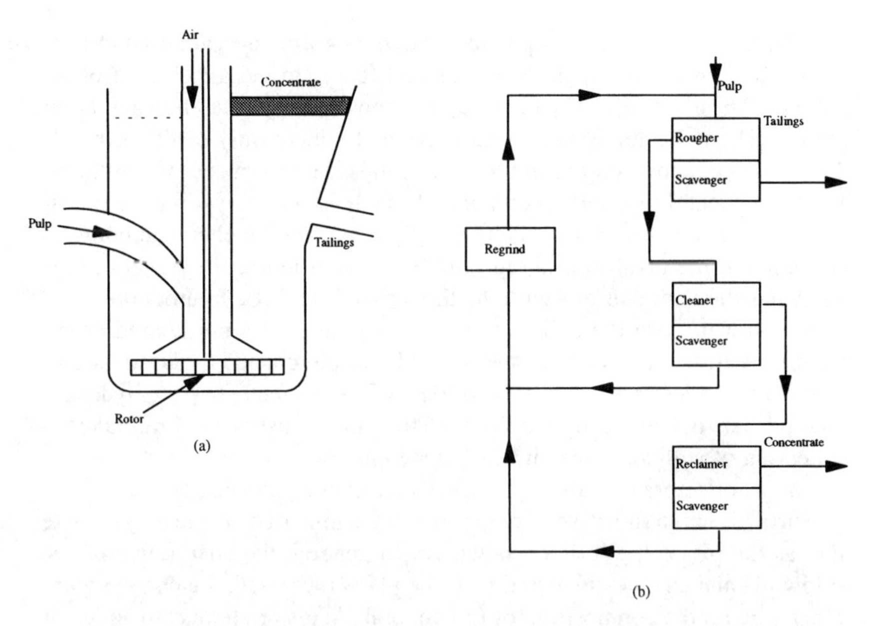

FIGURE 1.4. Flotation equipment. (a) Flotation cell, (b) typical arrangement of cells.

Ore particles are generally hydrophilic, which means that $[\gamma_{(s-g)} - \gamma_{(s-l)}]$ is greater than $[\gamma_{(l-g)}]$. A net charge is left on the ore particles when they are fractured during comminution. Water, which is a polar liquid, is attracted to the surface charge and the particles are readily wetted. Some chemical compounds are hydrophobic, with $\gamma_{(s-g)}$ greater than $\gamma_{(s-l)}$, so they are not wetted by water. If these compounds have polar groups which are attracted more strongly than water to the net surface charge on the solid particles, then the particle surface becomes coated with the reagent and is not wetted. If, now, a stream of air bubbles flows upward through the liquid, some bubbles will become attached to the coated solid, thus decreasing the apparent density. When the particle is small (e.g., 10 to 100 μm), relative to the diameter of the bubble, it is carried upward by the bubble to the gas-liquid surface. On arrival at the surface, however, the liquid film surrounding the bubble drains away and the bubble collapses, allowing the particle to descend again through the liquid. Drainage is prevented by adding a frothing agent such as pine oil or cresylic acid that stabilizes the liquid film covering the bubble and retains the solids in the surface froth. This is the basis of the froth flotation process.

The ore pulp is fed in at a point near the top of a cell that may have a capacity of over 50 m^3. Air bubbles rising from the bottom of the cell carry the minerals to the surface, where they are removed in the froth, while the unwanted particles descend through the water and are removed from the bottom of the cell, Figure 1.4a.

A variety of chemical compounds, known as **collectors,** are used to alter the surface properties of the particles and facilitate the attachment of air bubbles. These are typically reagents that consist of hydrocarbon and polar groups. The collectors must be selective and adhere only to the minerals that are to be recovered and not to other solids. In general, sulfide minerals are floated readily by adding compounds such as soluble sodium or potassium ethyl xanthates of the type $C_2H_5OS_2Na$. When the chemical contacts the mineral, the alkali ion is replaced by a metal ion in the mineral, thus attaching the xanthate molecule to the surface, and the hydrophobic end points outward into the solution where it can attach to the ascending air bubbles. Fatty acids (such as stearic acid and oleic acid) are used as collectors for oxides and carbonates, while chelating agents, (e.g., 2-hydroxy-5-nonyl-benzophenone oxime LIX 65N) can be used for simultaneous collection of sulfide, oxide and carbonate minerals.

Many collectors operate most effectively over a specific pH range since the surface charge (positive or negative) on the mineral particle may change sign as the pH of the solution is varied. In general, the charge is positive at low pH and changes to negative as the pH is increased, the change point depending on the composition of the mineral. Acids or alkalis are added to the pulp to adjust the acidity of the solution to the required range. These are known as **conditioners.** The composition dependence of the pH at which the sign of the surface charge changes can be exploited to achieve separate collection of different minerals from the pulp. Thus, in an ore containing Cu, Fe and Pb sulfides, the copper mineral, chalcopyrite, can be floated while PbS and FeS are retained in the solution by operating at high pH. After removal of the Cu, the Pb mineral is recovered while the FeS remains suspended in the solution by lowering the pH to the stage where the PbS acquires a positive charge but the FeS retains a negative charge.

Modifiers are chemicals that are added to improve the adhesion of the collector to the particles. Thus, ZnS is more readily floated by xanthates if a small amount of copper sulfate is added to the solution to form a thin layer of Cu over the mineral surface. Conversely, **depressants** render wettable a particle that would otherwise be recovered. PbS can be floated, while Cu, Fe and Zn sulfides are retained in the solution by adding sodium cyanide as a depressant. The cyanide does not affect the surface of PbS, but coats the other minerals and prevents flotation. After removal of lead, ZnS can be reactivated by adding $CuSO_4$ and using sodium butyl xanthate as the collector. It is not possible to separate the metallic elements in this way, however, when they exist in solid solution. Chalcopyrite is a solid solution of Cu and Fe sulfides. Any associated iron pyrites can be separated but the concentrate contains all the iron in the chalcopyrite, often resulting in roughly equal amounts of Fe and Cu in the product.

It is normal practice to operate flotation cells in series with at least the material rejected in the scavenger cells being recycled to reduce the loss in the tailings, Figure 1.4b. The recycled material may be ground to a finer size before reflotation to enhance mineral recovery. Problems arise, however, if the pulp contains very fine particles, as they interfere with particle movement within the cell and recovery of fine particles below about 20 μm diameter is poor. Attempts have been made to increase the recovery of fine particles by admitting air through a porous membrane, or by pressurizing the water to increase the solubility of air and then lowering the pressure to release the air as very small bubbles.[3]

With a sufficient degree of comminution to release all the mineral grains from the gangue and adequate flotation cells in series, it is theoretically possible to recover virtually all the mineral and reject the gangue. In practice, economics fix the limit of the concentration process. Beyond a certain stage, which varies from ore to ore and from one mineral to another, the yield from any additional flotation cells is not sufficient to justify the cost. Similarly, different minerals in the same ore can be collected separately, but a stage is reached beyond which further treatment becomes uneconomic. Much depends on the ability of the reduction plant to which the concentrate is dispatched to remove the residual gangue and other metals during the extraction process. A dried ZnS concentrate, for example, might contain up to 5% PbS, 1 to 2% CdS and 10% FeS, with the total weight of the metallic minerals other than zinc exceeding the weight of the gangue. It may then be less costly to separate the zinc from these other metallic elements during reduction of the ZnS than to attempt to increase further the purification of the mineral by flotation.

The efficiency of concentration of a mineral by froth flotation is markedly higher than by gravity methods, but it is also much more expensive. The amount of chemicals, other than conditioners, added to the pulp rarely exceed 1 kg t^{-1}, but they are not cheap. The rate of separation is also much slower than in classification, so more plant is required for a given throughput, although the capital cost is relatively low. The addition of chemical reagents to the water introduces ecological problems that do not arise with classification. The reagents absorbed by, or adherent on, the rejected ore constituents must be neutralized or removed by washing before the waste is discharged. All the water used in the process must also be treated before discharge to the environment. Particular care is required when noxious substances such as cyanides have been added to the pulp (see Chapter 8).

1.3.2.3. Electrostatic Concentration

Differences in the electrical properties of the ore constituents can be exploited to achieve separation. If a surface charge is induced on the particles, then conductive minerals such as the metallic sulfides, titanium minerals and magnetite quickly lose their charge, while it is retained on the nonconducting particles that comprise mainly the gangue materials. Hence,

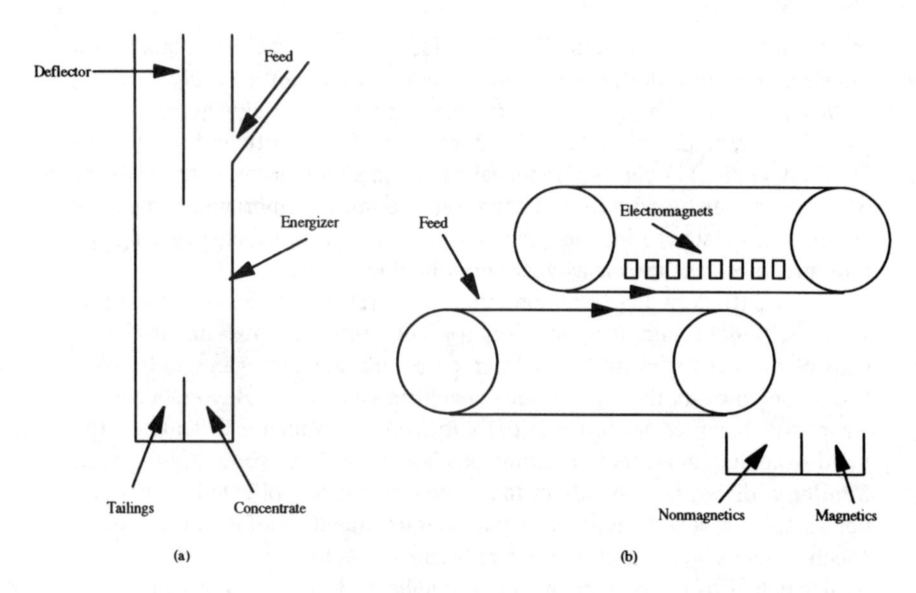

FIGURE 1.5. Electrostatic free fall (a) and magnetic (b) separators.

if a finely divided ore falls in a thin layer past an electrode charged at several thousand volts, the electrically conducting particles are attracted toward the electrode. The nonconductors are deflected and follow a different free fall path, Figure 1.5a. The separated materials are collected in a partitioned trough. In similar manner to the flotation process, the discharge collected from the middle portion of the trough may require recycling several times through the separator to obtain adequate recovery and concentration.

This mode of concentration is only effective if both the ore and the atmosphere are dry, since a coating of adsorbed moisture will convert nonconducting materials into conductors and separation is then impossible. The capital cost of the equipment is high and the capacity is low in comparison with the flotation process, so electrostatic separation is not often used for mineral dressing. It is applied more extensively to the removal of solids from the exhaust gases released in pyrometallurgical reduction.

1.3.2.4. Magnetic Concentration

A few minerals are either ferromagnetic (e.g., pyrrhotite-FeS or magnetite) or paramagnetic (e.g., chromite; ilmenite, $FeO{\cdot}TiO_2$; hematite; limonite, $Fe_2O_3{\cdot}xH_2O$; pyrolusite, MnO; or wolfram, $(FeMn)WO_4$), but most metallic minerals and gangue are nonmagnetic. This difference can also be used to effect separation. In one arrangement, a thin layer of fine ore is carried along a conveyer belt. A second conveyer belt, a short distance above the first belt, passes under a row of magnets. The magnetic particles are attracted by the magnets and carried on the underside of the upper belt to a discharge point beyond the end of the lower belt, Figure 1.5b. The particles adhering to the upper belt are caused to rotate during transit by alternating the polarity of successive magnets, thus aiding the separation by releasing any nonmagnetics that may have been elevated with the magnetic material. Magnetic attraction is not dependent on particle size, but the gap between the two belts is small to minimize the strength of the magnetic field required, thus placing a limit on the ore feed size. In an alternative arrangement, a magnetic field is used to deflect the magnetic particles during free fall, causing them to fall to one side of a partition while the undeflected diamagnetics accumulate on the other side of the partition.

Magnetic separation is not affected by water. Coarser particles are usually concentrated in the dry condition, but finely ground ore may be separated directly from the flowing pulp. As with the other concentration processes, it is often necessary to pass the material several times through the magnetic field to obtain a satisfactory level of beneficiation. The capital cost is high and the capacity is low in comparison with a flotation plant, and magnetic methods find limited use in mineral concentration. They are used more extensively for removing adventitious pieces of iron tools, etc. in the ore from the mine before feeding to the ore crusher and in the separation of ferrous metals from other metals in recycled scrap metal.

Both electrostatic and magnetic separation methods do not introduce any environmental hazards.

1.3.2.5. Economic Aspects of Beneficiation

The cost of beneficiation has an important bearing on the total cost of metal production. It is often cost effective to use more than one type of concentration process in the treatment of an ore. Thus, gravity methods may be used to obtain a crude separation of most of the gangue prior to flotation, or before grinding to a finer size suitable for flotation. The product from the flotation cells may be passed through a magnetic separator to reject ferrous minerals or to recover other minerals such as pyrrhotite or ilmenite. In some cases, magnetic separation precedes flotation.

The proportion of metallic minerals that can be recovered from an ore by concentration methods increases with increasing fineness of comminution; but the cost of grinding and the risk of mineral loss as dust or slimes rises as the particle size decreases. Similarly, the mineral loss in the tailings falls with an increase in the number of times the ore is recycled through the various concentration stages. However, the cost rises and throughput decreases as more material is recycled. The break-even point depends on the value of the refined metal and the mineral content of the ore. If the tailings contain, say, 0.1% of the mineral that is being recovered, this represents a loss of almost 10% of the value of an ore that contains 1.0% mineral, but only about 2% of an ore containing 5.0% mineral.

Typical mineral contents of material supplied by the concentrator are:

Copper ores	25–50% CuS
Lead ores	40–75% PbS
Tin ores	70–80% SnO_2
Zinc ores	50–60% ZnS

Theoretically, it is possible to produce a concentrate containing only one metallic mineral, but the cost would be prohibitive with most ores. On the other hand, the cost of reduction of the mineral to the metallic form and the removal of other elements by refining to obtain the required purity increases with decreasing purity of the concentrate. In practice, the level of concentration aimed for is based on evaluation of a number of factors. For example, it may not be economically viable to concentrate a low-grade lead ore beyond 45 to 50% PbS, whereas a high-grade ore may be treated to produce a 70 to 75% PbS concentrate. In both cases, the composition chosen is that which gives the maximum financial return to the operator, and the profit decreases if either a higher or a lower degree of concentration is achieved. In a similar manner, any other metallic minerals or noxious impurities are only separated to the extent that is financially viable. With a mixed ore of Pb and Zn sulfides, it is rarely profitable to produce a concentrate with less than about 2% Zn in the Pb concentrate, or 2% Pb in the Zn fraction. However, the customer (the reduction plant) may reduce the price it is prepared to pay if the concentrate contains large amounts of other elements that make the extraction process more complicated.

Since the density, wettability, and electrical and magnetic properties of bauxite ($Al_2O_3 \cdot 2H_2O$) fall within the range for the gangue materials, the beneficiation processes that have been described thus far are of limited use in the production of aluminum. In this case, the ore is purified by leaching (see chapter 6) to remove the impurities and produce pure alumina. Iron ores are amenable to concentration by any of the beneficiation processes. However, there is an abundance of ores containing at least 90% iron oxides and the cost of concentration, other than comminution and classification or washing to remove adherent clay from lower grade deposits, can rarely be justified. Both bauxite and iron ores are usually subjected only to crushing and coarse grinding. In the former case, the purpose is to increase the surface area for attack by the leach solution. Iron ores can be reduced in lump form, but it is usually advantageous to crush the ore to about −20 mm and then agglomerate the particles to improve the reducibility.

Thus, the cost of beneficiation is very low for Fe and Al ores. Concentration costs for other nonferrous minerals escalate rapidly as the metal content of the ore decreases and the fineness of dissemination of the mineral in the bedrock increases. The costs incurred at this stage can be a significant fraction of the total cost of production from low-grade ores in which the mineral is present as very small grains.

1.3.2.6. Dewatering

Drying the concentrate is usually the final stage of beneficiation. The product obtained from classification, flotation and wet magnetic concentration is a solid suspended in a large volume of water. Most of the water is removed by allowing the solids to settle in a shallow container. The rate of settling is dependent on the square of the particle diameter, as indicated by Stokes' Law and, hence, the time required for complete separation increases rapidly as the particle size decreases. Flocculents such as lime or polymers may be added to cause the fines to aggregate into larger apparent diameters and increase the rate of settling. Centrifuges achieve more rapid dewatering, but at higher cost. Vacuum or pressure (squeeze) filters can remove the remaining free water. The maximum aperture tolerated in the filter mesh is determined by the mineral particle size; thus, both the cost of the filter and the filtration time increase as the particle size is reduced.

The concentrate still contains about 5 to 10% of adsorbed and combined moisture after filtration. Some of this may be lost by evaporation during storage and in transit to the reduction plant, but heating is required to remove the remainder. Although the evaporation of moisture is endothermic and consumes heat in pyrometallurgical reduction, it is usually cheaper to remove the residual moisture at that stage than to attempt further drying at the concentrator. Where the concentrator and reduction plant are in close proximity, the ore is often filtered immediately before feeding into the extraction plant to reduce dust losses. Residual moisture is not a problem when the metal is extracted from the concentrate by leaching.

1.4. AGGLOMERATION

The dried powder form in which the material is supplied from the concentrator is suitable for immediate use in some of the extraction processes. In other cases, the powder form is not suitable. This is true particularly for blast furnace-type processes in which the ore descends down a shaft in countercurrent flow to a large volume of reducing gas. Fine material is readily entrained in the gas stream and large amounts are carried out of the furnace by the gas if raw concentrate is fed into the furnace. Ore fines readily stick to each other on heating to elevated temperatures, forming a mass that is less readily penetrated by the reducing gas. For such applications, the crushed ore or concentrate is first agglomerated into lumps of suitable size and strength.

1.4.1. Pelletizing

If only low strength is needed, the powder can be consolidated at ambient temperature by pressing into briquettes or forming into pellets with a suitable binder. Pellets can be formed by feeding the fines (< 1 mm dia.) onto a rotating disk, inclined slightly to the horizontal and with a lip around the periphery over which the pellets are discharged when they reach the required size. More commonly, the fines are fed to the higher end of a rotating drum that is inclined slightly to the axis of rotation. The output is screened and undersized pellets are recycled through the drum.

The particles can be made to agglomerate purely by the action of surface tension forces if just sufficient water is added to wet the particles and fill the voids in the growing pellet. However, loss of water during transportation or storage then results in partial or complete reversion to powder. Stronger bonding is obtained by adding a small amount of a suitable binder such as cement or bentonite, which coats the particles and forms bridges between them. For any given combination of ore, water and binder, the pellet size is determined primarily by the residence time on the disk or in the drum. This can be controlled by varying the speed of rotation or the angle of inclination. The preferred pellet diameter is usually in the range 10 to 20 mm.

Green pellets formed at room temperature are easily damaged and require careful handling. Greater strength can be conferred by heating them slowly to 1000 to 1100° C on traveling grates or in shaft kilns.

1.4.2. Sintering

Sintering is the other common method of agglomeration when high strength is required in the aggregate. If the ore is heated to a high temperature, but not sufficiently high to cause melting, the particles tend to sinter or weld together by incipient fusion at the points of contact. The necks thus formed between the particles have a negative radius of curvature and the surface of the neck is under tension. This tensile stress increases the number of vacant lattice sites in the surface, relative to an unstrained lattice. As a result, lattice vacancies tend to migrate by diffusion to any free surface, and atoms migrate in the reverse direction in an attempt to reduce the radius of the neck. The area of contact, and hence the strength of the bond between the particles, thus increases with an increase in the time at that temperature. The bonding which forms during the firing (induration) of pellets depends on the same mechanism.

The Dwight-Lloyd sintering machine consists of an endless chain of shallow pallets with perforated bases that travel horizontally, below a charging spout and leveling device, through the sintering zone and return beneath it. A coarser feed is preferred for sintering than for pelletizing and the ore is usually screened to −(15–20) mm. Excessive dust losses occur through the apertures in the pallets if very fine ore (<0.1 mm) is charged to the sinter strand, so material below this size produced by comminution or by a beneficiation process is pelletized. The pellets can be fired on the sintering machine.

1.4.2.1. Oxide Ores

Ores or concentrates in which the minerals are present as oxides are mixed in a drum with about 5% coal dust or coke breeze. Water is added to bring the moisture content up to about 10% to ensure adequate permeability through the sintering bed. The actual amount of water and fuel is adjusted continuously to control the maximum temperature attained. The ore is fed as a relatively thin layer onto the pallets as they enter the horizontal section of the machine. The charge then passes under an ignition hood where the fuel in the surface of the bed is ignited. A suction box underneath the pallets draws the combustion gases down through the ore bed. The combustion zone moves down progressively through the bed as the pallets traverse along the machine and reaches the bottom of the bed shortly before the pallets turn over at the end of the traverse to discharge the product. The sinter is cooled, crushed if necessary, sieved and the undersized material is recycled through the machine.

The ore is preheated and may be partially reduced by the combustion gases as they are drawn through the bed below the combustion zone. The air drawn through the bed above the combustion zone is preheated by the hot sinter and the ore is reoxidized by the air. Within the combustion zone, the ore particles are heated to a temperature at which some plastic deformation of the ore may occur and where atom transfer along the interface between contacting particles is relatively rapid. The maximum temperature may be reached slightly behind the combustion zone due to the exothermic heat generated by the reoxidation of the sinter.

A maximum temperature of 1200 to 1300° C may be attained in the sintering of iron ores. This is usually sufficient to cause partial fusion of some of the gangue, and the strength of the interparticle bonding is then increased by the glassy cement that forms when the melt solidifies as the combustion zone moves on. Excessive fusion of the gangue, however, results in complete coating of the ore particles with a fused layer that adversely affects the subsequent reducibility of the sinter and must be avoided. There is also a risk that chemical changes may occur that affect the reducibility. For example, silica may react with iron oxide to form the compound called "fayalite" ($2FeO \cdot SiO_2$), which melts at 1200° C and is much more difficult to reduce to metallic form than the iron oxides.

Other changes occur that are beneficial. Carbonates are decomposed with the release of CO_2 and the ore is completely dehydrated. Both of these reactions absorb heat (i.e., endothermic), so sintering reduces the thermal load in the reduction process. Some volatile species, such as arsenic and cadmium, that would adversely affect the metal properties are partially gasified during sintering. These chemical changes also occur during the induration of pellets. At first sight, it may appear that part of the thermal load and hence part of the fuel requirement is merely being transferred from the reduction process to the sintering unit; but the fuel used for sintering is finely divided coal or coke breeze that is generally unsuitable for use in the reduction process and would otherwise be rejected.

1.4.2.2. Sulfide Ores

A somewhat different practice is adopted for the sintering of sulfide ores because, during heating, the sulfur is removed as a gas (mainly SO_2) and the metallic mineral is oxidized:

$$MS_{(s)} + \tfrac{3}{2}O_{(g)} = MO_{(s)} + SO_{2(g)} \tag{1.5}$$

where M refers to the metal and the subscripts (s and g) refer to solid and gas, respectively. This reaction can be viewed as three consecutive steps:

$$2MS_{(s)} = 2M_{(s)} + S_{2(s)} \tag{1.6}$$

$$2M_{(s)} + O_{2(g)} = 2MO_{(s)} \tag{1.7}$$

$$S_{2(s)} + 2O_{2(g)} = 2SO_{2(g)} \tag{1.8}$$

For most minerals, the first step is mildly endothermic and the second step is exothermic, while the third step is strongly exothermic (see Chapter 4). The heat balance for the overall reaction (Equation 1.5) is a negative quantity and heat is released by the reaction. No coal or coke fuel is required for sintering if the ore is sufficiently rich in the sulfide mineral. In fact, it is often necessary to recirculate a significant portion of the sinter to avoid excessive temperatures being reached in the reaction zone.

A thin layer of ore is placed on the pallets and ignited with a flame by downdraft. A thicker layer of ore is then placed on top and the air flow is reversed to pass upward through the bed. The sintering mechanism is otherwise similar to that described for oxide ores. In addition to volatile species gasified from the ore, the exhaust gas contains large amounts of SO_2 and these species must be removed before the gas is discharged to the atmosphere. Environmental aspects are reviewed in Chapter 8.

1.4.2.3. Self-Fluxing Sinter

In the subsequent stage of pyrometallurgical reduction of the sinter, both the metal and the gangue are removed from the furnace in the molten condition. As indicated earlier, the minimum temperature at which the process can be operated is usually determined by the melting of the gangue. The opportunity is often taken, therefore, to incorporate the fluxes into the sinter that are required to lower the fusion temperature and produce a melt of the desired composition. These substances can be added in finer form and mixed more intimately with the gangue during sintering than is possible within the reduction furnace. The sinter is then described as "self-fluxing."

Some chemical reaction may occur between the gangue and the fluxes at the high temperatures occurring in the combustion zone, and care must be taken to ensure that an excessive amount of glassy bond is not formed. On the other hand, the fluxes may prevent the occurrence of adverse reactions between the gangue and the ore. Thus, the formation of fayalite in the sintering of iron ores can be prevented by the addition of lime as a flux to form calcium silicates.

1.4.3. Sintering versus Pelletizing

There is no clear consensus of opinion on whether sinter or pellets are a more suitable feed for a reduction process. It is often claimed that the porosity distribution is more favorable and enhances the reducibility of pellets. However, it is also claimed that the superior crushing strength of sinter results in a more open mass that allows better access of reducing gas within the furnace.

Pelletizing is a two-stage process in which the pellets are first formed and then indurated; whereas, sintering requires only the mixing of the materials before feeding to the sintering machine. However, the powder concentrate from a flotation plant, which is an ideal feed for pelletizing, cannot be fed directly to the sinter machine if it contains a significant fraction of particles less than 0.1 mm diameter.

Both pelletizing and sintering require greater initial capital outlay than is required for a classification or flotation plant; but the operating costs are relatively low and the cost of agglomeration is usually outweighed by the lowered costs for the reduction process resulting from the enhanced reducibility of the feed material. Agglomeration costs are avoided when the extraction process is capable of treating finely divided material.

1.5. REDUCTION AND REFINING

The ore is now ready for the final separation from the remaining gangue and for reduction to the metallic form. Ideally the metal would be produced with the required purity in a single step. In most cases, however, an impure metal is first produced in an extraction or smelting stage. The impurities are then removed by refining to produce a metal with the required composition. In some cases, as in the production of steel, other elements may be added during refining to produce alloy compositions. The types of processes used for this purpose can be classified under three headings:

Pyrometallurgy
Hydrometallurgy
Electrometallurgy

Each type will be considered subsequently in separate chapters, but it is convenient to consider the underlying principles at this stage.

1.5.1. Pyrometallurgy

As the name implies, pyrometallurgy involves treatment of the ore at elevated temperatures. In tonnage terms, it is the dominant method for the production of metals. Most commonly, the ore is heated and simultaneously reduced by solid carbon or by gases such as CO or hydrogen:

$$C + MO = M + CO \tag{1.9}$$

$$CO + MO = M + CO_2 \tag{1.10}$$

Most sulfide minerals are not readily reduced by these reactions and it is necessary first to roast the ore in an oxidizing atmosphere at an elevated temperature, but below the melting temperature of the constituents, to convert the sulfides to oxides. This transformation is achieved as part of the aggregation process when the ores are sintered.

Carbon is the cheapest available reducing agent. Hydrogen is much more expensive. However, neither of these agents are suitable for the reduction of metals that form very stable oxides. They are also not suitable for the reduction of metals such as Be, Cr, Ti, U and Zr, which readily form carbides or hydrides with consequent embrittlement of the metal. Other techniques are then used. For example, the mineral can be reduced by heating it with some other metal that forms a more stable oxide (metallothermic reduction). This is an expensive method, since the metal used to effect the reduction must first be reduced from the ore; thus two reduction stages are involved. Or, the metal may first be volatilized as a halide (e.g., MCl or MF) by heating in the presence of a halogen gas and subsequently recovered by decomposition of the halide.

Reduction by CO or H_2 may be accomplished below the melting point of the metal, as in the so-called direct reduction of iron ores, in which iron ore fines or pellets are reduced in the temperature range 700 to 900° C. This is well below the fusion temperature for both the metal and the oxides, but the temperature must be controlled to prevent the risk of the particles sintering together during reduction. The product is an intimate mixture of metal and gangue that contains some unreduced oxide. Separation of metal and gangue, and completion of reduction, must then be accomplished during the refining stage.

More commonly, the ore is heated to above the fusion temperatures of both the metal and the gangue in the reduction furnace. The metallic oxides are usually miscible with the molten gangue, but are relatively immiscible when reduced to the metallic form. Since the molten gangue has a lower density than the metal, it forms a separate layer called a *slag* and floats on top of the metal.

1.5.1.1. Slags

In addition to the effect on the minimum operating temperature for the process, the composition and viscosity of the slag is very important in other respects. The composition must be selected to ensure a low solubility of the metal in the slag. This becomes increasingly important as the metal content of the ore or concentrate charged to the furnace decreases. Thus, if the ore contains, say, 90% of the metallic mineral and only 10% of gangue, the volume of slag formed is small relative to the volume of the metal. If no fluxes are added and the slag contains 10% of the metallic mineral, this represents a loss of only about 1% of the metal in the slag. But if the charge material contains only 40% of the metallic mineral, the volume of the slag may be greater than the volume of the metal. If 10% of the metallic mineral is again retained in the slag, this now represents a serious loss of metal. In fact, the slag may contain more metal than the ore that was extracted from the ground. The addition of fluxes to lower the fusion temperature of the gangue results in the formation of a larger volume of slag and increases the metallic losses unless, at the same time, the solubility of the metal or the mineral in the slag is lowered by change in composition resulting from the flux additions.

The slag is frequently required to act as a receiver for unwanted elements that are present in the ore or in the concentrate. For example, Cu concentrates often contain roughly equal amounts of Cu and Fe sulfides. The process is then designed to convert the FeS into FeO, which combines with silica in the slag to form fayalite while retaining the Cu as a separate (matte) phase in the sulfide form:

$$2FeS_2 + SO_2 = 2FeO + 4SO_2 \tag{1.11}$$

$$2FeO + SiO_2 = 2FeO{\cdot}SiO_2 \tag{1.12}$$

The loss of some copper into the slag is unavoidable and this is very important when it is remembered that the average Cu content of ores from the mine is only about 1.0%. The Cu content of the slag is raised but the volume of slag is lowered with increasing Cu content in the concentrate; thus, the amount of Cu lost in the slag actually decreases with richer concentrates. There is an incentive, therefore, to increase the Cu content as far as possible in the beneficiation stages. When the value of the metal retained in the slag is high, it is common practice to recycle the slag from the smelter through a beneficiation stage. Similarly, the slags produced during refining may be recycled through the smelter. The viscosity of the slag is important with regard to the separation of the slag from the metal. Droplets of metal may be formed by reduction of mineral particles suspended within the slag layer, or may be ejected into the slag by gases emerging from the metal. The time taken for the droplets to separate from the slag and join the bulk metal rises with increasing slag viscosity. The metal loss is higher if insufficient time is allowed for complete separation.

At the end of the process, the metal and slag have to be removed (tapped) from the furnace and separated from each other. It is more difficult to completely drain a very viscous slag from the furnace and separate it from the metal than a slag that is more fluid. In general, the viscosity of molten sulfides is similar to the viscosity of molten metals. Oxide slags have a higher viscosity, which increases with increasing polymerization, or ionic complexity, of the melt. Silica and, to a lesser extent, alumina are the principal polymer formers in the slag and, if these are dominant in the gangue, basic oxides such as lime must be added as fluxes to reduce the viscosity. However, this raises the slag volume. Since viscosity decreases with increasing temperature above the fusion point, the alternative is to increase the operating temperature to a range where the viscosity is acceptable; but more energy is then consumed in raising the temperature.

The disposal of the slag remaining after the metals have been separated is a problem where slag heaps are banned and costs are incurred where it is used for landfill; but all the slags are not necessarily waste materials. They may be sold for a variety of applications (see Chapter 8). The composition of the slag may then have to be modified slightly, relative to that which is most desirable from a pyrometallurgical viewpoint, to obtain a sellable product.

1.5.1.2. Refining

Pyrometallurgical refining is conducted after separation of the metal from the smelter slag. A wide variety of techniques is employed at this stage. A new slag is often formed by addition of suitable compounds to act as a sink for elements that are removed selectively from the metal. Or, an element may be stirred into the metal to react with other elements that are dissolved in it and form an immiscible scum or **dross** that can be removed from the surface. In some cases, the metal is separated from the impurities by forming a volatile compound of either the metal or the impurity. Separation can also be achieved by cooling the metal after most of the refining has been completed. This technique, which is called **liquation,** depends on the decrease in the solubility of the impurity as the temperature of the molten metal is decreased. These and other methods are considered in detail in Chapter 5.

1.5.1.3. Refractories

The molten metal must be contained by materials that are stable at the high temperatures involved and that can protect the vessel holding the melt from attack by the slag and the metal. This protection is usually achieved by lining the vessel with an oxide or a mixture of oxides that has a higher melting point than the operating temperature of the furnace. The refractory oxides most commonly used for this purpose and their fusion temperatures are listed in Table 1.3.

Beryllia, thoria and zirconia are very expensive and are only used for special applications. In bulk metal production, the other oxides in the list, or mixtures of them, are used for the lining. Monoliths of the pure oxides

TABLE 1.3
Refractory Oxides and Their Melting Temperatures

Oxide	Melting temperature, °C
Al_2O_3	2050
BeO	2570
CaO	2600
MgO	2800
SiO_2	1710
ThO_2	3050
ZrO_2	2700

have to be prepared by fusion or sintering, the cost of which can only be justified for very special applications. Less-pure oxides are generally used and small additions of other oxides are made to produce a small amount of liquid during firing of the refractory. This subsequently solidifies as a cement in a similar way to the bonding of the ore particles during sintering. In consequence, these refractories begin to soften at temperatures well below the fusion temperatures of the pure oxides and cannot be used to contain the melt at temperatures above about 1800° C. However, the cost is only about one quarter of the cost of fused, pure oxide refractories.

When an oxide refractory is in contact with an oxide slag, there is a risk that the refractory will be attacked by the slag and dissolve some of the lining. Both the volume and the viscosity of the slag may then be increased. This attack is minimized by choosing a refractory that is as compatible as possible with the slag. Thus, a silica-rich slag is contained in a silica refractory (acid practice), while a lime-rich slag is contained in a magnesia or a dolomite lining (basic practice).

Carbon can also withstand very high temperatures and is used as a refractory, for example, to contain the molten iron in the blast furnace, and to contain the molten salt and the metal in the electrowinning of aluminum. It is readily gasified by oxygen, however, and its use is limited to conditions where the partial pressure of oxygen is very low. It cannot be used in contact with metals that readily form carbides. It is usable in the blast furnace only because the iron is almost saturated with carbon.

1.5.1.4. The Economic Aspects of Pyrometallurgy

The capital cost of a pyrometallurgical furnace and the associated plant is very high. If coke is used as the fuel and/or the reductant, there is also a high cost for construction of the coke ovens and the gas handling equipment. Large amounts of energy are consumed in heating the charge to the required temperature and in the endothermic reduction reactions. More energy is consumed during the refining stage. This results in the copious evolution of the "greenhouse" gas, CO_2, when carbon fuels are combusted to provide the heat.

Although the refractory lining is selected to be as compatible as possible with the slag, corrosion and erosion cannot be prevented completely; so the lining requires frequent patching and periodic renewal. Cost is also incurred in the acquisition of fluxes with the requisite purity and in the disposal of the large amounts of slag that are produced. Environmental constraints on the emission of carbonaceous and noxious gases, dust and other pollutants also add to the total cost (see Chapter 8).

On the positive side, the chemical reactions proceed more rapidly at high temperature and the pyrometallurgical extraction and refining of the metals is achieved much more quickly than by other techniques that operate at or about ambient temperature. The processes are more amenable to large-scale production and can produce more metal per man-hour than is achieved by other methods. Consequently, the lowest production costs are usually obtained by this mode of production. In most cases, a metal must be melted at some stage in manufacture in order to produce castings of the required shape or to consolidate it by rolling, forging or other forms of mechanical working. The metal is molten at the end of the pyrorefining stage and can be cast directly into the required shape without the supply of any additional energy. Most of the sensible heat in the exhaust gases from the furnaces can also be recovered and utilized for space heating, electricity generation, etc. so that the total amount of energy consumed in metal production by pyrometallurgical methods is no greater and is often considerably less than is required for production by other means.

Pyrometallurgical refining is sometimes not capable of producing the metal in a sufficiently pure form. Care is also required to prevent or minimize contamination of the metal from other sources. Thus, the coal or coke used as an energy source or as a reductant contains variable amounts of sulfur. This may dissolve in a metal that contains very little of this element derived from the ore and raise the total sulfur content to an unacceptable level (e.g., in iron production in the blast furnace). Additional refining steps are then required to remove the extraneous contamination. The ash remaining after combustion of the coal or coke dissolves in the slag and must be taken into account when determining the fluxes that must be added to obtain the required slag composition.

1.5.2. Hydrometallurgy

It is possible to extract most metals from the ore using a commercial adaptation of the natural processes of chemical weathering that occur in the formation of ore deposits. Rainwater percolating through the ore left in the ground, or stacked in heaps on the surface, can dissolve or **leach** the metallic minerals, which are then recovered from the effluent. This is an attractive way of recovering metals from low-grade deposits that are not sufficiently rich to justify pyrometallurgical treatment, since the cost of beneficiation is avoided.[4] However, it is a very slow process and the concentration of the minerals in the effluent is also very low, making recovery difficult. The reactions can be accelerated by percolating a solution containing a strong acid or alkali through the ore. Further acceleration can be obtained by the presence of certain naturally occurring bacteria—*thiobacillus ferrooxidans*—in the solution. Theoretically, it is possible to recover all the metallic values from the ore if the leach solution is left in contact with the ore for a sufficient length of time.

Capital and operating costs are very low, but the rate of extraction is also very low in comparison with pyrometallurgical methods. Some minerals are not readily leached. Thus, sulfide minerals may be roasted to convert the sulfide to a sulfate (which may be water soluble) or to an oxide, prior to leaching. It is sometimes cheaper, however, to recover a metal from a low-grade concentrate by leaching than by pyrometallurgy. One of the common applications of the technique is the recovery of small amounts of metals and minerals from slags, flue dust and other wastes and effluents. A higher metal recovery from the ore can usually be obtained by leaching rather than by other methods. For example, copper oxide ores are not easily treated by smelting and refining because large amounts of Cu are lost in the slag; thus, leaching is preferred for these ores.

The rate of a chemical reaction increases as the temperature is raised, so the leaching rate can be increased by heating. However, at atmospheric pressure, the maximum temperature of aqueous solutions is limited to less than 100° C to prevent loss of water by boiling. Higher temperatures are possible if the ore is leached under pressure in an autoclave, and this is the standard practice for the extraction of pure alumina from bauxite. The cost of treatment is higher, however, and low-grade ore must first be prepared as a high-grade concentrate to avoid heating and pressurizing large amounts of gangue.

The solvents used for leaching are water soluble and are usually not expensive, but there are exceptions. Gold, for example, can only be leached completely in cyanide solutions. This highlights the problems of toxicity that can occur with the solutions. When toxic chemicals are used, they must be washed out completely from the ore residue before it is discharged to waste, producing large quantities of dilute leach solution from which the metals must be recovered. Reagent costs are conserved by recycling the solutions, but all metals must be removed before the spent liquor is finally discharged.

Leaching is not very selective. Any mineral present in the ore and soluble in the particular acid or alkaline solution used will dissolve, and care must also be exercised to prevent the dissolution of some of the gangue materials. Selective separation can be obtained more readily during the recovery of the metals from the leach solutions.

A wide variety of techniques are used for recovery. The simplest—and cheapest—is known as cementation. Thus, copper can be extracted from an oxide ore with a sulfuric acid solution and the Cu deposited as a loose granular deposit if the solution is then brought into contact with solid steel scrap metal:

$$CuO + H_2SO_4 = CuSO_4 + H_2O \tag{1.13}$$

$$CuSO_4 + Fe = Cu + FeSO_4 \tag{1.14}$$

Other methods include precipitation by changing the pH of the solution or by adding other chemicals, gaseous reduction, solvent extraction and ion exchange (see Chapter 6). The last two methods can be highly selective, but are very expensive. Their use is limited mainly to the recovery of the more reactive metals that are produced in relatively small quantities and that command a high selling price. All of these methods produce a powder requiring heating to consolidate it, either by melting and casting or by a powder metallurgy route. The metal can be recovered in a consolidated form by electrolysis of a concentrated leach solution (see Chapter 7), but it still requires melting to shape it and to obtain good mechanical properties.

1.5.3. Electrometallurgy

Although it is economical to use electricity as the energy source for melting metals, its use for the extraction and refining of metals by electrometallurgical techniques is very expensive. There is a high energy loss in the conversion of fossil fuels into electrical energy. Despite major advances in the generation of electric power, the thermal efficiency of generation from solid or gaseous fuels is still less than 40% (although efficiencies of over 50% may be achieved with the development of a combined cycle using gas or oil firing and steam generation from the waste heat). Further losses arise due to low efficiency in the electrochemical reactions. Thus, the processes tend to be located in countries where abundant supplies of cheap hydroelectric power are available.

The process is essentially the same as that which occurs during the charging of a lead-acid accumulator. Passage of an electrical current between two electrodes immersed in a solution containing the metal causes the metal to be deposited on the negative electrode (cathode) in the cell. The metal is usually contained in an aqueous solution, which may have been produced in a hydrometallurgical operation; but it may also be contained in a mixture of fused salt, as in aluminum and magnesium extraction. In electrorefining, the impure metal forms the anode, which dissolves in the solution, and the pure metal is plated out on the cathode.

The high cost and consumption of electricity results in restriction of electrowinning to the extraction of the very reactive metals that are difficult to produce by other methods. However, in both electrolytic extraction and refining, it is possible to obtain metals in a very pure condition and it is common practice to electrorefine metals produced by pyrometallurgy for the more demanding applications.

Achievement of high purity often requires the purification of the electrolyte to remove other metals that could be coprecipitated on the cathode. The treatment required varies from one metal to another. The elements can be arranged in a sequence (the electrochemical series, see Table 2.1) according to the potential difference required between the electrodes to cause deposition of the metal from a specific solution. Thus, of the common engineering metals, Cu requires a smaller potential difference than either Fe or Zn, such that Cu can be readily refined from all traces of these elements. But Fe and Cu must be removed completely from a solution containing these species before pure Zn can be deposited.

Metals are most commonly electrorefined from a sulfate solution. Gold, silver, the platinum group metals and most nonmetallic elements are insoluble or only sparingly soluble in such solutions at the pH used for refining. Hence, as the anode dissolves, these elements accumulate in a sludge at the bottom of the plating tank. This sludge is a major source of the precious metals, and the values obtained from treatment of the sludge often compensate for the cost of electrorefining.

Where hydrometallurgical extraction and electrorefining are practiced on the same site, the spent electrolyte from electrolysis can be recycled to the extraction plant. Otherwise, the electrolyte must be purified before discharge from the plant.

In contrast to metals produced by pyrometallurgy, the electrorefined product contains virtually no dissolved nitrogen or oxygen gas. However, hydrogen may be generated during electrolysis and the slabs often require remelting to remove dissolved hydrogen. The capital cost per unit of production is high, requiring power generation, conversion to a high amperage, low-voltage DC supply and distribution via low resistance cables and bus bars to the electrolysis cells. Operating costs are also high. When fused salt electrolysis is used for extraction, the salts are very aggressive to the refractory lining, due to both their composition and low viscosity. It is customary to chill a layer of the salt over the lining where it is exposed to the melt, thus protecting the refractory but increasing the thermal losses.

1.6. CHOICE OF PROCESS

It is difficult, if not impossible, to produce the more reactive metals with sufficient purity via a pyrometallurgy route. The remaining metals could be produced by any of the methods that have been described. The choice of route is then determined by the cost of production relative to the market price for the metal. It is influenced by the type and concentration of the metallic mineral in the ore or in the concentrate, the other elements that are present, the purity required in the end product and the scale of the operation.

Reduction and refining by pyrometallurgy, using carbon, CO or H_2 as the fuel and as the reductant, is usually the cheapest route and is best suited to large-scale operation. The high capital cost is offset to some extent in the modern integrated plant, typical of the iron and steel industry, in which the metal is extracted and refined, then continuously cast from the furnace into bars, slab or sheet. Energy available in the exhaust gas from the extraction and refining furnaces is recovered elsewhere in the fabrication processes. The metal leaves the works in semifinished form. Plants of this type are suited to large-scale operation and are not economical when only small quantities of metal are produced.

The major disadvantage of pyrometallurgy is usually the relatively low purity of the product, although the purities that can be obtained are adequate for the majority of applications. It also requires the availability of a suitable fuel and reductant at a reasonable price. Major improvements in the energy efficiency of the processes over the last 2 decades have decreased the energy requirements by as much as 30%; but world supplies of suitable fossil fuels are diminishing, and processes dependent on them must inevitably increase progressively in cost. The recent introduction by many countries of stringent environmental legislation has also created problems for the operator, involving additional costs to remove noxious substances such as SO_2, NO_x, dust and metal fumes from the exhaust gases before discharge to the atmosphere. These additional costs may eventually result in a financial incentive to change to alternative hydro- and electrometallurgical processes for the production of some nonferrous metals.

Many metallic minerals can be dissolved in a cheap solvent, but hydrometallurgy is inherently a slow process and requires vastly more space for a similar throughput. It can be operated economically on a much smaller scale, however, and is better suited to the production of some of the less common metals that are only required in small quantities. The cost of setting up a small-scale operation is relatively low. Environmental restraints are less of a problem. This route may be favored, for example, when low-grade sulfide ores are being processed, since the sulfur can ultimately be converted into sulfuric acid, precipitated as sulfur or neutralized as CaS. It is difficult and costly to remove low concentrations of SO_2 from the exhaust gas when these ores are processed by pyrometallurgy.

Hydrometallurgy is not suitable when the ore contains precious metals in amounts that justify recovery but are not sufficient to justify their extraction by a separate solvation process after the principal metal has been extracted from the ore. With pyrometallurgical processing, these valuable elements are dissolved in the metal that is extracted and can be recovered either by treatment of the molten metal [e.g., desilvering of Pb (see Chapter 5)] or during electrorefining.

Most metals could be produced by electrowinning from fused salts, but this is the most energy-intensive route and the cost can be justified only for the more reactive metals. The process is suitable for relatively large-scale production, as instanced by the production of aluminum and, with closely controlled operation, it is possible to produce metal with a purity of 99.99% in the one operation, eliminating the need for further refining.

Electrorefining is the most common route for treatment of metals extracted by hydrometallurgy. The high recovery of the metal from the ore without the need for extensive grinding and recirculating of a slag and the high purity of the product can result in similar cost for this route to that for the pyrometallurgy processes when low-grade ores are treated. Electrorefining of metal produced by pyrometallurgy is the primary method of production of very high purity metal.

The choice of production route is sometimes also influenced by political factors. A country may wish to establish, or continue, an extractive industry almost regardless of cost and provide subsidies to support operation of an otherwise uneconomical route. For example, 30 years ago the open-hearth furnace process was the major method used for the production of steel. It is very inefficient in terms of both the energy consumption and the rate of production in comparison with modern methods of steel manufacture, and the process is now obsolete in most parts of the world. But open-hearth furnaces are still being operated in the countries that constituted the former Soviet bloc and in a few other countries, and 15% of the total world steel production was still being made by this process in 1991. The cost of steel produced in this way could only compete with that produced by more modern methods with the aid of large state subsidies.

1.7. TYPICAL PRODUCTION ROUTES

Figure 1.6 shows typical routes for the production of the more common metals. These routes are described briefly here. The chemical reactions that occur are examined in detail in later chapters.

1.7.1. Aluminum

Aluminum is produced from bauxite, which is first purified by digesting with caustic soda at 150 to 200° C at a pressure of about 4 atm to separate the alumina from iron oxides, silica and titania (the principal impurities in the ore). Alumina is extracted as sodium aluminate, $NaAlO_2$, leaving a residue of red mud consisting primarily of iron oxides and aluminum silicate. The alumina is then recovered by hydrolysis of the solution:

$$NaAlO_2 + 2H_2O = NaOH + Al(OH)_3 \tag{1.15}$$

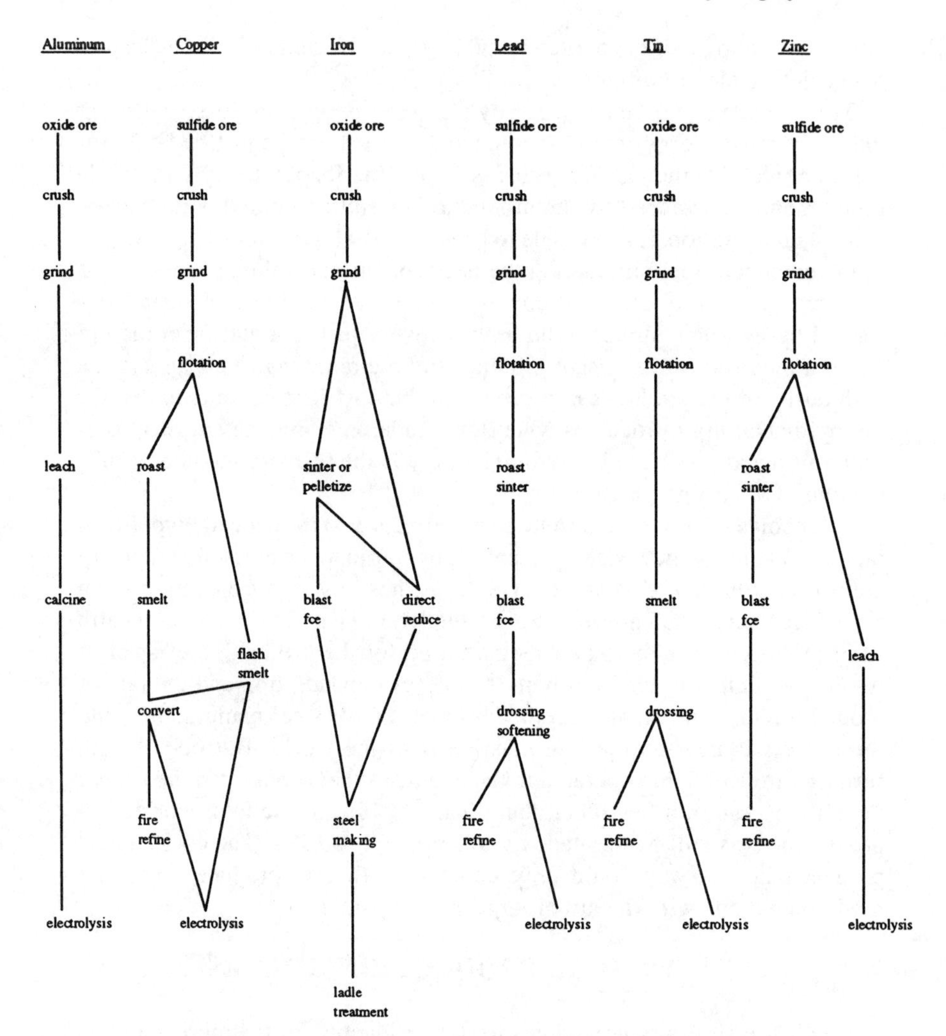

FIGURE 1.6. Principal production routes for the common metals.

The alumina precipitate is filtered off and calcined to expel the combined water. The caustic soda is recirculated after evaporation of the water added for hydrolysis, while the alumina, which now contains very little impurity, is fed to an electrowinning cell, Figure 1.7a. Low concentrations ($<10\%$) of alumina are dissolved in molten cryolite (Na_3AlF_6), to which small amounts of AlF_3 and CaF_2 are added to lower the melting point of the salt to about 900° C. The alumina is electrolyzed between carbon anodes immersed in the salt and a carbon cathode that forms the lining of the cell. Passage of a DC current results in accumulation of molten Al at the cathode, below the salt melt, and the release of oxygen that combines with the carbon

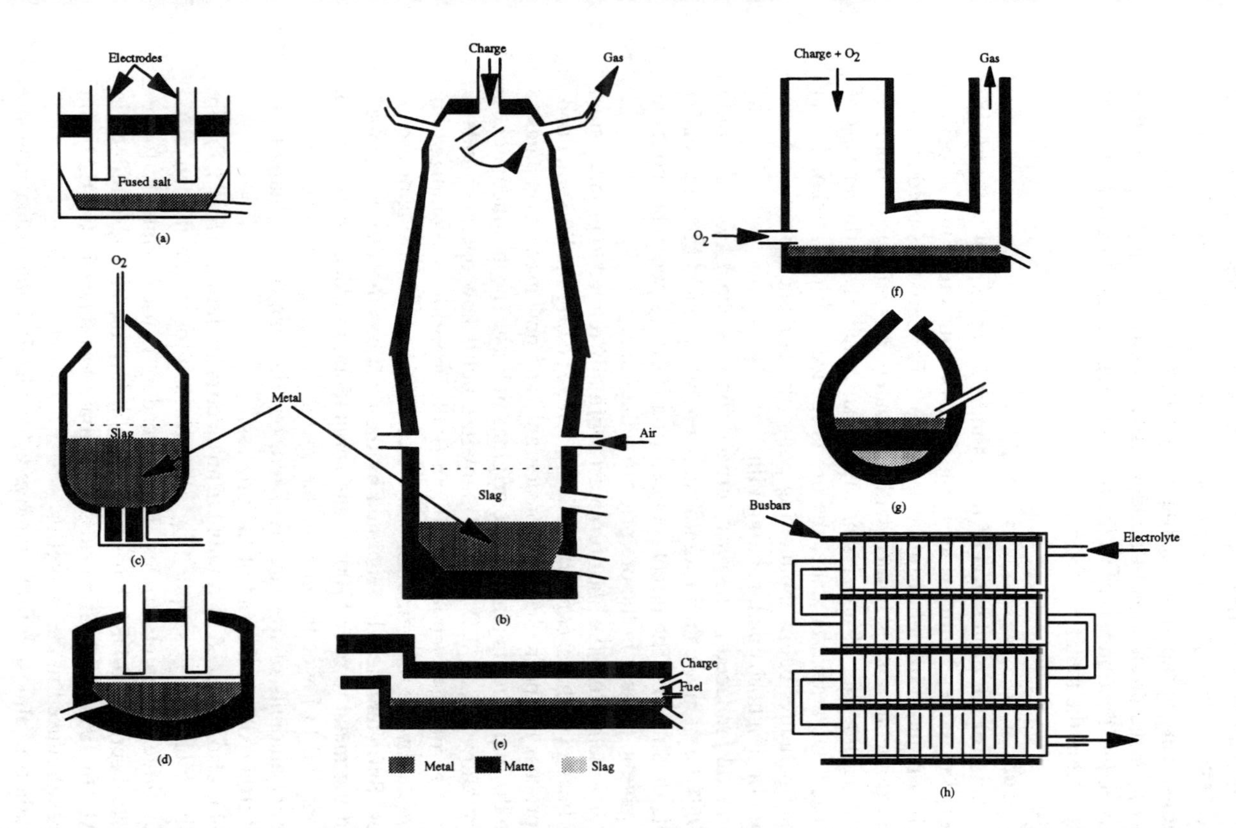

FIGURE 1.7. Schematic diagrams (not to scale) of extraction and refining units. (a) Electrowinning, (b) blast furnace, (c) BOS converter, (d) electric arc furnace, (e) reverberatory furnace, (f) flash smelter, (g) Pierce Smith converter, (h) electrorefining cells.

at the anode to escape as CO gas.[5] The anode is thus consumed and has to be renewed continuously or replaced periodically. The metal produced has a purity of 99.99%, requiring no further treatment for the most demanding operations. This is the simplest route (in terms of the number of stages involved) for the production of a metal. It is also unique in the sense that the bauxite purification serves as the refining stage, preceding reduction of the oxide from the metal.

1.7.2. Iron

The route for the production of iron is more complex, but still requires relatively few stages.[6] Iron oxides in the form of lump ore, sinter or pellets are mainly reduced in a blast furnace, Figure 1.7b. Carbon is used as both fuel and reductant and is burned with preheated, oxygen-enriched air injected through a ring of nozzles or tuyeres surrounding the lower part of the furnace. On a smaller scale, the ore is reduced with H_2 and/or CO at lower temperatures below the melting point (direct reduced iron) in a vertical kiln or in a horizontal, rotating kiln.

The metal produced in the blast furnace contains about 4% C and only about 95% Fe. It may contain significant amounts of Si and Mn and smaller amounts of S and P. The direct reduced iron (DRI) is purer, but contains all the gangue present in the ore prior to reduction.

The hot metal from the blast furnace is normally refined to produce steel in a vertical, cylindrical converter, Figure 1.7c. The original Bessemer converter process has been changed considerably and now uses oxygen gas to oxidize the impurities in place of the original air blast. The modified process was first known as the LD (Lintz-Donowitz), but is now operated under a variety of acronyms such as QBOP and BOS. Impurities are oxidized out of the metal by the oxygen blast and either absorbed into a slag or escape as a gas. Sufficient heat is generated by the reactions to raise the temperature of the metal to over 1700° C and scrap metal is added, both to act as a coolant and to lower the cost of the metal. Blowing is stopped before significant amounts of iron are oxidized, but the slag may contain 10 to 20% of iron oxide at the end of the process.

Steel is also produced by refining in an electric arc furnace (EAF), Figure 1.7d, in which heat is supplied by arcs struck through the molten metal between three carbon electrodes. This procedure is used primarily for refining steel scrap and DRI, but some blast furnace metal is also refined in the EAF. In 1991, 28% of world steel output was made in the EAF and 56% was produced in the converter.

If carbon is almost entirely removed, the metal contains unacceptable amounts of oxygen and elements such as Al, Mn and Si (which form more stable oxides than Fe) are added in small amounts to deoxidize the metal. It is possible in this way to produce steel containing about 99.5% Fe. For higher purity, the metal is transferred while still molten to a ladle treatment station where, under closely controlled conditions, new slags are formed and additions made to remove more S and/or P. Agitation in a partial vacuum removes hydrogen and oxygen from the metal. These treatments can produce 99.9% Fe. For very high purity with greater freedom from nonmetallic inclusions, and particularly when large amounts of readily oxidizable alloy elements are added, the metal may be remelted either in a vacuum furnace or in contact with yet another slag in the electro-slag refining process.

1.7.3. Copper

A still more complex route is followed for the production of copper.[7] The principal path for the processing of sulfide ores is indicated in Figure 1.6. Since Cu is now produced from ores containing an average of less than 1% of the metal, the ore must first be concentrated by flotation to produce a feed suitable for the reduction processes. The concentrate may contain as much, or even more Fe than Cu.

Sulfide concentrates are first roasted in a multiple-hearth furnace or in a fluidized bed to remove part of the sulfur. The ore is then charged to a reverberatory or electric smelting furnace, Figure 1.7e, in which the iron is partially oxidized to FeO and combined with silica to form a molten slag that also contains the gangue materials, while the copper is retained together with some of the iron in a sulfide phase called a **matte.** Conditions are closely controlled to minimize Cu loss to the slag. More modern plants combine the roasting and smelting operations in a single-stage, flash smelting process, Figure 1.7f, in which the concentrate is injected with oxygen gas into the furnace. Provided that the temperature in the smelter is high enough to initiate the reaction, the exothermic oxidation of the FeS provides all the heat required to raise the temperature of the charge to about 1250° C. The oxidation reactions are spontaneous (hence the name of the process) and are completed before the particles are entrained in the slag and matte layers. The production rate with flash smelting is markedly higher than with the traditional route, but the process is more difficult to control and consequently the slag tends to contain more Cu.

The matte is next converted to metallic Cu in a Pierce Smith converter, Figure 1.7g, a horizontal, cylindrical vessel with submerged gas inlets (tuyeres) and an open mouth that serves as the gas outlet. Oxygen is blown through the matte to oxidize the sulfur, the reaction heat supplying all the thermal energy required. When oxidation is complete, the metal contains 98 to 99% Cu, 0.8 to 1.0% O and smaller amounts of sulfur and other elements. This is called "blister copper."

In a separate, fire refining furnace, the residual S and other impurities such as Pb and Zn are carefully oxidized into a slag or into the atmosphere. Oxygen dissolved in the Cu is next removed by bubbling a hydrocarbon gas through the bath to lower the oxygen content to about 0.05%. The product is called "tough pitch copper." Alternatively, phosphorus is added to form P_2O_5 in the slag. A small amount of P then remains in the metal, which is sold as phosphorus-deoxidized copper. For the more demanding applications (e.g., for high electrical conductivity) and particularly when the ore contains As and/or Sb, the blister Cu is cast into slabs and electrorefined, Figure 1.7h. Copper ores often contain small quantities of Ag, Au and the precious metals, and the recovery of these elements from the electrolyte sludge partially compensates for the cost of electrorefining.

Oxide ores are leached with H_2SO_4. The extract is treated with a chelating agent (see Chapter 6) to increase the concentration of $CuSO_4$ in the solution to a level suitable for recovery of the Cu by electrolysis.

1.7.4. Lead

Lead production is an example of a procedure in which impurities are removed in a series of discrete stages.[8] Extraction again commences with concentration of the ore, but the metal content of the concentrate is usually much higher than in Cu concentrates. Sulfide ores are sintered to aggregate the particles and convert the sulfides to oxides. Fluxes are incorporated in the sinter to lower the melting temperature of the slag formed during reduction in the Pb blast furnace.

The blast furnace is small in comparison with the Fe blast furnace and the operating temperature is lower; thus, the upper portions can be formed from water-cooled steel panels, thereby reducing the refractory costs. Carbon is used as both fuel and reductant. The atmosphere is reducing with respect to Pb, but oxidizing to Fe and Zn, so that these impurities accumulate in the slag. If the ore contains arsenic, a separate layer of **speiss** (mainly iron arsenide) may form between the metal and the slag.

Lead and its compounds create severe environmental problems (*cf.* the move to unleaded petrol). The risk of Pb release into the atmosphere by volatilization during sintering has led to the development of alternative processes such as flash smelting, in which the concentrate is oxidized and then reduced in one furnace.

After separation from the slag, the metal is cooled slowly to just above the melting point. The solubility of impurities such as As, Fe, Sb and Sn in the metal decreases as the temperature falls and, since these elements have a lower density than Pb, the excess amounts rejected from solution float to the metal surface. Copper is also removed as Cu_2S by adding small amounts of sulfur. During this treatment, the metal is exposed to air and some Pb is oxidized, forming a scum or dross containing 50 to 60% Pb in which the impurities are contained. The operation is called **drossing.**

The metal is next transferred to a furnace with a large surface area per unit volume and reheated to about 700° C for softening. Compressed air is blown into the metal to oxidize As, Sb and Sn. The Sb content can be reduced to less than 0.02% by this treatment.

The slag is separated from the metal and the temperature is adjusted to 450 to 480° C for recovery of any Ag dissolved in the metal by adding Zn dust to form an Ag-Zn dross, which floats to the surface of the metal (Parkes process). More silver can be recovered by lowering the temperature of the metal to just above the melting point to decrease the solubility of Ag in the Pb. Any gold in the ore is recovered at the same time. After removal of the Ag from the bath, the lead is heated to about 600° C under high vacuum. The Zn in solution is volatilized and condensed on chill plates, from where it is recycled; or it is removed as a chloride by the addition of $PbCl_2$. Small amounts (0.1 wt%) of NaOH and $NaNO_3$ are then stirred into the melt to scavenge remaining traces of Sb and Zn if the intended use requires very low concentrations of these elements. The product is 99.99% Pb.

The slag or dross removed from the metal at each of the refining stages contains significant amounts of Pb in either the metallic or oxide form. The Pb is recovered by recycling the slag to an earlier stage, usually after recovery by separate processes of any other valuable elements contained therein. For special applications, or when bismuth is present in the ore, the Pb is electrorefined in a fluosilicate electrolyte in which Bi is insoluble.

On a smaller scale, the ore is leached with $FeCl_3$ or with ammoniacal NH_2SO_4 after an oxidizing roast, and is then electrolytically recovered.

1.7.5. Tin

The extraction of tin is relatively simple in comparison with the production of Cu and Pb, but the high price of the metal justifies extensive treatment of the slags which are formed to recover all the values.[9] The principal sources of Sn are the placer deposits of cassiterite which are found in the Far East. These are readily processed to yield concentrates containing about 70% Sn and relatively few impurities other than Fe. Some concentrates produced from mined ores require a preliminary roasting to remove As_2O_3 by volatilization.

The concentrate is smelted with coal as the reductant in a reverberatory or an electric furnace. The slag that separates contains large amounts of Sn and Fe, so it is removed and resmelted with additions of coal and lime to produce an Sn-Fe alloy (hard head) containing 4 to 8% Fe. The slag is removed, silica is added and the Fe is oxidized to form a fayalite slag. Alternatively, the Fe can be removed by adding ferrosilicon, relying upon the wide area of liquid immiscibility in the Fe-Si-Sn ternary system, which results in the Sn separating from the Fe-Si layer.

For many applications, the only refining required is liquation, in which the impure metal is placed at the top of a sloping hearth and slowly heated to over 900° C. The Sn melts and drains to the bottom of the hearth, leaving a residue of Fe containing some Sn. Air may be blown through the Sn to preferentially oxidize any Fe still present.

Ores recovered by underground mining usually contain other elements that are extracted in the Sn and necessitate more extensive refining treatments. Thus, Cu or Pb can be removed by adding S and Cl_2, respectively, to form insoluble compounds that float in a dross. For higher purity, and particularly if Bi is present in the ore, the metal is electrorefined.

1.7.6. Zinc

The production of metals having a high vapor pressure at elevated temperatures can be illustrated by consideration of the extraction of zinc.[10] Zinc ores frequently contain large amounts of Pb, and both metals can be recovered simultaneously in the Zn blast furnace. The ore is concentrated and sintered to agglomerate the particles and eliminate sulfur. The sinter and flux are then charged to the blast furnace, again using coke as both fuel and reductant. Lead is reduced preferentially in the upper part of the furnace and trickles through the charge to accumulate in the hearth. Zinc is not reduced until the ore arrives at the hottest zone of the furnace. The Zn volatilizes as rapidly as it is reduced, and the vapor rises with the ascending reducing gases to the top of the furnace. The gases pass through sprays of molten Pb in a condensing chamber. The Zn dissolves in the Pb and is recovered by liquation as the temperature of the Pb is decreased in a separate chamber. The Zn is skimmed off and the Pb is recycled to the condensing chamber. Very close control is necessary to prevent reoxidation of the Zn vapor before it is condensed.

The product contains some Pb and Fe, but these impurities are often acceptable if the metal is used for galvanizing. If cadmium is present in the ore, it is recovered in the metal and can be removed by vacuum distillation since, at any temperature, the vapor pressure for Cd is higher than for Zn.

Alternatively, the concentrates are calcined and leached with H_2SO_4. The $ZnSO_4$ solution is neutralized with ZnO to precipitate Fe and Sb. Zn dust is added to precipitate Cd and Cu. After filtration to remove the solids and concentration to increase the Zn content, the metal is recovered by electrolysis to produce Zn with a purity of 99.9%. This is now the principal route for the extraction of Zn from ores containing little or no Pb.

1.7.7. Other Metals

Some other metals are produced by routes similar to those that have been described. Thus, magnesium is extracted from seawater as $Mg(OH)_2$, which is chlorinated to form $MgCl_2$ and the chloride is reduced by fused salt electrolysis, in similar manner to the production of Al.[11] One major difference is that Al has a higher density than the molten salt and sinks to the hearth, where it is protected from reoxidation by the salt layer. Mg has a lower density than the salt and floats on top of the bath, so the atmosphere above the melt must be controlled to prevent reoxidation of the metal. Magnesium can also be produced from the mineral dolomite ($CaO \cdot MgO$).

Manganese ores can be reduced to form an Fe-Mn ferroalloy in the blast furnace. However, the temperature that can be attained in the furnace is inadequate for the production of the pure metal, and higher purity is obtained by electrolysis of a leach solution.

Nickel is often associated with Fe, Cu and small amounts of precious metals in the ore and the production route is similar to that for Cu up to the matte stage.[12] The iron is oxidized completely to form a fayalite ($2FeO{\cdot}SiO_2$) slag, and the matte is overblown to yield Cu_2S, Ni_3S_2 and a small amount of metallic phase into which are partitioned the precious metals in the ore. The matte is then cooled slowly to allow time for a coarse separation of crystals of Cu_2S and Ni_3S_2. The sulfides are separated by comminution and flotation, the metallic phase is recovered magnetically, and the nickel-rich sulfide fraction is sintered to form an oxide agglomerate that is smelted to produce a crude Ni product. Cu, As and other impurities must be removed before the pure metal is produced by electrorefining; or, the Ni is extracted from the matte by a chlorine leach and the solution is purified by precipitation of As, Fe and Pb and solvent extraction of Cu before electrodeposition.[13]

Chromium can be extracted as an Fe-Cr ferroalloy by smelting with carbon in an EAF, or by metallothermic reduction with silicon when low-C ferrochromium is required. For higher purity, the ore is roasted with sodium carbonate to form sodium chromate, which is then recovered by leaching and dehydration. After purification, the chromate is heated with sulfur to convert the sodium to Na_2SO_4 and the Cr to an oxide. The oxide is reduced to metal either by the thermite process or by electrolysis from a chloride solution.

Titanium and zirconium are obtained by heating the ore concentrates with carbon in an EAF to form the metal carbides. The carbides are reacted with chlorine to form the chlorides $TiCl_4$ and $ZrCl_4$, which are then reduced to the metal by reaction with Mg to form a more stable chloride (Kroll process). $MgCl_2$ and excess Mg are removed by leaching to leave a metal sponge that is consolidated by vacuum melting or by a powder metallurgy route.

1.8. FACTORS AFFECTING THE PRICE OF METALS

1.8.1. Cost

Most of the important factors that determine the relative costs for the production of metals have now been identified. Thus, with reference to ore deposits, the high metal content of Al and Fe ores would be expected to lead to a lower metal cost than for, say, Cu which is produced from very lean ores. Whereas abundant supplies of rich Al and Fe ores remain to be exploited, the proven reserves of Cu ores are more limited and may not last for more than about 50 years at the present rate of consumption.[14] Hence,

TABLE 1.4
Typical Total Energy Requirements for the Production of Metal from the Ore

Metal	GJ tonne^{-1}
Al	220–250
Cu	95–110
Fe	25–30
Pb	20–25
Ni	140–150
Sn	145–155
Zn	60–70

the average grade of Cu ores will decrease more rapidly with time than the average grade of Al and Fe ores, and the price differential should increase correspondingly.

As the metal content of an ore decreases, it becomes more important to recover as much of the mineral from the ore as possible, with consequent increase in the cost of beneficiation. The costs incurred do not depend solely on the metal content, however, because they rise with increasing dispersion of the mineral in the bedrock and with increasing resistance to fracture of the rock. The capital and operating costs for comminution can represent three quarters and more than half, respectively, of the total cost for beneficiation when the ore has to be ground to less than about 100 μm to release the minerals.

The range and complexity of the extraction and refining routes used for the production of the metals has been indicated. Each additional stage in the manufacturing route increases the capital cost for the provision of the processing equipment and for the transfer of the metal between the stages. The operating costs are also raised through change in a number of factors, including increases in the number of operatives required, in the fuel and/or chemical consumption (and in the consequent pollutant and effluent problems), and in the number of production units that have to be maintained. Production costs thus rise with increases in the number of stages, except when the total production time is decreased and hence the productivity is increased sufficiently to compensate for the cost of separation of the process into discrete stages.

The energy consumed in recovering a metal from the ore varies over a wide range from one metal to another, depending on the relative stability of the metallic minerals (see Chapter 2) and the processes used for extraction and refining. Typical values for good practice with ores of average composition and currently being processed for the recovery of each metal are given in Table 1.4.

The energy consumption is increased when a lower grade ore, or an ore containing a higher concentration of noxious impurities, is processed. It is also increased when the metal is extracted from less suitable minerals. Thus, the energy required for the production of Ni from oxide ores may be at least twice the value given in the table for processing the more common sulfide ore.[15]

Aluminum production shows the highest energy consumption, which is a consequence of the energy required for extraction of a metal from a very stable oxide by electrowinning. In marked contrast, the production of steel requires only about one tenth this amount of energy. This difference is partly due to the very large size of the iron and steel manufacturing units, but mainly because the oxides of Fe are much less stable than alumina and can be reduced and refined by more simple pyrometallurgical techniques. Thus, although both metals are extracted from ores with not too dissimilar metal content and more stages are involved in the production of steel than of Al, the cost of the production of steel is less than half the cost of Al.

The energy consumed in Cu production is roughly half that required for Al. However, the very lean ores from which Cu is extracted, with the consequent high cost of beneficiation, and the multistage route required for the preparation of the pure metal result in a production cost that is nearly twice that of Al. Nickel consumes more energy for production than is required for Cu, Ni ores are scarce, and the manufacturing route is even more complex, resulting in a further marked increase in production cost.

Tin is extracted primarily from very lean placer deposits. These contain few impurities but they require extensive work to produce a rich concentrate. The metal can be extracted by simple pyrometallurgical techniques, but extensive treatment of residues and slags to recover metal values is justified by the very low mineral content in the ore. The total energy consumption is inflated by the energy consumed in beneficiation and recycling and is similar to that for Ni production. World production of Sn is very low in comparison with the other nonferrous metals listed in Table 1.4, and the production units have a small annual capacity. In consequence, Ni and Sn have the highest production costs in this group.

Lead and zinc ores have higher metal contents than those used for Cu, Ni and Sn production, reducing the beneficiation costs, and the production route is of intermediate complexity. Both metals often occur in the same ore and can be extracted simultaneously in the blast furnace. Zn forms a more stable oxide than Pb, and the energy required for Zn extraction is more than twice the amount required for the production of Pb. There is an added bonus for Pb in the value of the silver bullion recovered. Both metals are produced on a small scale when compared with Fe. As a result, Zn costs more than Pb, but both metals cost more to produce than Fe and less than Cu.

The cost of production of any metal rises as the purity required is increased. It increases rapidly when electrorefining is necessary to lower the residual concentration of harmful impurities to only a few parts per million (ppm).

Metals can be recycled when their useful life in manufactured products has been exhausted. Since it is already present in metallic form, the energy required to regenerate the metal from the scrap is less (and often significantly less) than for the production of new metal from the ore (see Chapter 8). Recycling thus conserves the stock of ores in the ground, resulting in less rapid deterioration in the average metal content of the ores that are mined than otherwise would be the case, and provides more opportunity for improvements in manufacturing efficiency to compensate for the cost increase arising from a worsening average ore grade. Costs may be lowered when some recycled scrap metal can be incorporated in the charge for the refining of new metal.

The actual cost of any one unit operation or process (e.g., Cu smelting or electrorefining) may vary markedly from one company to another. It depends on the quality, cost and composition of the input materials, the quality and composition of the product, the scale and the efficiency of the operation, local labor costs and many other factors.

The total cost of production in any enterprise can be divided into fixed and variable costs. The former includes the cost of buildings, capital equipment and other fixed assets that must be provided before the operations can be conducted. The variable costs include the cost of the raw materials and energy consumed in the process. Labor is charged under the ''variable'' heading in a market economy (i.e., one with minimal state interference) since the size of the labor force can be increased or decreased to respond to changes in the demand for the product. In many countries, however, legislation prohibits a reduction in manpower, and salaries are then part of the fixed cost. Inevitably, when a weak demand results in a restricted output, production costs per tonne of product increase more rapidly in these countries than in others that allow a market economy to operate. Production can then be maintained only with the aid of massive state subsidies. Higher charges must be levied (and/or wage levels restrained) during periods of full production to recoup the subsidy.

Other similar factors, which are not related directly to the extraction and refining operations, may be included in the total cost. Hence, it is difficult to obtain a true indication of the actual operating costs for any process.

1.8.2. Price

The order in which the costs of metal production vary has now been explained. Additional factors affect the selling price. The price of most nonferrous metals is determined daily by trading on Metal Exchanges such as the London Metal Exchange and is published in financial newspapers. Metal is offered for sale at the Exchange on behalf of the manufacturers and is sold at a price customers are prepared to pay. The price for each metal varies from day to day in response to variations in supply and demand.

It has been claimed[16,17] that there is an inverse relationship between the price and the consumption of a metal. This dependence is sometimes used by the metal producers to exert some degree of control over the metal price. When a metal is supplied mainly or entirely by a small number of producers, it is feasible for them to inflate the price by limiting production, or to stabilize the price by holding part of the production in reserve (i.e., buffer stock) if the demand falls and to release the stock when the demand rises.

The latter practice is effective over a relatively short time span, but can fail when major changes occur in supply or demand. Demand can be curtailed by an industrial recession as in the early 1990s, resulting in a fall in price. Supplies can be severely disrupted by unforeseen events, such as a prolonged strike or political disturbance in one of the major centers where the ore is mined or the metal is extracted, resulting in a price rise. Thus, the buffer stock of Sn proved to be inadequate when supply was disrupted in the 1970s and the metal price rose rapidly. This encouraged exploitation of smaller and lower grade deposits that were uneconomical to work at the previous controlled price. The reopening of the Cornish tin mines was one consequence of this price rise. More Sn became available for sale after the new sources had been brought into production. Meanwhile, the users of the metal were actively seeking ways to reduce their requirement for Sn. The major outlet for the metal is in the manufacture of tinplate. Can manufacturers were induced by the high price to develop a thinner coating of Sn that gave corrosion protection to the cans equivalent to that obtained previously with much thicker coatings. For some applications, Sn was replaced completely by a coating of chromic oxide. As a result, the amount of metal available for sale eventually exceeded the reduced demand, and the market price fell rapidly. Many of the new producers, including the Cornish tin mines, were forced into liquidation.

A similar cyclic pattern has occurred in the price of nickel, production of which is controlled by a small number of major producers. Nickel is classed as a strategic metal and various governments built up stockpiles of the metal to protect against the interruption of supplies at the time of the Korean War. This caused an increase in price, followed by the development of more ore deposits and attempts by the user to substitute other metals for Ni (e.g., the use of Mn in stainless steel). These changes led to a fall in the market price for the metal, which accelerated when the stockpiles were released after the war. Disruption of supply from one of the major production plants in the late 1960s again led to a marked increase in price. The cycle was repeated with the development of fresh sources and the substitution of other elements, or simply the deletion of Ni from the alloy composition, followed by a rapid fall in price when full output was restored. The major market for Ni is for the manufacture of austenitic stainless steel and world production of this material increased almost twofold during the 1980s. Rising demand for Ni resulted in a rapid price rise, again stimulating the development of new production capacity, and this was followed by a fall in price as industry moved into recession at the end of the decade.

It is more difficult to achieve control over the market when a metal is produced by many independent companies located in different countries, but it is not impossible. Thus, the export prices for steel were controlled by the International Steel Cartel during the 1930s.[18] The cartel collapsed at the outbreak of the Second World War and has not been reformed but, for a period during the 1980s, the price of steel in Western Europe was fixed by the European Commission.

Political action can also cause a significant change in the price of a metal. Following the demise of the Soviet Republic and the ending of the Cold War, the CIS released large stocks of aluminum, gold and titanium onto the markets to obtain hard currencies for other purchases. This resulted in oversupply and a significant fall in the price of those metals during the early 1990s.

It is apparent, then, that the price of metals, relative to each other, may vary markedly over a short time span and for a variety of reasons. A more meaningful picture of the price changes is obtained by averaging the price of each metal over successive periods of, say, 5 years for a span of about 50 years. When the resultant data are corrected for inflation, it is evident that the real price for any metal has either remained constant or has fallen steadily throughout the 20th century (see, for example, Reference 19).

This trend is the cumulative result of changes in many factors, but is primarily due to improved efficiency in the concentration processes, producing richer grades of concentrate with lesser amounts of undesirable elements, and to improved efficiency of the extraction and refining routes. The latter has arisen partly from an increase in the capacity of the production units with a reduction in the man-hours per tonne of metal produced and a decrease in energy consumption.

A major factor affecting the trend has been the development of a scientific understanding of the variables controlling the rate and extent of the chemical reactions that occur during processing and how the variables can be changed to achieve the required objectives more rapidly and consistently. In the first half of this century, control of a process was exercised by "rule of thumb," based on the accumulated experience of the operatives. The situation has changed dramatically over the past few decades until, today, reliable predictions can be made, for example, of the changes in the control parameters that must be imposed and the time required to achieve a specific change in metal composition with minimum loss of metal in the slag. Sensors have been developed that measure changes as they occur in the operating variables and provide data for updating the predictions as the reactions progress. Some operations are now controlled completely by computer, removing the need for continuous assessment and adjustment by the operator. This change has been aided by the development of sensors that can provide almost instantaneous readouts of the state of some of the controlling parameters (see Chapter 9). Automatic control is developing rapidly, but there is a need for more extensive and reliable data to use as input to the computer programs and for a more extensive range of sensors to monitor continuously a wider range of variables.

FURTHER READING

Burt, R. O. and Mills, C., *Gravity Concentration Technology,* Elsevier, Amsterdam, 1984.

Fuerstinau, D. W., Ed., *Flotation of Sulphide Minerals,* Elsevier, Amsterdam, 1985.

Goldschmidt, V. M., *Geochemistry,* Clarendon Press, Oxford, 1958.

Hayes, P. C., *Process Selection in Extractive Metallurgy,* Hayes Publishing, Sherwood, Queensland, Australia, 1985.

Wills, B. A., *Mineral Processing Technology,* 2nd ed., Pergamon, Oxford, 1981.

Chapter 2

THEORETICAL PRINCIPLES—1 THERMODYNAMICS

The preparation and concentration of ores, prior to reduction to the metallic form, is dependent mainly on the physical and mechanical properties of the minerals. In contrast, the reduction of the minerals and the purification of the metal produced are mainly chemical processes involving chemical reactions. Two concepts are very useful in characterizing those reactions.

Thermodynamics can be used as an aid in the selection of the specific chemical reactions that can achieve the desired objective. This approach shows whether or not a particular reaction can occur under any chosen set of imposed conditions and can identify the concentrations of the various species that coexist in each phase when equilibrium is attained (i.e., the extent of a reaction). However, it does not provide information from which it is possible to directly deduce the rate at which a reaction will proceed toward the equilibrium condition.

The study of the **kinetics** of reactions can provide the latter information. This can be roughly subdivided into **chemical kinetics,** which provides information on the rate of formation and rupture of the bonds between atoms or molecules, and **mass transfer kinetics,** which identifies the rate at which the various species move around in the system under consideration. In general, reactions are slow between solid substances, but are much more rapid when the substances are molten or gasified. In many cases, the equilibrium condition is approached quite rapidly when molten reactants are brought into contact at elevated temperatures. Flash smelting is an example of almost instantaneous attainment of near-equilibrium. But there are other examples, as in the reduction of silica from the slag in the iron blast furnace to produce Si dissolved in the molten Fe, where equilibrium is not fully achieved even after lengthy periods of contact between the molten phases.

Hence, it is necessary to consider both the thermodynamic and kinetic aspects of the various reactions in order to manipulate and control the extraction and refining operations.

2.1. THERMODYNAMIC RELATIONS

Many textbooks explain in detail the principles of chemical thermodynamics (see, for example References 20 to 24), so only a brief résumé of those aspects which are important for an understanding of extraction metallurgy is presented here.

2.1.1. Thermodynamic Systems

It is conventional to consider the conditions that exist and the changes that occur within the relevant portion of the universe, which is called the "thermodynamic system." The system is enclosed by real or imaginary boundaries across which some form of transfer of mass or energy may or may not be permitted. A homogeneous system contains only one phase, such as a gas or a liquid or solid solution, whereas a heterogeneous system contains more than one phase.

The components of a system are the minimum number of pure substances from which the system can be formed. The **state** of the system is defined completely by fixing the values of an appropriate number of the variables of state. Thus, the state of a homogeneous system is defined when the four state variables—temperature (T), pressure (P), volume (V) and composition—are fixed. Composition may be expressed in various ways, including weight percent, mole fraction [$N_A = n_A/(n_A + n_B)$], molal concentration (m_A = number of gram-atoms, gram-ions, etc. of component A per 1000 g solution), and molarity (C_A = number of moles of A dissolved in 1 liter of solution).

The state of the system is changed when one or more of the variables is changed. The system may be transferred from one state to another by some kind of reaction. From the thermodynamic viewpoint, the progress of the reaction is regarded as either reversible or irreversible. In a reversible reaction, the system remains virtually at equilibrium throughout the process and can be restored to the initial state without any permanent change (e.g., in the heat content) in the surroundings. In contrast, a spontaneous process is irreversible and restoration to the initial state requires an exchange of heat and/or work with the surroundings. Since the principal application of thermodynamics in extraction and refining is the determination of the distribution of various species between the phases when equilibrium is approached, we shall be concerned primarily with reversible conditions.

For convenience in the consideration of changes occurring in a system, it is useful to define thermodynamic terms dependent on the state variables. These should be mathematically exact differential terms that are dependent only on the initial and final states of the system and independent of the path chosen to achieve the final state. For further simplification, it is assumed that either P or V in the system is held constant. In the majority of extraction and refining operations, the pressure within the system is at, or close to, atmospheric pressure and constant pressure relations are then adequate to describe the state of the system.

2.1.2. Free Energy, Enthalpy and Entropy

At constant pressure, the thermodynamic driving force for a reaction is defined by the sum of the Gibbs free energy, or available energy, G, for each of the participating species. G is defined by the relation:

$$G = H - TS \tag{2.1}$$

where H is the enthalpy or heat content of the species and S is the entropy or unavailable energy. The enthalpy, in turn, is dependent on the internal energy (U), P and V:

$$H = U + PV \tag{2.2}$$

This is a mathematical consequence of the **First Law of Thermodynamics,** which states simply that energy can neither be created nor destroyed. If the change in enthalpy accompanying a change in a system is regarded as a measure of the energy produced by the change, then the entropy change can be regarded as a measure of that part of the energy which is consumed by internal rearrangement within the system and is not available for doing useful (i.e., external) work.

The absolute value of the internal energy cannot be measured, so it is necessary to define a reference state for the enthalpy, to which an arbitrary numerical value can be assigned. By convention, this is chosen as zero for each pure substance in the form in which it is stable at 25° C (298.16 K). The enthalpy at any other temperature can then be calculated using data for the heat capacity (C_p) of the substance at constant pressure:

$$H_{(T_2 - T_1)} = \int_{T_1}^{T_2} C_p \mathrm{d}t \tag{2.3}$$

Likewise, the enthalpy change for a reaction can be determined from the summation of the heat capacities of the products and the reactants **(Kirchhoff Equation):**

$$\Delta C_p = \Sigma C_{p(\text{products})} - \Sigma C_{p(\text{reactants})} \tag{2.4}$$

and

$$\Delta H = \Sigma C_p \mathrm{d}T \tag{2.5}$$

If one of the components of the reaction undergoes a change of phase (e.g., from solid to liquid or from liquid to gas between the reference temperature and the temperature under consideration), it is necessary to add (for the products) or subtract (for the reactants) the latent heat of transformation (L_t) at the transformation temperature, T_t:

$$\Delta H_{T_2} = (H_{T_2} - H_{T_1}) = \int_{T_1}^{T_t} \Delta C_{p(T_1 \rightarrow T_t)} dT \pm L_t + \int_{T_t}^{T_2} \Delta C_{p(T_t \rightarrow T_2)} dT \tag{2.6}$$

These equations can be used for heat balance calculations, as illustrated in Worked Examples 1, 2 and 3.

The enthalpy change for a reaction is equal to the heat change accompanying the reaction at constant pressure or, more simply, the **heat of reaction.** When H is positive, the reaction is endothermic and heat must be supplied if the reaction is to proceed. A negative value signifies an exothermic reaction that may occur spontaneously. However, there are many examples of exothermic reactions that are not spontaneous. The simple explanation for this behavior is that the enthalpy change is caused by the system trying to minimize its potential energy, and this is opposed by the system attempting simultaneously to minimize its kinetic energy. The latter is measured by the entropy term in Equation 2.1.

The natural tendency of a system is to attempt to achieve the maximum degree of randomness or disorder consistent with the imposed conditions; but this may be prevented by the accompanying enthalpy change. For example, when two gaseous species are in contact, the molecules of the two species mix together in random order and do not remain separated as discrete entities. Entropy can be regarded as a measure of the randomness that exists in the system. Disorder increases during a spontaneous process and the **Second Law of Thermodynamics** can be expressed in a form that shows the consequent change in the entropy:

> "The entropy increases when a spontaneous change occurs in a system which is isolated from its surroundings and entropy is a maximum at equilibrium."

In a more generalized form, this means that the entropy of the universe is constant at equilibrium and is greater than zero for a spontaneous process. The entropy change in the universe is equal to the difference between the entropy changes in the system and in the surroundings; i.e.,

$$\Delta S_{(\text{universe})} = \Delta S_{(\text{surroundings}} - \Delta S_{(\text{system})} \geq 0$$

Thus, the entropy change for a reaction is a measure of the amount of energy consumed in achieving the final state of disorder of the atoms or molecules participating in the reaction. This energy cannot be converted into useful work, so it is a measure of the unavailable energy.

The entropy change accompanying a change of temperature for a component or for a reaction can be calculated in analogous manner to the change in the enthalpy. In contrast to enthalpy, however, there is a fixed zero level for entropy, which is defined by the **Third Law of Thermodynamics:**

> "The entropy of any substance which is in complete internal equilibrium is zero at the absolute zero of temperature (0 K)."

In consequence, the entropy of a substance at a temperature T is given by the relation:

$$S_T = S_o + \int_{T_o}^{T_t} \frac{C_{p_1}}{T} \, \mathrm{d}T \pm \frac{L_t}{T_t} + \int_{T_t}^{T_T} \frac{C_{p_2}}{T} \mathrm{d}T \tag{2.7}$$

where L_t and T_t again signify the latent heat of transformation at the transformation temperature and $S_0 = 0$, signifying complete order of the atoms or molecules at 0 K.

The free energy of a system was expressed in terms of H and S in Equation 2.1. It follows that the temperature dependence of G can be expressed in terms of C_p. For a component that undergoes no phase change in the temperature range considered:

$$\Delta G = G_{T_2} - G_{T_1} = \int_{T_1}^{T_2} C_p \mathrm{d}T - T \int_{T_1}^{T_2} \frac{C_p}{T} \mathrm{d}T \tag{2.8}$$

It is readily shown that G decreases when a reaction occurs spontaneously and equals zero when the system arrives at the equilibrium state. This is expressed in the form:

$$\Delta G \leq 0 \tag{2.9}$$

which signifies that the change in G is less than zero (i.e., a negative quantity) for a spontaneous process.

A similar set of relations to Equations 2.1 through 2.9 can be written in terms of the heat capacity at constant volume (C_v) for reactions that occur under constant volume conditions.

G, H and *S* are described as *extensive* state properties, which means that their values are additive **(Hess' Law)**. Thus, for the reaction:

$$Pb + \tfrac{1}{2} O_2 = PbO \tag{2.10}$$

the change in a state property is given by:

$$X = X_{PbO} - X_{Pb} - \tfrac{1}{2} X_{O_2} \tag{2.11}$$

where X represents *G, H* or *S*. Note that the values obtained relate to the formation of 1 mol PbO from 1 mol Pb and $\tfrac{1}{2}$ mol oxygen gas. The free energy, etc. of formation of the oxide per mole of O_2 is equal to $2X$. Now, the reactants and products in this reaction could be pure or they could be mixed with other substances. The values of the extensive properties are dependent on the state in which each of the participants exists, so it is convenient to identify a reference or standard state for the components. By convention, the standard state is usually defined as the most stable form of the pure substance that can exist at the particular temperature and pressure under consideration. The values of the state properties for substances in their standard states are designated as the standard free energy, etc. and are identified by the symbols G°, H° and S°.

Heat capacity data for pure substances are often tabulated in the form:

$$C_p = a + bT + cT^{-2} \tag{2.12}$$

When relations of this form are inserted into Equation 2.6, it is apparent that the variation of H with temperature is nonlinear. Since Equation 2.12 is also used for the calculation of G and S, these terms also vary nonlinearly with temperature. C_p equations are rarely accurate to better than ±0.5%, however, and the error band is often much larger. For our purposes, therefore, all three state properties can usually be expressed as linear relations, without significant error, between the temperatures at which phase changes occur. Extensive tabulations of critically assessed data for C_p and for standard enthalpies, entropies and free energies of elements and compounds, as a function of the temperature, are available in the literature.[25–30] Selected data are also available from computer banks that are frequently updated, such as the MTDATA source at the National Physical Laboratory, Teddington, England. Other databases have been described in a recent review paper.[31]

2.2 SOLUTIONS

The standard state properties apply only to pure elements and compounds. In extraction and refining operations, the objective is to produce a metal with the required purity, but the metal first produced from the ore is usually very impure and several other elements may be dissolved in it. The purity is increased by refining, but only approaches 100% toward the end of the treatment. The slags formed during pyrometallurgical operations always contain a number of components. Hence, additional relationships are required to fully describe these complex solutions.

2.2.1. Partial Molar Free Energy

The total free energy, G', of a multicomponent solution containing n_A, n_B, n_C . . . n_i moles of components A,B,C, etc. can be expressed as:

$$G' = f(P,T,n_A,n_B,n_A \ldots n_i) \tag{2.13}$$

If a very small quantity, ∂n_A, of component A is added to the system at constant P and T, the total free energy is changed by an amount $\partial G'$. The ratio of these two quantities is called the **partial molar free energy** of component A ($\overline{G}_A$) in the solution:

$$\overline{G}_A = \left[\frac{\partial G'}{\partial n_A}\right]_{(T,P,n_B,n_C \ldots n_i)} \tag{2.14}$$

This quantity is also called the "Gibbs chemical potential" and represented by the symbol μ_A.

At constant P and T, the total free energy of the system is equal to the product of the number of moles and the partial molar free energy for each component:

$$G' = \Sigma n_i \overline{G}_i \tag{2.15}$$

and the free energy per mole of the system is:

$$G = \Sigma N_i \overline{G}_i \tag{2.16}$$

where N_i is the mole fraction of component i and G is the integral molar free energy of the system. If Equation 2.15 is differentiated without holding the composition constant, then subtracted from Equation 2.14, and finally divided through by $(n_A + n_B \ldots n_i)$, the product is the **Gibbs-Duhem** equation:

$$N_A \, d\overline{G}_A + N_B \, dG_B \ldots + N_i \, d\overline{G}_i = 0 \tag{2.17}$$

When the variation with composition of the integral molar free energy of a system is known, the partial molar free energies of the constituents can be evaluated from this relation, either mathematically or using a graphical method. Consider a binary (two-component) system. Complete differentiation of Equation 2.16 and subtraction of Equation 2.17 for a system with components A and B yields:

$$dG = \overline{G}_A \, dN_A + \overline{G}_B \, dN_B \tag{2.18}$$

Multiplying throughout by N_A/dN_B $(= -N_A/dN_A)$ and adding Equation 2.16 gives:

$$\overline{G}_A = G + (1 - N_A)\frac{dG}{dN_A} \tag{2.19}$$

$$\overline{G}_B = G + (1 - N_B)\frac{dG}{dN_B} \tag{2.20}$$

These equations describe the tangent to the G-N_B or G-N_A curve at the composition $(1–N_B)$ and $(1–N_A)$. The intercepts of the tangents at $N_A = 1$ and $N_B = 1$ give the values of $\overline{G}_A$ and $\overline{G}_B$ for the solution containing N_A mole fraction of component A, Figure 2.1.

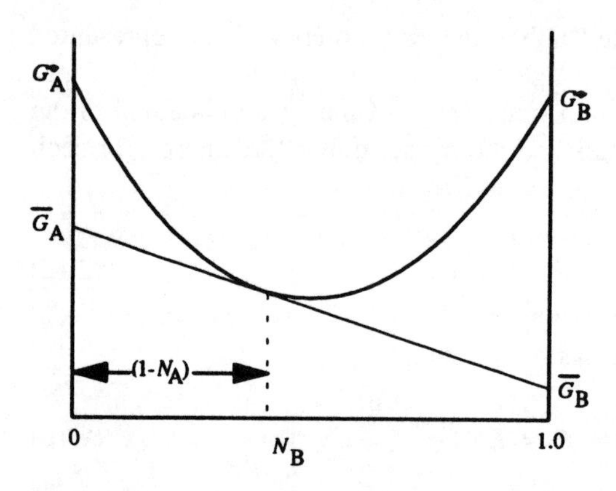

FIGURE 2.1. Tangent intercept method for the determination of the partial molar free energies of the constituents of a binary solution.

Similar relationships can be derived for the partial molar enthalpy and entropy of the components. They are related to $\overline{G}_i$ by the analogous form of Equation 2.1:

$$\overline{G}_i = \overline{H}_i - T\overline{S}_i \tag{2.21}$$

2.2.1. Fugacity

An ideal gas is defined as one that obeys the equation of state, $PV = RT$, where R is the gas constant (8.3143 J K^{-1} mol^{-1}). This relationship is used in the determination of the free energy of a gas. It can be shown that the free energy change for a system which does only reversible mechanical work is given by:

$$dG = VdP - SdT \tag{2.22}$$

If the work is done isothermally by the expansion of 1 mole of an ideal gas:

$$dG = VdP = RT\,dP/P = RT\,d\ln P \tag{2.23}$$

But no real gas obeys the ideal gas relation at all temperatures and pressures. A quantity called the fugacity, f, is therefore defined such that $fV = RT$ under all conditions and

$$dG = RT\,d\ln f \tag{2.24}$$

where ln signifies logarithm to the base e. On integration at constant temperature,

$$G = RT\ln f + I \tag{2.25}$$

where I is an integration constant, the value of which is chosen to make $f \rightarrow P$ as $P \rightarrow 0$.

The free energy change accompanying the expansion of a gas from the standard pressure of 1 atm to any other pressure is equal to the change in the partial molar free energy of the gas. This is shown by integrating equation 2.24 between the two pressure limits to obtain:

$$\Delta \bar{G} = (\bar{G} - G^{\ominus}) = RT \ln \left[\frac{f}{f^{\ominus}} \right] \tag{2.26}$$

2.2.3. Activity

All condensed substances have an equilibrium vapor pressure that changes with variations in T, P and composition. In similar manner to the way in which the behavior of a real gas is compared to the behavior of an ideal gas, it is useful to consider the actual behavior of a component in a solution relative to the way that it would behave in an ideal solution. It is convenient to make this comparison in terms of the vapor pressure of the substance. Since the vapor is a gas that may not exhibit ideal gas behavior, however, the actual comparison is made in terms of the fugacity of the vapor.

An ideal liquid or solid solution is defined as one in which the fugacity of the vapor of each substance is directly proportional to the concentration (atomic or mole fraction) of that substance:

$$f_A = f_A^{\ominus} N_A \tag{2.27}$$

where $f_A^{\ominus}$ is the fugacity of the pure substance at the temperature considered. This is a statement of **Raoult's Law.** The ratio $f_A/f_A^{\ominus}$ is called the activity (a) of the substance such that in an ideal solution:

$$a_A = N_A \tag{2.28}$$

In real solutions, the fugacity of the vapor and hence the activity of a component may be greater or less than that for an ideal solution. These are classed, respectively, as positive and negative deviations from ideal behavior. A measure of the deviation is given by the **activity coefficient** (γ), which is defined as:

$$\gamma_A = \frac{a_A}{N_A} \tag{2.29}$$

Thus, $\gamma = 1$ for an ideal solution, is greater than unity for a positive deviation from ideality, and is less than unity for a negative deviation. Both a

and γ can be regarded as correction factors that relate the actual to the ideal behavior of a species in solution.

Substituting Equation 2.27 into Equation 2.26, for component A in a solution,

$$\Delta\overline{G}_A = (\overline{G}_A - G_A^{\circ}) = RT \ln a_A \tag{2.30}$$

which relates the activity to the partial molar free energy. It follows that as $\overline{G}_A = G_A^{\circ}$ for pure A, the activity is unity when the component is present in its standard state.

The left-hand side of Equation 2.30 is called the **partial molar free energy of mixing,** or the relative partial molar free energy of the component and is designated as $\overline{G}_A^M$. The integral molar free energy of mixing for a binary solution of A and B is given by:

$$\begin{aligned} G^M &= \overline{G}_A^M N_A + \overline{G}_B^M N_B \\ &= (\overline{G}_A N_A + \overline{G}_B N_B) - (G_A^{\circ} N_A + G_B^{\circ} N_B) \end{aligned} \tag{2.31}$$

Similar relationships hold for the relative partial molar and integral molar entropies and enthalpies of mixing.

2.2.4. The Equilibrium Constant

The free energy change for the formation of a pure oxide, MO, from a pure metal, M, and oxygen gas at 1 atm pressure, i.e.,

$$2M + O_2 = 2MO \tag{2.32}$$

is given by:

$$\Delta G^{\circ} = 2G_{MO}^{\circ} - 2G_M^{\circ} - G_{O_2}^{\circ} \tag{2.33}$$

because all the constituents are present in their standard states. If the metal and the oxide are not pure, but are dissolved respectively in other metals and oxides and if the pressure of oxygen is not unity, then the free energy change is:

$$\Delta G = 2\overline{G}_{MO} - 2\overline{G}_M - \overline{G}_{O_2} \tag{2.34}$$

Subtracting Equation 2.33 from Equation 2.34 yields Equation 2.35, since $\Delta G = 0$ at equilibrium.

$$-\Delta G^{\circ} = 2(\overline{G}_{MO} - G_{MO}^{\circ}) - 2(\overline{G}_M - G_M^{\circ}) - (\overline{G}_{O_2} - G_{O_2}^{\circ}) \tag{2.35}$$

Substituting the appropriate forms of Equation 2.30 for the first two terms and Equation 2.21 for the third term on the right-hand side yields:

$$\Delta G = \Delta G^{\ominus} + RT \ln\left(\frac{a_{MO}^2}{a_M^2 \cdot f_{O_2}}\right) = \Delta G^{\ominus} + RT \ln\left(\frac{a_{MO}^2}{a_M^2 \cdot p_{O_2}}\right) \quad (2.36)$$

assuming ideal behavior of the gas.

Since $\Delta G^{\ominus}$ has a precise numerical value at any given temperature and pressure, it follows that the term in brackets is also constant and describes the activities and fugacities, or the partial pressures for ideal gases, of the constituents of the reaction at that temperature and total pressure. It is designated accordingly as the thermodynamic equilibrium constant, *K*, for the reaction; and, since $\Delta G = 0$ at equilibrium, it follows that:

$$\Delta G^{\ominus} = -RT \ln K \quad (2.37)$$

The equilibrium constant is temperature dependent. From Equation 2.22, at constant temperature and pressure, one obtains:

$$\left(\frac{\partial G^{\ominus}}{\partial T}\right)_p = -\Delta S^{\ominus} \quad (2.38)$$

and since $G^{\ominus} = H^{\ominus} - TS^{\ominus}$, it follows that:

$$\Delta G^{\ominus} = \Delta H^{\ominus} + T\left(\frac{\partial \Delta G^{\ominus}}{\partial T}\right)_p \quad (2.39)$$

Dividing throughout by $1/T^2$ and rearranging:

$$\left(\frac{\partial(\Delta G^{\ominus}/T)}{\partial T}\right)_p = -\frac{\Delta H^{\ominus}}{T^2} \quad (2.40)$$

Substitution of Equation 2.37 then gives the temperature dependence of *K* (the **van't Hoff Isochore**):

$$\left(\frac{\partial \ln K}{\partial T}\right)_p = \frac{\Delta H^{\ominus}}{RT^2} \quad (2.41)$$

$\Delta H^{\ominus}$ usually does not vary rapidly with temperature and, over a small range, it is often adequate to assume that it is temperature independent. Then,

$$\ln\left(\frac{K_{T_1}}{K_{T_2}}\right) = \frac{\Delta H_T^{\ominus}}{R}\left(\frac{1}{T_2} - \frac{1}{T_1}\right) \tag{2.42}$$

or

$$\Delta H_T^{\ominus} = \frac{RT_1T_2(\ln K_{T_1} - \ln K_{T_2})}{T_1 - T_2} \tag{2.43}$$

where $\Delta H_T^{\ominus}$ is the standard enthalpy change in the temperature range under consideration.

A reaction is highly feasible when K is greater than 1 (i.e., $\Delta G^{\ominus}$ is negative). Conditions are much less favorable when $\Delta G^{\ominus}$ is positive and $K < 1$. Some reaction may still occur, but the extent of reaction is usually very low. From Equation 2.41, it is evident that K increases with temperature for endothermic reactions, favoring the forward reaction. The converse applies for exothermic reactions.

2.25. Solution Models

It is necessary to know the relationship between the activity and the concentration for each of the participants in order to solve Equation 2.36 and determine the equilibrium concentrations of the components of a reaction. Many such relationships have been determined experimentally. However, there are gaps remaining in our knowledge and, in these cases, recourse can be made to models of solutions that allow estimates to be made of the values.

2.2.5.1. The Ideal Solution

This model simply assumes that the activity of each species is equal to the atom or mole fraction concentration of the species in the solution. Hence, the activity is independent of the temperature. Furthermore, from Equation 2.30,

$$\overline{G}_A^M = \overline{G}_A - G_A^{\ominus} = RT \ln N_A \tag{2.44}$$

thus, it follows from Equation 2.40, expressed in terms of the free energy and enthalpy of mixing, that $\overline{H}_A^M$ is zero. Hence,

$$\overline{S}_A^M = -\frac{\overline{G}_A^M}{T} = -R \ln N_A \tag{2.45}$$

so the integral molar entropy of mixing is given by:

$$S^M = (N_A \overline{S}_A^M + N_B \overline{S}_B^M + \ldots N_i \overline{S}_i^M) \tag{2.46}$$

$$= -R(N_A \ln N_A + N_B \ln N_B + \ldots N_i \ln N_i) \tag{2.47}$$

and the integral molar free energy of mixing is:

$$G^M = RT(N_A \ln N_A + N_B \ln N_B + \ldots N_i \ln N_i) \tag{2.48}$$

2.2.5.2 *The Regular Solution Model*

This model gives a better fit to many metallic systems. Very few real solutions approximate ideal behavior. Many systems show approximately ideal S^M, but exhibit small, finite values of H^M. Hildebrand[32] introduced the regular solution model to describe these systems. For a binary solution, from Equations 2.30 and 2.31,

$$G^M = RT(N_A \ln a_A + N_B \ln a_B) \tag{2.49}$$

However, if the entropy of mixing is ideal, then:

$$S^M = -R(N_A \ln N_A + N_B \ln N_B) \tag{2.50}$$

Therefore,

$$H^M = RT(N_A \ln a_A + N_B \ln a_B) - RT(N_A \ln N_A + N_B \ln N_B) \tag{2.51}$$

$$= RT(N_A \ln \gamma_A + N_B \ln \gamma_B \tag{2.52}$$

and

$$\overline{H}_A^M = RT \ln \gamma_A; \quad \overline{H}_B^M = RT \ln \gamma_B \tag{2.53}$$

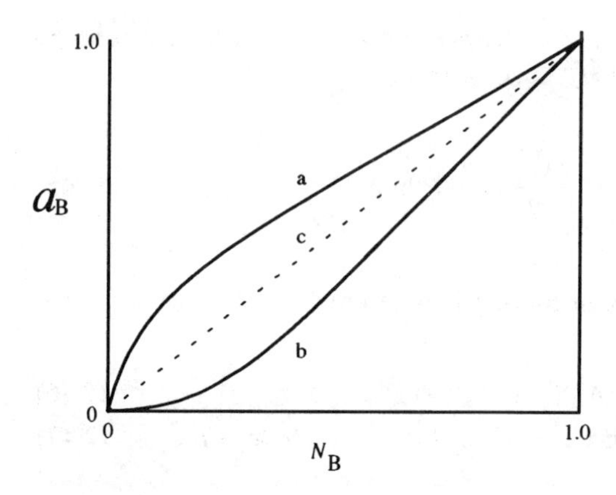

FIGURE 2.2. Activity-concentration curves showing (a) positive and (b) negative departure from (c) ideal behavior.

Thus, the value of the activity coefficient (and hence the activity) for each component of a regular solution can be determined from the relative partial molar heats of mixing.

In a regular binary solution of A and B atoms or molecules, the mixing enthalpy arises from the difference between the mutual attraction of the two species. If there is no difference between the forces of attraction to form A-A, B-B and A-B groupings, H^M is zero, the solution exhibits ideal behavior and the activity varies linearly with concentration, Figure 2.2. The species tend to segregate into separate clusters when the A-A and B-B attraction is greater than A-B. The atoms or molecules can then escape more readily into the vapor phase than from a random mixture. As a result, the vapor pressures and hence the activities of the species show a positive deviation from ideality (curve a in Figure 2.2). The activity coefficients are greater than unity and, from Equation 2.53, the mixing enthalpy is positive, so formation of the solution is endothermic. Conversely, a negative deviation from ideality (curve b in Figure 2.2) is caused by a preferential association of unlike species to form A-B pairs. The activity coefficients are then less than unity, the mixing enthalpy is a negative quantity and solution formation is exothermic.

Preferential association, however, to form either A-A and B-B or A-B pairs reduces the randomness of the solution and hence lowers S^M. Thus, the regular solution model, which assumes a purely random value for S^M, is an approximation that becomes increasingly invalid as the deviation from ideality is increased. It is a useful concept that can often be used to obtain at least an indication of the direction in which the activity of a component of a metallic solution may be expected to deviate from ideal behavior. It is less useful for estimation of activities in molten oxides (e.g., slags), in which large ring or chain molecules may form and modify the intermolecular attractions.

2.2.5.3 Excess Quantities

These are used to evaluate the difference between real and ideal solutions. For any component in a solution, the difference between its free energy in the actual solution and in an ideal solution is equal to the difference between Equations 2.30 and 2.44. This difference is called the **excess partial molar free energy, $\overline{G_A^E}$,** of the component:

$$\overline{G_A^E} = (G_A^{\ominus} + RT \ln a_A) - (G_A^{\ominus} + RT \ln N_A) \quad (2.54)$$

$$= RT_A \ln \gamma_A \quad (2.55)$$

Hence, from Equation 2.38, expressed in terms of excess partial molar quantities:

$$\left(\frac{\partial \overline{G_A^E}}{\partial T}\right)_p = -\overline{S_A^E} \quad (2.56)$$

where $\overline{S_A^E}$ is the excess partial molar entropy. However, the entropy is ideal for a regular solution and $\overline{S_A^E} = 0$. It follows that $RT \ln \gamma_A$ is independent of temperature and:

$$RT \ln \gamma_A = \overline{H_A^M} = \alpha_A N_B^2 \quad (2.57)$$

$$RT \ln \gamma_B = \overline{H_B^M} = \alpha_B N_A^2 \quad (2.58)$$

where α is a constant that is also often independent of temperature. When this is so, $\alpha_A = \alpha_B$.

These equations can be used to extend experimental measurements made over a limited composition range to a wider range if the solution conforms to regular behavior. They can also be used to calculate activities at other temperatures from measurements made at only one temperature. This practice is adopted frequently in the absence of more extensive data, even when there is no evidence that the solution is regular.

2.2.6. Dilute Solutions

Frequently in extraction processes and particularly toward the end of the refining operations, the solutes dissolved in the solvent metal are present at only low concentrations. A useful approximation can then be applied. At very high dilution, the vapor pressure of a solute, B, is directly proportional to its concentration in the solution:

$$p_B = \kappa N_B \quad (2.59)$$

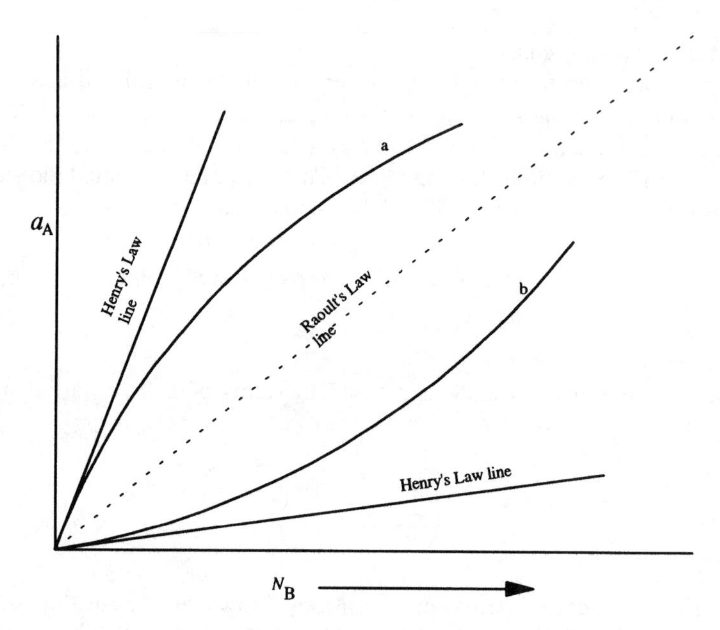

FIGURE 2.3. Henry's law behavior of solutes showing (a) positive and (b) negative deviation from ideal behavior.

where κ is a constant for any particular solution. This is a statement of **Henry's Law.** From Equation 2.27, it follows that:

$$a_B = \kappa N_B \tag{2.60}$$

at high dilution. The constant, κ, is equal to the slope of the tangent to the curve of activity vs. composition at infinite dilution, Figure 2.3. However, the ratio a_B/N_B is equal to γ_B. At infinite dilution, this is designated as γ_B^o.

Henry's Law is a limiting condition, which is approached by all solutes at extreme dilution. In practice, it is sometimes found that $\gamma_B = \gamma_B^o$ up to at least 1 atom percent concentration. One activity measurement within this range is then sufficient to define the activity-composition relationship for the whole of the range. Using the Gibbs-Duhem Equation (Equation 2.17), it is readily shown for a binary solution that the solvent obeys Raoult's Law (i.e., shows ideal behavior) over the composition range in which the solute obeys Henry's Law. At higher concentrations, when the solute activity deviates from the Henry's Law line, a positive deviation from Henry's Law behavior occurs when the solute shows negative deviation from ideal behavior and vice versa.

A useful corollary of this law is **Sievert's Law,** which relates the solubility of a gas (e.g., H_2, N_2 or O_2) in a liquid or a solid solution to the partial pressure of that species in the gas phase, assuming ideal gas behavior:

$$N_X = \kappa \, n \sqrt{P_{X_n}} \tag{2.61}$$

where n is the number of atoms per mole of gas. Thus, the solubility of diatomic nitrogen gas in a metal is given by:

$$N_N = \kappa \, P_{N_2}^{1/2} \tag{2.62}$$

Generally, gases are only sparingly soluble in liquid metals and even less soluble in the solid state, and Sievert's Law adequately describes the solubility.

2.2.7. Alternative Standard States

Thus far, the standard state in which the activity of a species is unity has been chosen conventionally as the pure component in its most stable state at the temperature and pressure under consideration. Quite frequently, this standard state has no real significance when, for example, a pure substance would normally exist either as a solid or as a gas at a particular temperature and pressure, but forms a liquid solution with a solvent under the same conditions (e.g., Cu and S in molten Pb). In any case, when considering solutes in dilute solution, it is often more convenient to select some other standard state such that the activity is unity within the concentration range of interest. Two alternative standard states are commonly used for this purpose, both of which are based on Henry's Law.

2.2.7.1. The Infinitely Dilute Atom Fraction Standard State

This state is defined such that the activity of the solute, B, approaches the atom fraction concentration of the solute as the concentration tends to zero:

$$a_B \rightarrow N_{B(\text{limit}, \, N_B \rightarrow 0)} \tag{2.63}$$

However, the activity of the solute is given by:

$$a_B = \frac{f_B}{f_B^0} = \kappa f_B \tag{2.64}$$

This is effectively a statement of Henry's Law and it follows that $a_B = N_B$ up to the limit where Henry's Law is obeyed by the solute when the activity scale is defined by this alternative standard state. Within this range, the atom fraction can be substituted for the activity of the solute.

2.2.7.2. *The Infinitely Dilute Weight Percent Standard State*

This is analogous to the preceding case, the difference being a change in the concentration scale. Here, the activity is set equal to the weight percent concentration of solute up to the limit to which Henry's Law applies and is equal to unity at 1 wt% of the solute.

2.2.7.3. *The Activity Coefficient*

This must be redefined for the alternative standard states. Any deviation from the equality of activity and concentration (i.e., from Henry's Law) in both alternative standard states is measured in terms of a Henry's Law activity coefficient, *h*, i.e.,

$$h_B = a_B/N_B \text{ or } h_B = a_B/\text{wt\% B} \tag{2.65}$$

depending on the chosen alternative standard state. The ratio of the activities of the solute in the pure substance and the dilute standard states is constant over the concentration range in which Henry's Law applies. At constant composition, for the atom fraction scale:

$$\left[\frac{a_{B(\text{pure})}}{a_{B(\text{dilute})}} = \frac{\gamma_B}{h_B}\right]_{N_B \text{ constant}} \tag{2.66}$$

where the subscripts, pure and dilute, refer to the solute activities in the two standard states. However, in the concentration range where Henry's Law applies, $h_B = 1$ and $\gamma_B = \gamma_B^0$. Hence,

$$\left[\frac{a_{B(\text{pure})}}{a_{B(\text{dilute})}} = \gamma_B^0\right]_{N_B \text{ constant}} \tag{2.67}$$

Similarly, on the wt% scale:

$$\left[\frac{a_{B(\text{pure})}}{a_{B(\text{dilute})}} = \gamma_B^0 \frac{N_B}{\text{wt\%B}}\right]_{N_B \text{ constant}} \tag{2.68}$$

2.2.7.4. *The Standard Free Energy*

For a reaction, the standard free energy is also altered by a change in the standard state chosen for one or more of the constituents. The activities expressed in terms of the alternative standard states are useful only if they can be inserted in the equilibrium relation for a reaction (e.g., Equation 2.36); but $\Delta G^{\ominus}$ has been defined thus far only with reference to the pure stable substance, as in Equation 2.35. If alternative standard states are used, the $G^{\ominus}$ values for the solutes must be changed to correspond to the chosen standard state.

The free energy change accompanying the transfer of 1 mol solute from the pure substance to the infinitely dilute atom fraction standard state, i.e.,

$$B_{(pure)} \rightarrow \mathrm{B}_{(dilute)}$$

is expressed as:

$$\Delta G^{\ominus}_{\mathrm{B}} = G^{\ominus}_{\mathrm{B(dilute)}} - G^{\ominus}_{\mathrm{B(pure)}} \tag{2.69}$$

However, the partial molar free energy of the solute is independent of the chosen standard state; thus,

$$\Delta G^{\ominus}_{\mathrm{B}} = RT \ln a_{\mathrm{B(pure)}} - RT \ln a_{\mathrm{B(dilute)}} \tag{2.70}$$

$$= RT \ln \gamma^{0}_{\mathrm{B}} \text{ (from Equation 2.67)} \tag{2.71}$$

$$= 0 \text{ for an ideal solution, where } \gamma^{0}_{\mathrm{B}} = 1 \tag{2.72}$$

Hence,

$$G^{\ominus}_{\mathrm{B(dilute)}} = G^{\ominus}_{\mathrm{B(pure)}} + RT \ln \gamma^{0}_{\mathrm{B}} \tag{2.73}$$

Likewise, in terms of the dilute wt% standard state:

$$G^{\ominus}_{\mathrm{B(dilute)}} = G^{\ominus}_{\mathrm{B(pure)}} + RT \ln \left[\gamma^{0} \frac{N_{\mathrm{B}}}{\mathrm{wt\%\ B}} \right] \tag{2.74}$$

The ratio (N_{B}/wt% B) can be expressed as a simple number for a dilute solute. In a binary solution of elements A and B, the atomic fraction of the solute (N_{B}) is:

$$N_{\mathrm{B}} = \left[\frac{\mathrm{wt\%\ B/M_B}}{\dfrac{\mathrm{wt\%\ B}}{M_{\mathrm{B}}} + \dfrac{(100 - \mathrm{wt\%\ B})}{M_{\mathrm{A}}}} \right] \tag{2.75}$$

where M_{A} and M_{B} are the atomic weights of the solvent and the solute, respectively. For a dilute solute, the first term in the denominator is small compared to the second term, and the relation can be approximated to:

$$N_{\mathrm{B}} = \mathrm{wt\%\ B \cdot M_A}/100\, M_{\mathrm{B}} \tag{2.76}$$

or

$$N_{\mathrm{B}}/\mathrm{wt\%\ B} = M_{\mathrm{A}}/100\, M_{\mathrm{B}} \tag{2.77}$$

If a solute in a liquid solution normally exists as a solid or as a gas at the same temperature and pressure, the free energy change accompanying the change of state of the pure component must be added to Equation 2.73 or 2.74 to obtain the free energy value appropriate to the chosen standard state. Thus, at the melting point of an element, $G_S^{\ominus} = G_l^{\ominus}$ and the free energy change is zero. Therefore,

$$S_f = L_f/T_f \tag{2.78}$$

where S_f and L_f are the entropy and enthalpy (latent heat) of fusion, respectively, at the melting point, T_f. Assuming that S_f and L_f are independent of temperature, the hypothetical free energy change accompanying the change of a pure substance from solid to liquid at any other temperature is given by:

$$G_f = L_f - L_f/T_f \tag{2.79}$$

2.2.8. Multicomponent Solutions

Calculation of the equilibrium distribution of a solute between a dilute solution of that element in a molten metal and a molten slag or gas would be relatively simple if only binary metallic solutions had to be considered. Unfortunately, the metal usually contains several solutes and each of these may affect the interatomic forces between the solvent and the solute under consideration. In so doing, they change the activity of that solute, even when the other solutes are present only in very low concentrations. Experimental determination of the activity of each solute in any particular solvent in the presence of all probable combinations and concentrations of other solutes and covering the full temperature range of interest in extraction and refining operations would be an almost impossible task. Approximate values of activities in multicomponent solutions can be obtained, however, by making a few simplifying assumptions. The alternative standard states are generally used for this purpose.

Consider a ternary solution containing two solutes, X and Y, in dilute solution. The activity coefficient, h_X, for element X is a function of the concentrations of both elements X and Y. This can be expressed as:

$$h_X = h_X^X + h_X^Y \tag{2.80}$$

where h_X^X is the activity coefficient for X at any specific concentration in a binary solution of the solvent with element X, and h_X is the activity of X at the same concentration of X in the ternary solution. The factor, h_X^Y, is a measure of the effect of Y on h_X in the solution. It is commonly found that, within the limits of experimental error, h_X^Y is a logarithmic function of the concentration of Y, but is independent of the concentration of X. Hence, the concentration dependence of this term can be expressed as:

$$\frac{\partial \ln h_X^Y}{\partial N_Y} = \varepsilon_X^Y \tag{2.81}$$

and

$$\frac{\partial \log h_X^Y}{\partial \text{wt\% Y}} = e_X^Y \tag{2.82}$$

The interaction coefficients, ε_X^Y, and e_X^Y, identify the activity coefficient of X relative to the infinitely dilute atom fraction and to the wt% standard states, respectively.

A similar set of relationships could be defined for h_X in another ternary solution with the same solvent but with some other solute, Z, to give the interaction coefficients, ε_X^Z and e_X^Z. If it is assumed that each of the interaction coefficients is not affected by the presence of other solutes, they can be combined to determine h_X in the multicomponent solution:[33]

$$\begin{aligned} \ln h_X = N_X \frac{\partial \ln h_X}{\partial N_X} &+ \left[N_Y \frac{\partial \ln h_X}{\partial N_Y} + N_Z \frac{\partial \ln h_X}{\partial N_Z} \right] \\ &+ \frac{1}{2} \left[N_X^2 \frac{\partial^2 \ln h_X}{\partial N_X^2} + N_X N_Y \frac{\partial^2 \ln h_X}{\partial N_X \partial N_Y} \right] + \ldots \end{aligned} \tag{2.83}$$

The second and higher-order derivatives become very small at very low concentrations and can be disregarded. Equation 2.81 or 2.82 can be substituted for the first derivatives to give:

$$\ln h_X = N_X \varepsilon_X^X + N_Y \varepsilon_X^Y + N_Z \varepsilon_X^Z \ \ldots \tag{2.84}$$

with respect to the atom fraction standard state, or:

$$\log h_X = \text{wt\% X}\ e_X^X + \text{wt\% Y}\ e_X^Y + \text{wt\% Z}\ e_X^Z \ \ldots \tag{2.85}$$

for the wt% scale. Any number of interaction coefficients can be combined in this way.

Equations 2.84 and 2.85 are strictly valid only at very high dilution, but in practice they are frequently used to compute solute activities at concentrations up to about 1 atom percent or 1 wt%. The errors arising from neglecting second and higher-order derivatives become increasingly important as both the deviation from ideality and the concentration increase, but the uncertainty in the experimental measurements rarely justifies the use of the higher-order terms.

The values of interaction coefficients have been determined most extensively for solutes dissolved in iron,[30,34,35] but compilations for solutes in aluminum,[36] cobalt and copper[37] and nickel[38] are also available.

2.3. GRAPHICAL REPRESENTATION OF THERMODYNAMIC DATA

There are several ways in which the variation of thermodynamic properties with change in the state variables of a system can be portrayed in graphical form. They are usually two-dimensional diagrams in which two of the state variables—T, P and composition, N—or some derivative of them are allowed to vary while the third of these and the fourth state variable, V, are held constant. The representations can be used for visual comparison of data, for rapid evaluation of the conditions required to achieve a chosen objective, and sometimes for the estimation of unknown thermodynamic properties.

2.3.1. Phase Diagrams

A phase diagram is a map which shows how, at equilibrium, coexisting phase boundaries are moved by changes usually in the state variables, N and T, with pressure and volume held constant. The positions of the phase boundaries are determined by the thermodynamic properties of the system: thus conversely, knowledge of a diagram can be used to evaluate, or at least estimate, thermodynamic data. This will be demonstrated with reference to binary phase diagrams, but the same principles apply to systems containing more than two components.

Solving Equation 2.47 for an ideal binary solution, it is found that S^M is symmetrical about a maximum value of 5.8 J mol^{-1} at $N_A = N_B = 0.5$, falling to zero at N_A and $N_B = 1$. From Equations 2.49 and 2.50, $G^M = -TS^M$ and the total free energy of an ideal solution is given by:

$$G = [N_A G_A^\ominus + N_B G_B^\ominus] - TS^M \qquad (2.86)$$

These quantities are represented as a function of composition at constant temperature in Figure 2.4. The intercepts in Figure 2.4c at N_A and $N_B = 1$ correspond to the values of $G^\ominus$ for the two components, referred to the pure substance standard state, at the temperature considered.

2.3.1.1. Complete Liquid and Solid Solubility

This is found in systems that conform fully or closely to ideal behavior. Let the curve in Figure 2.4c represent the free energy for solid mixtures of A and B. A similar curve can be drawn to represent the liquid mixtures.

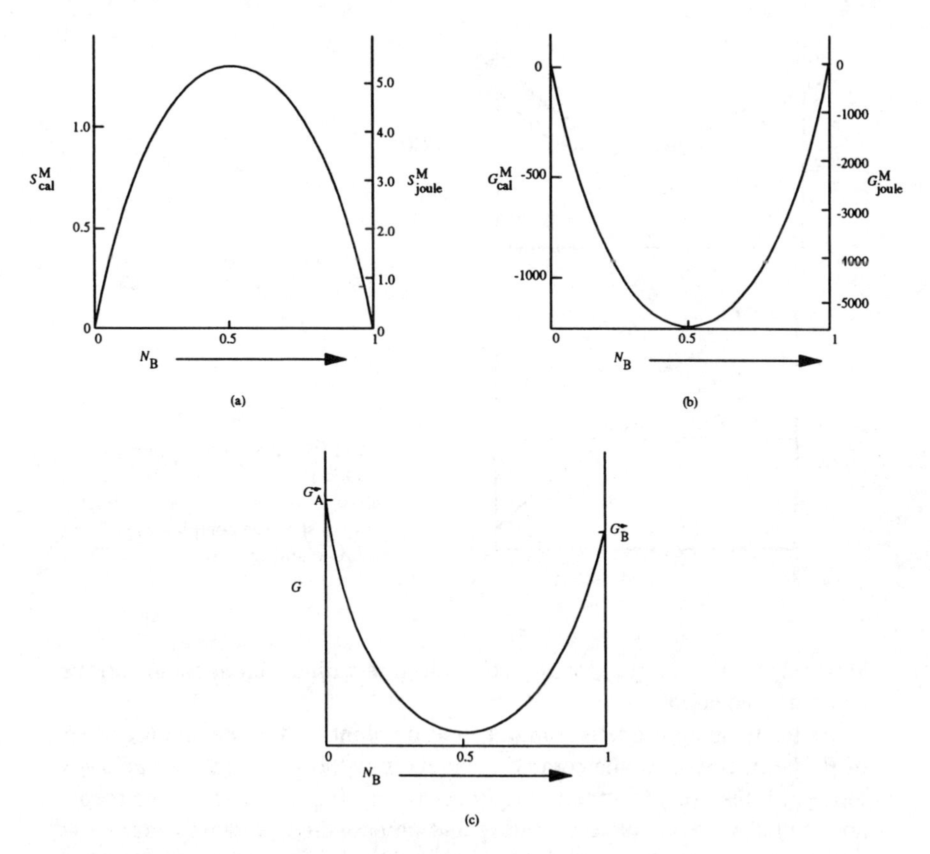

FIGURE 2.4. Variation with composition of (a) entropy of mixing, (b) free energy of mixing and (c) total free energy for an ideal solution.

Thermodynamically, the most stable state is the one with the lowest (i.e., most negative) free energy under the imposed conditions. It follows that at temperatures above the melting points of the two constituents, $G^\ominus_l$ is more negative than $G^\ominus_s$, and the G-N_B curve for the liquid mixture lies below the corresponding curve for the solid at all compositions. Below the melting points of both constituents, the situation is reversed and the G-N_B curve for the solid lies below the curve for the liquid. If B melts at a lower temperature than A (or vice versa), the solid and liquid curves must cross in the temperature range between the two melting points, as shown in Figure 2.5a. To the left of the diagram, the solid is more stable than the liquid, while the liquid is more stable than the solid at the right-hand side. There is an intermediate region, between x and y, where a common tangent touches both curves. Between these points, a mixture of solid with composition x and liquid with composition y has a lower free energy than a mixture of

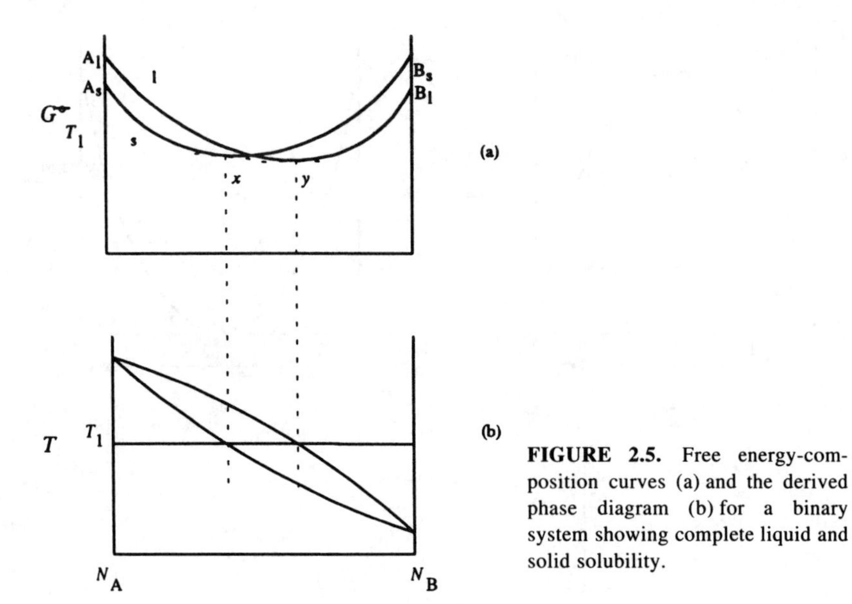

FIGURE 2.5. Free energy-composition curves (a) and the derived phase diagram (b) for a binary system showing complete liquid and solid solubility.

either all solid, all liquid, or any other mixture with different compositions of liquid and solid.

As the temperature falls from the melting point of A to the melting point of B, the intercepts of the common tangent with the liquid and solid curves move progressively from left to right across the diagram. If these intercepts are plotted in terms of temperature and composition, a phase diagram is obtained showing complete liquid and solid solubility, Figure 2.5b. This type of diagram is found only in those systems that do not deviate markedly from Raoult's Law. Hence, in the absence of experimental data, the assumption of ideal behavior (i.e., activity = atom fraction) for systems showing this type of behavior should not produce major errors.

2.3.1.2. Eutectic and Peritectic Diagrams

These diagrams are typical of systems that show positive deviations from ideal behavior. The S^M-composition curve for nonideal systems has different values to those shown in Figure 2.4a and the maximum in the curve may be displaced from $N_A = N_B = 0.5$, but TS^M remains a positive quantity. The enthalpy of mixing is now finite and contributes to the total free energy. If mixing of the components is exothermic, H^M is negative and the G-N_B curve is displaced downward, although the origins at N_A and $N_B = 1$ are unchanged. Conversely, endothermic mixing gives positive values of H^M and displaces the G-N_B curve upward to less negative values.

If H^M is sufficiently large and positive, the G^M-N_B curve may show an inflection at intermediate compositions, where the enthalpy outweighs the

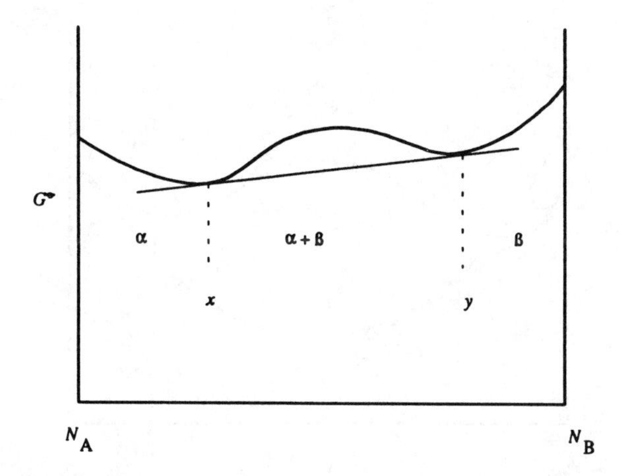

FIGURE 2.6. Free energy-composition curve for a binary system with a positive heat of mixing.

entropy contribution to the total free energy, Figure 2.6. If the curve represents the total free energy of the solid at some temperature below the melting points of A and B, then a solid solution (e.g., α) is stable to the left of point *x;* and a second solid solution (e.g., β) is more stable to the right of point *y,* where *x* and *y* define the points of contact of the common tangent to the free energy curve. Between these points, a phase mixture of (α + β) with compositions *x* and *y* has a lower free energy than any other mixture of the α and β phases. Similarly, if the curve represents the free energy of the liquid at a temperature above the melting points, *x* and *y* define the limits of a region of liquid immiscibility (i.e., a mixture of two liquids).

Since enthalpy and entropy do not change rapidly with a change in temperature, it is evident from the definitional relationship, $G = H - TS$, that the enthalpy contribution to the free energy becomes less significant as the temperature increases. It follows that the width of the miscibility gap decreases as the temperature is raised and is eventually closed if the temperature is raised sufficiently.

The situation shown in Figure 2.7 results when the total free energy of the liquid is represented by a smooth curve, while the free energy of the solid shows an inflection, but the melting points of A and B are similar. Figure 2.7a describes the situation at some temperature T_1, just below the melting points of the constituents, while Figure 2.7b shows the condition at T_2 where α, β and liquid are all in equilibrium with each other. On a temperature-composition plot, these intercepts describe a eutectic diagram. If the melting points of A and B differ markedly, the free energy curves are more skewed and a peritectic diagram results. Hence, both eutectic and

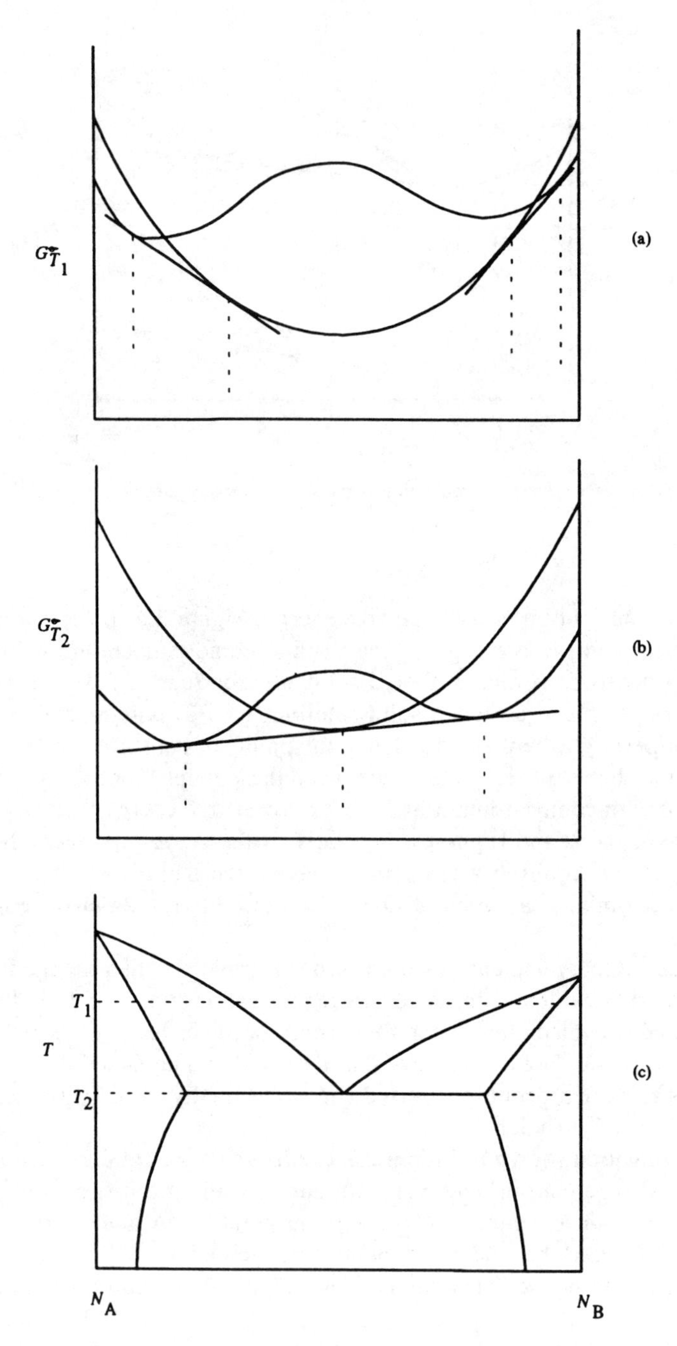

FIGURE 2.7. Free energy of the liquid and solid phases for a binary system that forms a eutectic phase diagram.

peritectic types of phase diagrams suggest that H^M is positive, and the activities of the components will show a positive deviation from ideal behavior. In a similar way, it can be shown that systems forming a stable compound usually have a negative value of H^M, and the activities show negative deviation from ideality.

2.3.1.3. Thermodynamic Data From Phase Diagrams

These data can sometimes be used for the evaluation of activities, etc. in the absence of experimental data. With reference to Figure 2.1, it has been shown that the intercepts at $N_A = N_B = 1$ of the tangent to the free energy curve at any composition x give the partial molar free energies of the two components in a solution of composition x. It follows that, when a tangent is common to two free energy curves, the partial molar free energy of each component is the same in the two phases at the compositions defined by the common tangent. For example, at any given temperature in the two-phase region of Figure 2.5b:

$$\overline{G}_{A(\text{s at solidus})} = \overline{G}_{B(\text{l at liquidus})} \tag{2.87}$$

where the subscripts, s and l, refer to solid and liquid, respectively. From Equation 2.30, however, one obtains:

$$\overline{G}_{A(\text{s at solidus})} - G^{\ominus}_{A(\text{s})} = RT \ln a_{(\text{s at solidus})} \tag{2.88}$$

$$\overline{G}_{A(\text{l at liquidus})} - G^{\ominus}_{A(\text{s})A(\text{l})} = RT \ln a_{(\text{l at liquidus})} \tag{2.89}$$

Hence,

$$a_{A(\text{s at solidus})} = a_{A(\text{l at liquidus})} \tag{2.90}$$

when both activities are referred to pure solid A as the standard state. If the solid solubility of B in A is negligible, then $a_{A(\text{s at solidus})}$ is close to unity and it follows that the activity of A at the liquidus is also approximately unity (relative to the pure solid standard state). More commonly, the activity in the liquid, relative to the pure supercooled liquid standard state, is required. It can readily be shown that this can be determined using the relationship:

$$\ln a_{A(\text{l at liquidus})} = \int \frac{L_f}{RT^2} dT \tag{2.91}$$

where L_f (the latent heat of fusion) is often regarded as a constant that is independent of temperature. When the temperature considered is significantly below the melting point, however, it is more appropriate to calculate L_f as a function of temperature.[39]

When the component B shows significant solid solubility in A, the activities at the liquidus can only be calculated in this way if the activity of A at the solidus is known. In the absence of experimental data, it is often assumed that the activities in the solid conform either to ideal or to regular solution behavior. After the activities along the liquidus line have been calculated, they can be extrapolated to any higher temperature in the liquid phase by assuming regular solution behavior and using Equation 2.58. The activities of the solute at the same temperature can then be calculated using the Gibbs-Duhem Equation 2.17.

It must be recognized that the activities thus determined may not be very accurate. From consideration of Figure 2.6, it is evident that a common tangent can be drawn to the liquidus and solidus curves even at very high dilution of the solute, such that zero solubility of a solute is not possible thermodynamically at any temperature other than the melting point of the solvent. Hence, the assumption of unit activity of the solvent at the solidus is an approximation, even in those systems that show apparently zero solid solubility. Calculation of the activities at the solidus by assuming ideal or regular solution behavior is also an approximation. Some uncertainty is attached to the experimentally determined positions of the liquidus and solidus lines, since any impurities in the metals used to determine the phase diagrams affect the temperatures of the phase changes and it is difficult to avoid some undercooling or superheating in detection of the arrest temperatures. Each of these factors contributes to the uncertainties in the values of the calculated activities. In the absence of experimental values, however, approximate values calculated in this way are often a useful guide and are likely to be more accurate than the alternative assumption of ideal behavior for the complete range of liquid solutions.

2.3.2. Free Energy-Temperature Diagrams

A plot showing the variation with temperature of the free energy of formation of compounds of a similar type is a convenient way of illustrating the relative stabilities of those compounds. This form of representation was first introduced by Ellingham[40] and was expanded by Richardson and Jeffes[41] to summarize data for the free energy of formation of pure metal oxides from the pure metal and oxygen gas at a pressure of 1 atm. The graphs are often referred to as Ellingham or Oxygen Potential Diagrams. The latter name is not adequate to describe all diagrams of this type, for the representation has been expanded to illustrate the stabilities of metal carbides, nitrides, sulfides, etc. that do not involve reaction with oxygen gas.

2.3.2.1. The Oxygen Potential Diagram

The diagram shown in Figure 2.8 is a free-energy temperature diagram for the formation of some metal oxides that are important in extraction

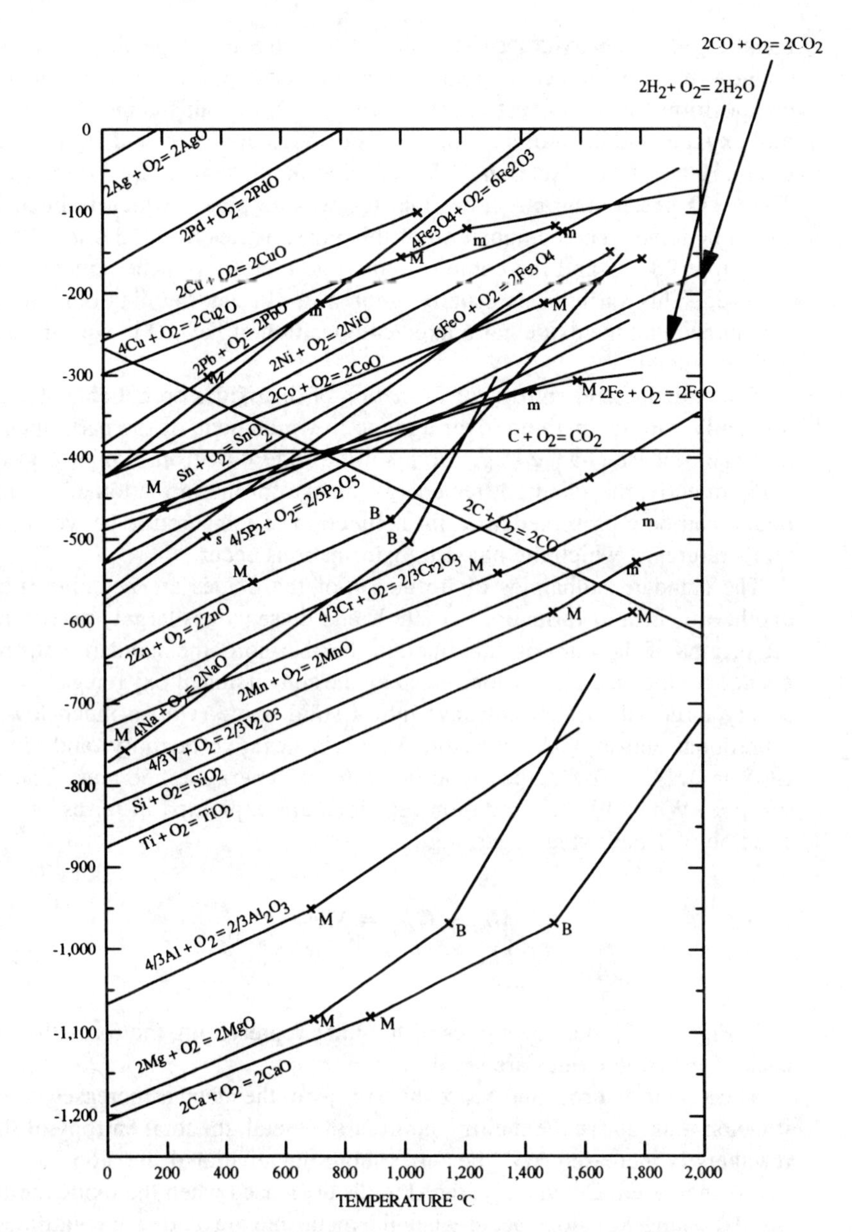

FIGURE 2.8. The oxygen potential diagram for the formation of oxides. M and B indicate the melting and boiling points, with capital symbols for the metals and lower-case symbols for the oxides.

metallurgy. Some oxides exist over a concentration range that may not include the stoichiometric composition. For example, the stoichiometric composition for the lowest oxide of iron (FeO) contains 1 mol Fe and 1 mol oxygen, and the atomic weights of Fe and O are 55.85 and 16, respectively. Hence, from Equation 2.75, the oxide should contain 22.3% oxygen. The Fe-O phase diagram shows that the wustite phase, which is FeO, is cation deficient. The composition of this phase varies from 23.6 to 24.5% oxygen at 900 °C and the composition range increases as the temperature is raised. This variation is usually ignored in plotting the diagram, and it is assumed that the oxide has a fixed composition of $Fe_{0.95}O$ in equilibrium with the metal.

The enthalpy and entropy of formation of the oxides do not change significantly with temperature over a range in which none of the participants undergo a phase change (e.g., from solid to liquid or from liquid to gas). Consequently, the standard free energy change for the formation of a compound can be represented as a linear function of temperature between the temperatures at which the phase transformations occur.

The standard enthalpies of formation of the oxides are negative (i.e., exothermic heat of formation) at 298 K and these values largely determine the origins of the lines on the diagram. Furthermore, the standard entropy change for the reactions is dominated by the conversion of oxygen gas (with a very large vibrational entropy) into a solid oxide (with a much lower vibrational entropy). As a result, $\Delta S^{\ominus}$ is a negative quantity, and since $\Delta G^{\ominus} = \Delta H^{\ominus} - T\Delta S^{\ominus}$, the slope of each free energy-temperature line is positive. When all the formation reactions are expressed in terms of the reaction of 1 mol oxygen gas, e.g.,

$$M_{(s)} + O_{2(g)} = MO_{(s)} \tag{2.92}$$

$$\frac{3}{2}M_{(s)} + O_{2(g)} = \frac{1}{2}M_3O_{4(s)} \tag{2.93}$$

as in Figure 2.8, then the slopes of the lines representing the formation of each of the solid oxides are similar.

The atomic disorder and hence the entropy of the metal is increased when it melts; thus, above the melting point of the metal, the total entropy of the reactants is increased, $\Delta S^{\ominus}$ becomes more negative, and the slope of the line is increased. Conversely, the slope is decreased when the oxide melts. Similar changes in slope occur when the metal and the oxide are volatilized. The entropies of fusion and volatilization are roughly 10 and 92 J mol^{-1}, respectively **(Trouton's rule)**. Hence, the change of slope is more marked when vaporization occurs than with melting. When a change of phase occurs, the free energies of the two phases are equal at the phase change temperature. Thus, the lines show only a change of slope and no other discontinuity.

Oxygen potentials are readily evaluated qualitatively from the diagram. The positive slope to the free energy lines signifies that the stability of the metal oxides decreases with increasing temperature (i.e., ΔG° becomes less negative). This can be demonstrated by consideration of the equilibrium constant for the formation of an oxide. From Equation 2.36, the equilibrium constant for the reaction:

$$2M + O_2 = 2MO$$

is given by:

$$\Delta G = \Delta G^{\circ} + RT \ln \left[\frac{a_{MO}^2}{a_M^2 \cdot p_{O_2}} \right] \tag{2.94}$$

If oxygen gas at less than 1 atm pressure is in equilibrium with a pure metal and its pure oxide, then $\Delta G = 0$ and the activities of the condensed phases are unity. Hence,

$$\Delta G^{\circ} = + RT \ln p_{O_2} \tag{2.95}$$

(See Worked Example 6.) The quantity, $RT \ln p_{O_2}$, is called the oxygen potential of the system, and the equality with ΔG° gives rise to the name for this form of representation of oxide formation. Thus, the diagram is, effectively, a *P-T* plot.

When $\Delta G^{\circ} = 0$, the oxygen pressure is equal to 1 atm. The line for the formation of silver oxide, Ag_2O_3, intersects the horizontal line at $\Delta G^{\circ} = 0$ at 190 °C. This signifies that, at equilibrium, silver oxide dissociates into metallic silver and oxygen gas if the oxide is heated to above that temperature in pure oxygen. If it is heated in air ($p_{O_2} = 0.21$ atm), the oxygen potential is lower and the oxide will tend to dissociate at an even lower temperature. The oxides of the precious metals (e.g., Au, Pd) also dissociate at quite low temperatures. However, inspection of Figure 2.8 shows that much higher temperatures are required to cause the simple dissociation of oxides such as Cu_2O and PbO. The dissociation temperature increases as the value of ΔG°_{298} (or the oxygen potential) for the formation of the oxide becomes more negative and is in excess of 2000 °C for the majority of the common metals. It is apparent, then, that the lower the line is located on the diagram for the formation of an oxide, the greater is the stability of the oxide. It follows from this that, in principle, any metal is capable of reducing the oxide of a metal located at a higher position on the diagram.

2.3.2.1. The Role of Carbon and Hydrogen as Reducing Agents

This role can also be assessed from the diagram. Two lines are shown on Figure 2.8 for the oxidation of carbon. One of these lines represents the reaction:

$$C_{(s)} + O_{2(g)} = CO_{2(g)} \tag{2.96}$$

Here, 1 mol oxygen gas is consumed by reaction with 1 mol of solid carbon to form 1 mol of CO_2 gas. There is no change in the number of gas moles, so the overall entropy change has only a small positive value and the line on the diagram is almost horizontal. In the other reaction:

$$2C_{(s)} + O_{2(g)} = 2CO_{(g)} \tag{2.97}$$

there is an increase in the number of moles of gas, giving rise to a large positive value for the entropy change and a negative slope to the line.

The lines for these two reactions cross at 710 °C. CO_2 is more stable at lower temperatures and CO is more stable at higher temperatures. Hence, it is evident from the diagram that the oxides of Co, Cu, Ni and Pb can be reduced to the metallic form by reaction with carbon to form CO_2, i.e.,

$$2MO + C = 2M + CO_2 \tag{2.98}$$

Reduction by carbon of some of the more stable oxides, such as FeO and ZnO, is only feasible at reasonably attainable temperatures because the gaseous reaction product is CO and the stability of CO increases as the temperature rises.

Gaseous reductants can also be used to reduce some of the metal oxides. Lines are shown on the diagram for the oxidation of hydrogen and carbon monoxide:

$$2H_{2(g)} + O_{2(g)} = 2H_2O_{(g)} \tag{2.99}$$
$$2CO_{(g)} + O_{2(g)} = 2CO_{2(g)} \tag{2.100}$$

Both lines have positive slopes, similar to the slopes for the metal-metal-oxide reactions, since 3 mol of gas are consumed in each of the reactions to generate 2 mol of a different gas with an overall reduction of 1 mol of gas. From the position of the lines on the diagram, it appears that CO is the more efficient reductant at low temperatures, and hydrogen is more efficient at high temperatures. It must be remembered, however, that this deduction relates only to the extent and not to the rate of the reactions. It is evident from the diagram that neither CO nor H_2 is as effective as solid carbon as a reducing agent at high temperatures.

The lines drawn for Equations 2.99 and 2.100 represent the equilibrium condition when each of the species is present in the standard state at 1 atm pressure. However, the equilibrium constant for Equation 2.99 is given by:

$$\Delta G^{\ominus} = -RT \ln \frac{p^2_{CO_2}}{p^2_{CO} \cdot p_{O_2}} \tag{2.101}$$

It is evident from this that ΔG° is not changed if the pressures of CO and CO_2 are greater or less than unity, provided that the ratio, p_{CO_2}/p_{CO}, remains equal to unity. A similar relaxation applies to Equation 2.100.

In practice, the metal ore is exposed to an atmosphere of reducing gas, and reaction with the oxide may result in complete or only partial oxidation of the reductant. The effect of the degree of oxidation of CO, for example, can be seen from consideration of Equation 2.101. If p_{CO_2} is less than p_{CO}, the numerical value of the right hand side of the equation is lowered, ΔG° is more negative, and the position of the line on the diagram describing Equation 2.99 is rotated downward about a point corresponding to the origin of the line at a temperature of 0 K. Conversely, the line is rotated upward if p_{CO_2} is greater than p_{CO}. The H_2-H_2O line responds similarly to changes in the relative partial pressures of H_2 and H_2O.

2.3.2.2. Impure Metals and Oxides

These exhibit activities of less than unity. The free energy-temperature relationships shown in Figure 2.8 for the formation of the metal oxides are not valid when the metal or the oxide are not present in their standard states (i.e., as pure substances). From Equation 2.36, it is apparent that the value of the activity quotient is smaller and ΔG is more negative if the oxide activity is less than unity. But when equilibrium is attained, $\Delta G = 0$ and ΔG° is not affected by the activities of the participants in the reaction. Thus, the partial pressure of oxygen must decrease correspondingly to restore the activity quotient to its original value. Effectively, this means that the position of the line representing the formation of the oxide is moved downward on the diagram and the oxide is more difficult to reduce.

The activity of a metal oxide is less than unity when it is dissolved, for example, in a molten phase. In a batch-type process in which the oxide is not continuously replenished, the concentration of the oxide in the molten phase decreases continuously as the oxide is reduced to the metal. Since $a = \gamma N$, it is probable that the activity of the oxide also falls continuously and the equilibrium oxygen potential becomes progressively more negative as reduction proceeds. Conditions that are more strongly reducing than are suggested by the position of the appropriate free energy-temperature line must then be maintained in the reactor to ensure that the oxide is fully reduced to the metal. As a result, other oxides that appear from the diagram to be more stable than the one under consideration may also be reduced if they are present in the molten oxide phase. Conversely, an oxide may only be reduced at a higher oxygen potential than suggested by the diagram if the activity of the metal product is lowered by solution in another metal.

Such changes in the activities of the metals and the oxides can cause quite large changes in the relative stabilities of the oxides. For example, the diagram indicates that MnO and SiO_2 are much more stable than FeO, but both of these oxides are reduced in the iron blast furnace to form dilute solutions of Mn and Si in the molten iron.

To summarize, standard free energy-temperature diagrams are a very useful way of illustrating the relative stability of compounds, but care is required in their interpretation. More detailed consideration in terms of the appropriate equilibrium constants is required in order to determine the extent of competing reactions.

2.3.3. Predominance Area Diagrams

A form of presentation was first proposed by Kellog[42] to portray the changes that occur during the roasting of sulfide ores. The plots are often referred to as Kellog or Predominance Area Diagrams.

There are three constituents—metal, oxygen and sulfur—in a system comprised of the metal and the compounds which it forms with oxygen and sulfur. According to the phase rule:

$$F = C - p + T + P \tag{2.102}$$

where F, C and p represent the number of degrees of freedom, the number of components, and the number of phases that can coexist, respectively. At constant pressure, P, when $C = 3$ the maximum value of F is three and, accordingly, a three-dimensional plot is required to show the changes in the equilibria when the temperature, T, and two of the composition variables are changed. For simplicity, therefore, it is customary to show only the effects caused by change in the composition variables on an isothermal, two-dimensional plot. Conventionally, the axes show the partial pressures of the oxygen and sulfur species in the equilibrium gas phase, plotted on a logarithmic base, so that the diagram is a form of a *P-N* plot.

Sulfur may exist in the gas in a number of forms, including S_2, S_4, SO, SO_2 and SO_3. At equilibrium the concentrations of all these species are interrelated through reactions such as:

$$S_2 + 2O_2 = 2SO_2 \tag{2.103}$$

$$2SO_2 + O_2 = 2SO_3 \tag{2.104}$$

Thus, the partial pressures of all the sulfur species at any chosen temperature can be determined from knowledge of the partial pressure of any one of them and the values of the relevant equilibrium constants. Any one of the sulfur gases could be selected as the variable for the plot. Normally SO_2 is chosen, since this is usually the predominant species.

2.3.3.1. Construction of the Diagram

Construction commences with a listing of all known compounds formed by the metallic element with sulfur and oxygen. The standard free energies

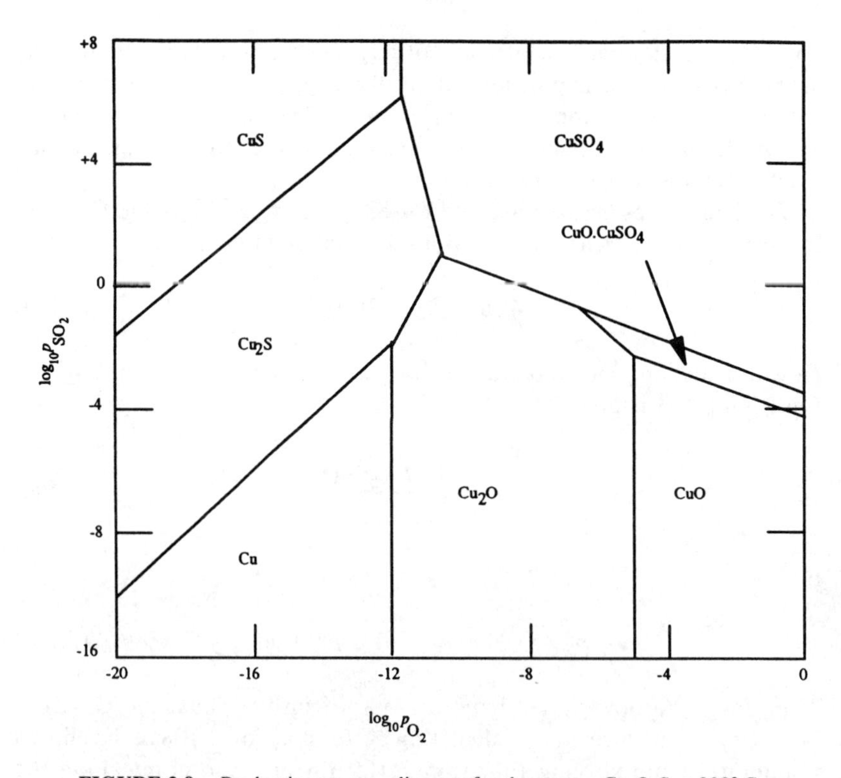

FIGURE 2.9. Predominance area diagram for the system Cu-O-S at 900° C.

of formation of each of the compounds and the free energies of reaction between all possible pairs of compounds are then calculated at the chosen temperature to determine which reactions are probable (i.e., have the most negative free energies) at that temperature. In practice, the construction can be simplified, as illustrated by consideration of the Cu-O-S diagram shown in Figure 2.9. The diagram is subdivided by straight lines bounding phase fields in which the element or compound named is the predominant species.

Consider the boundaries between the fields labeled Cu, Cu_2O and CuO. These represent the equilibrium conditions at the chosen temperature for the reactions:

$$2Cu + \tfrac{1}{2}O_2 = Cu_2O \tag{2.105}$$

$$Cu_2O + \tfrac{1}{2}O_2 = 2CuO \tag{2.106}$$

If the condensed substances are present as pure phases with unit thermodynamic activities, the equilibrium constants for both reactions simplify to:

$$K = \frac{1}{p_{O_2}} \tag{2.107}$$

Thus, the reactions are independent of p_{SO_2} and the boundaries are parallel to the SO_2 axis. Their positions along the oxygen abscissa are fixed accordingly by calculation of the equilibrium value for p_{CO_2} for the two reactions from known values for the standard free energy change for the reactions at the temperature considered.

The boundaries between Cu and Cu_2S, and between Cu_2S and CuS, are independent of p_{O_2} when the reactions are written in the form:

$$4Cu + S_2 = 2Cu_2S \tag{2.108}$$

However, the ordinate is plotted as log p_{SO_2}. The equilibrium constant for Equation 2.103 is given by:

$$K = \left(\frac{p^2_{SO_2}}{p_{S_2} \cdot p^2_{O_2}}\right) \tag{2.109}$$

or

$$\log p_{SO_2} = \tfrac{1}{2}\log K + \tfrac{1}{2}\log p_{S_2} + \log p_{O_2} \tag{2.110}$$

Therefore, the boundaries have a slope of unity. Thus, the Cu-Cu_2S boundary can be located by calculating p_{S_2} for reaction 2.108 at the chosen temperature and solving Equation 2.109 for any one point along the boundary. The line can then be drawn with the required slope through that point.

The line separating Cu_2S and $CuSO_4$ describes the equilibrium condition for the reaction:

$$Cu_2S + 3O_2 + SO_2 = 2CuSO_4 \tag{2.111}$$

for which the equilibrium constant is given by:

$$K = \left(\frac{1}{p^3_{O_2} \cdot p_{SO_2}}\right) \tag{2.112}$$

or

$$-\log p_{SO_2} = 3\log p_{O_2} + \log K \tag{2.113}$$

when the activity of the condensed substances is unity, so that the boundary between these phases has a negative slope of 3.

The slope and position of other boundaries can be determined in similar fashion, and the lines are then extended to intersect with the other boundary lines as appropriate. For accuracy in construction, it is important to ensure that the free energy data used to determine the values of the various equilibrium constants are consistent. That is, they should all be based on the same data for the standard free energies of the metal, oxygen and sulfur.

2.3.3.2. The Effect of Activity and Temperature Variations

This effect can also be assessed. A boundary represents the condition where both participating phases are at unit activity and, in consequence, the ratio of the activities is unity. At a point just below the Cu-Cu_2S boundary, Cu is the predominant phase; thus, the activity ratio is greater than unity. But a_{Cu} cannot be greater than unity, relative to the pure substance standard state, so the activity of Cu_2S must be less than unity at this point. The activity of Cu_2S falls with increasing distance from the boundary into the Cu phase field. Additional lines could be drawn within the phase field to represent the conditions corresponding to different values (e.g., 0.5, 0.1, 0.05) for the activity of Cu_2S. The position of these lines can be fixed by calculation of p_{S_2} in equilibrium with pure S_2 and Cu_2S at the chosen activity and then by calculation of p_{SO_2}, using Equation 2.109. Similar lines can be determined for the activities of the other phases bounding any particular phase field.

It is evident that the area of the diagram in which any particular phase can exist is increased in all directions if the activity of that phase is less than unity, while all other phases remain at unit activity. In marked contrast to a phase diagram, a boundary on a predominance area diagram does not define the limits of existence of a phase, nor does a phase field signify the existence of only one phase. It merely indicates which phase is present as the major constituent. This explains the name of the diagrams; they show the conditions under which each of the phases is the predominant species in a mixture of phases.

The positions of the boundaries are also changed by change in the temperature. The free energy-temperature diagram (Figure 2.8) shows that p_{O_2} in equilibrium with Cu and Cu_2O and with Cu_2O and CuO is increased as the temperature is raised and, correspondingly, the vertical boundaries in Figure 2.9 move to the right with increasing temperature. Similarly, p_{S_2} (and hence p_{SO_2}) in equilibrium with Cu and Cu_2S, etc. is increased and all the sloping boundaries move upward to higher values of p_{SO_2} as the temperature is raised. Calculation of a series of diagrams to cover a range of temperatures is fairly simple, but laborious. Programs have been written, therefore, to facilitate computer calculation of the diagrams (see, for example, References 43 and 44).

2.3.3.3. Application

Application of this type of diagram can yield useful information. Thus, Figure 2.9 shows that $CuSO_4$ is only stable at 900 K at high values of

p_{SO_2} and p_{O_2}. Since copper oxides and sulfides are insoluble and the basic sulfate, $CuO{\cdot}CuSO_4$, is only sparingly soluble in water, the gas atmosphere required to produce the soluble $CuSO_4$ during roasting must lie within the boundaries of the top right-hand phase field. However, the sulfate phase becomes unstable at this temperature at $p_{SO_2} = 1$ atm when p_{O_2} is less than 10^{-8} atm. Since the lower boundary of the sulfate phase moves to lower p_{SO_2} as the temperature is lowered, it is evident that roasting to produce the soluble sulfate should be carried out at the lowest possible temperature at which the reaction can be completed in reasonable time. Conversely, a high temperature is required when the ore is roasted, with the intention of removing all the sulfur from the mineral.

In comparing predominance area diagrams for several metal-O-S systems, gas atmospheres can be identified that will result in the conversion of one sulfide (e.g., Co or Cu) into a soluble sulfate while another sulfide (e.g., Fe) is left as an insoluble oxide. Computer calculations facilitate the combination of diagrams for two or more metals in one diagram.

2.3.4. Pourbaix Diagrams

The Pourbaix diagram, named for the originator, is used for consideration of reactions that occur in aqueous solutions in a similar way to the use of predominance area diagrams for examination of gas-solid reactions. Before considering the construction of the diagram, it is first necessary to outline the thermodynamic relations applicable to aqueous solutions.

2.3.4.1. Electrochemistry

This is the collective term for these considerations. Elements or compounds are present as ions when they dissolve in aqueous solution; i.e.,

$$M_{(s)} = M^{n+}_{(\text{in solution})} + ne^- \qquad (2.114)$$

where n is the valence of the species and e^- is an electron. Thus, a reaction occurring in a solution can be regarded as a transfer of electrons from one type of ion to another, or it could result from the application of an electrical potential to the solution to cause, for example, Reaction 2.114 to move either in a forward or reverse direction. The electrical potential required to achieve this change is related to the free energy change for the reaction.

When the equilibrium condition in a system is changed isothermally and under reversible conditions, the accompanying change in the free energy is a measure of the maximum amount of work that can be obtained from the system. It is also a measure of the amount of work that must be done to cause the reaction to proceed reversibly in the opposite direction. If the change occurs reversibly and at constant P and T in a galvanic cell (i.e., negatively charged anions flow toward one electrode and positively charged cations flow in the reverse direction, with a balancing flow of electrons through an external circuit between the electrodes), then ΔG is equal to the maximum value of the electrical work that can be obtained from the cell. The latter is equal to the product of the reversible potential (emf), E, and the amount of electrical charge transferred.

The transfer of one gram-electron through the cell requires the expenditure of 1 Faraday (96,487 C) energy. If the reaction involves the transfer of z gram-electrons (equal to z gram-moles of substance), then,

$$\Delta G = -zFE \tag{2.115}$$

When the reactants and products are present in their standard states, then,

$$\Delta G^{\ominus} = -zFE^{\ominus} \tag{2.116}$$

where $E^{\ominus}$ is called the standard emf of the cell.

The free energy-temperature diagrams can readily be translated into standard emf-temperature diagrams using Equation 2.116. Thus, 1 mol oxygen is consumed with the transfer of four electrons in the reactions shown on the oxygen potential diagram. Hence, $\Delta E^{\ominus} = -\Delta G^{\ominus}/(4 \times 96{,}487)$ V, which is equal to $-2.59 \times 10^{-6}\,\Delta G^{\ominus}$, and the ordinate can be converted to an $E^{\ominus}$ scale by use of this factor.

The enthalpy and entropy changes can also be expressed in terms of the electrical work, *viz:*

$$\Delta S^{\ominus} = zF\left(\frac{\partial E^{\ominus}}{\partial T}\right)_p \tag{2.117}$$

$$\Delta H^{\ominus} = -zF\left(E^{\ominus} - T\frac{\partial E^{\ominus}}{\partial T}\right) \tag{2.118}$$

The standard state for a species in aqueous solution is usually defined as a solution containing 1 mol of that species per kilogram solution (i.e., $a = 1$ at $C = 1$, where C = molality). The reference state used to measure the potentials of other reactions is normally selected as the standard hydrogen electrode:

$$2H^+ + 2e^- = H_{2(g)} \tag{2.119}$$

By convention, reactions of this type are written with the electrons on the left-hand side of the equation. The change in the standard electrode potential, $\Delta E^{\ominus}$, is set at zero for equilibrium between H_2 gas at 1 atm pressure and a solution containing 1 mol kg^{-1} hydrogen ions.

Equation 2.119 describes the reduction of hydrogen ions to hydrogen gas. The standard reduction potentials of metal ions can be defined in a

TABLE 2.1
Standard Electrode Potentials (in Volts) for 1 N Solutions at 25° C

Reaction	Potential
$Au^{3+} + 2e^- = Au$	1.50
$Cl_2 + 2e^- = 2Cl$	1.36
$2H^+ + \frac{1}{2}O_2 + 2e^- = H_2O$	1.23
$Pt^{2+} + 2e^- = Pt$	1.20
$Ag^+ + e^- = Ag$	0.80
$Fe^{3+} + e^- = Fe^{2+}$	0.77
$Cu^+ + e^- = Cu$	0.52
$Cu^{2+} + 2e^- = Cu$	0.34
$As^{3+} + 3e^- = As$	0.15
$Sb^{3+} + 3e^- = Sb$	0.10
$H^+ + e^- = \frac{1}{2}H^2$	0.0
$Pb^{2+} + 2e^- = Pb$	−0.13
$Sn^{2+} + 2e^- = Sn$	−0.14
$Ni^{2+} + 2e^- = Ni$	−0.25
$Co^{2+} + 2e^- = Co$	−0.28
$Cd^{2+} + 2e^- = Cd$	−0.40
$Fe^{2+} + 2e^- = Fe$	−0.44
$2S + 2e^- = S^{2-}$	−0.48
$\frac{1}{2}O_2 + 2e^- = O^{2-}$	−0.56
$Zn^{2+} + 2e^- = Zn$	−0.76
$Mn^{2+} + 2e^- = Mn$	−1.05
$Al^{3+} + 3e^- = Al$	−1.67
$Ti^{2+} + 2e^- = Ti$	−1.75
$Mg^{2+} + 2e^- = Mg$	−2.40
$Na^+ + e^- = Na$	−2.71
$Ca^{2+} + 2e^- = Ca$	−2.85

similar way and are listed in Table 2.1, relative to the standard hydrogen electrode. Each of the reactions listed is termed a ''half-cell reaction.'' If a reaction involving any two of the half-cells occurs in a cell in which the metals and the metal ions are present at unit activity, then the metal with the most positive or least negative potential is able to reduce the other metal from solution. Since

$$\Delta G^{\circ} = -zFE^{\circ} \quad \text{(from Equation 2.117)}$$
$$= -RT \ln K \quad \text{(from Equation 2.37)}$$

then,

$$E^{\circ} = \frac{RT}{zF} \ln K \tag{2.120}$$

The electrode potential for a reaction such as:

$$Zn^{2+} H_{2(g)} = Zn + 2H^+ \tag{2.121}$$

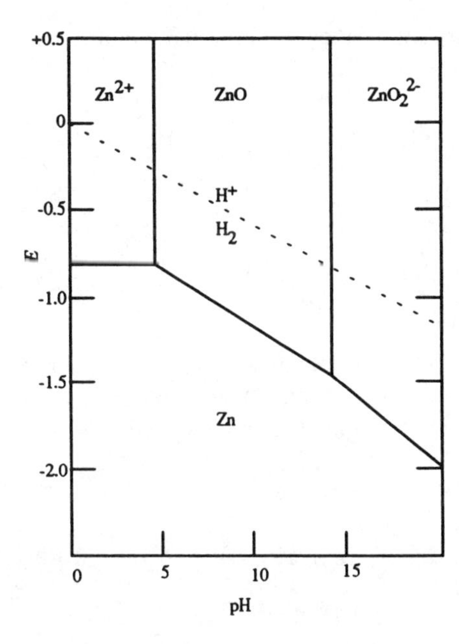

FIGURE 2.10. Pourbaix diagram for the zinc-water system.

under nonstandard conditions is given by:

$$E = E^{\ominus} - \frac{RT}{zF}\ln\left[\frac{a_{Zn} \cdot a_{H_2}^2}{a_{Zn^{2+}}^2 \cdot p_{H_2}}\right] \qquad (2.122)$$

This is the Nernst equation. At 25° C (298 K), the quantity $(RT/F) \cdot \ln x = 0.0591 \log x$; and since $z = 2$ for Equation 2.121, $(RT/zF) \cdot \ln x = 0.0296 \times \log x$, where x is the activity quotient. Furthermore, by definition, the pH of the solution is equal to $-\log a_{H^+}$.

2.3.4.2. Construction of the Diagram

The construction follows similar principles to those applied in deriving a predominance area diagram. A Pourbaix diagram is a graphical representation of the relative stabilities of a solid metal and its solid compounds and the ions produced therefrom in an aqueous solution, as a function of their electrode potentials and the pH of the solution. The diagrams can be quite complex. Taking a metal-metal oxide-water system as an example, the metal may form a number of different oxides, hydroxides and ions if it exhibits multiple valence. All possible reactions between each possible pair of solids, and between each of the solids and each of the ions, must be considered, the equilibrium relationships evaluated and the results plotted against the coordinates. The method of construction can be illustrated with reference to a simplified form of the zinc-water diagram, Figure 2.10.

The value of the equilibrium constant for the dissolution of metallic zinc in water to produce zinc ions in solution:

$$Zn^{2+} + 2e^- = Zn \tag{2.123}$$

for which $E^{\ominus} = -0.76$ V is given by:

$$E = -0.76 - \frac{0.0591}{2}\log\left(\frac{1}{a_{Zn^{2+}}}\right) \tag{2.124}$$

when a_{Zn} is unity. If the activity of the zinc ions is also unity (i.e., a 1 m solution), then $E = -0.76$ V. This value is independent of the pH of the solution, so the boundary between the two phases is represented by a horizontal line on the diagram.

The reaction between Zn and ZnO can be written as:

$$ZnO + 2H^+ + 2e^- = Zn + H_2O \tag{2.125}$$

for which $E^{\ominus} = -0.50$V. Hence,

$$\begin{aligned} E &= -0.50 - \frac{0.0591}{2}\log\left(\frac{1}{a_{H^+}^2}\right) \\ &= -0.50 - 0.0591\text{pH} \end{aligned} \tag{2.126}$$

when the activities of Zn, Zn ions and the water solvent are all equal to unity. Thus, this boundary has a negative slope of 0.0591 and passes through the point, $E = -0.50$ at pH $= 0$. Similarly, the boundary between Zn and the zinc oxide anion is determined from:

$$ZnO_2^{2-} + 4H^+ + 4e^- = Zn + 2H_2O \tag{2.127}$$

$$\begin{aligned} E &= E^{\ominus} - \frac{0.0591}{2}\log\left(\frac{1}{a_{H^+}^4}\right) \\ &= +0.90 - 0.1182\text{pH} \end{aligned} \tag{2.128}$$

The reactions between Zn^{2+} ions and ZnO, and between ZnO and the oxide anion, do not involve transfer of an electron charge:

$$ZnO + 2H^+ = Zn^{2+} + H_2O \tag{2.129}$$

$$ZnO_2^{2-} + 2H^+ = ZnO + H_2O \tag{2.130}$$

and, consequently, the boundaries representing these two reactions are independent of the electrical potential. According to the phase rule, in a three-component system (e.g., Zn-O_2-H_2) at constant T and P, three phases can only meet at a point. Hence, these boundaries appear in Figure 2.10 as vertical lines that start at the intersections of two associated boundaries. (Note that the equations are written to exclude the formation of hydroxyl ions, which would otherwise complicate the construction of the diagram.)

2.3.4.3. Application of the Diagrams

This can be demonstrated with reference to Figure 2.10. The diagram shows that ZnO is stable over a wide pH range, from mildly acidic to moderately strong alkaline solutions. Thus, a strong acid solution is required to extract Zn as Zn^{2+} ions, and the oxide anion, ZnO_2^{2-}, is only predominant in strongly alkaline solutions. Pure solid Zn can only be deposited from a solution of Zn^{2+} ions if the emf is less than -0.76V. Table 2.1 shows that the horizontal line between the metal and the metal ion lies at more positive potentials for most of the common metals and, accordingly, they can be recovered more readily than Zn by electrorefining.

The solid substances may be at unit activity in aqueous extraction and refining processes; but the concentration of the ions in solution rarely approaches the standard state of a 1 m solution and the activity of the ions is then less than 1. In the description of the predominance area diagrams, it was shown that the boundaries of any particular phase move outward if the activity of that phase is reduced, while the activity of the phases occupying the adjacent regions remains at one. A similar condition applies to Pourbaix diagrams. Thus, the boundary between ZnO and Zn^{2+} moves to higher pH values as the activity of the ions is lowered, and most of the Zn can be recovered from solution by electrodeposition even in neutral solutions. At the same time, the boundary between Zn and Zn^{2+} ions moves downward to even more negative potentials as the activity of the ions is lowered and recovery of the metal from solution becomes more difficult. The change in the position of the boundary is readily determined by substituting the value of the ion activity in the appropriate equilibrium constant and recalculating the value for E.

A diagonal broken line is also shown in Figure 2.10. This represents the boundary between hydrogen ions in solution and hydrogen gas when both species are present at unit activity. Hydrogen ions may be produced by reversal of reactions of the type described by Equations 2.125, 2.127, 2.129 and 2.130 (i.e., by reaction from right to left). Small amounts are also produced by partial dissociation of the water solvent into hydrogen and hydroxyl ions. The H^{2+} ions may, in turn, be released as H_2 gas according to the reaction:

$$2H^+ + 2e^- = H_{2(gas)} \tag{2.131}$$

The position of the boundary line for this reaction is given by $E = -0.0591\text{pH}$ since, by convention, $E^{\circ} = 0$ when p_{H_2} is equal to 1 atm at 25° C. The line moves upward on the diagram if p_{H_2} is lowered to <1 atm.

Since the hydrogen line lies at less negative potentials than the Zn-Zn^{2+} boundary, it is evident that it is easier to reduce H^+ ions to hydrogen gas than to reduce Zn^{2+} to the metal, and precautions have to be taken to prevent this from happening during electrodeposition (see Chapter 7).

Use of the diagrams is not restricted to aqueous solutions. They are also applied in consideration of electrowinning from fused salts, as in the production of aluminum.

2.3.5. Other Diagrams and Applications

The most commonly used forms of graphical representation of thermodynamic data have been described, but other forms are also used. For example, data for the formation of metal compounds can be plotted to show the variation with temperature of the equilibrium constants for the formation reactions. This form of plot clearly distinguishes those reactions that are favored by an increase in temperature (i.e., a positive slope) from those that occur less readily when the temperature is raised (negative slope). Or, a predominance area diagram may plot log p_{CO}/p_{CO_2} along the abscissa to show directly the effect of the actual gas atmospheres on the boundaries of the species.

The graphical representations also have wider use outside the fields of extraction and refining. Thus, free energy-temperature and predominance area diagrams are useful in studies of hot corrosion (e.g., dry oxidation, sulfidation, etc.), while Pourbaix diagrams are used in connection with wet corrosion and electroplating.

One major limitation of the diagrams should always be remembered. They indicate only the extent to which the various reactions can proceed and give no indication of the rate of the reaction or, indeed, if equilibrium can be achieved. One example can be used to illustrate this point. The oxygen potential diagram shows that iron will oxidize if the partial pressure of oxygen is greater than 10^{-43} at 25° C, and it is well known that unalloyed iron does oxidize slowly under these conditions. The same diagram also shows that aluminum should oxidize at the same temperature at very much lower partial pressures of oxygen. On this basis, it would be expected that aluminum would oxidize at a very much faster rate than iron. In fact, Al oxidizes spontaneously to form a very thin layer of alumina, but thereafter it oxidizes at a very much lower rate at ambient temperature and only exhibits a higher rate at elevated temperatures. Hence, the kinetics must also be considered for a full understanding of the reactions.

Chapter 3

THEORETICAL PRINCIPLES—2 KINETICS

The objective in extraction metallurgy is to achieve a change in the chemical composition to convert a metallic mineral into a metal with the purity required for the intended use by the cheapest practical route. Kinetics determine the time taken to achieve a desired end point (e.g., metal composition) and thus affect the productivity of any process. In general, as the productivity is raised the capital costs, the energy consumption, the man-hours, etc. per tonne of production are all decreased and the profitability is increased. Hence, control of the reaction kinetics may make the difference between a viable and a nonviable production route.

There is a very important difference between the application of thermodynamic and kinetic data in extractive metallurgy. The thermodynamic quantities, G, H and S, are state properties whose values are dependent only on the initial and final states of the system. They are not affected by how a reaction is achieved, whether by a single step or by a multiplicity of steps, or by the total mass of the participants. Consequently, the value of an equilibrium constant, for example, determined with a small quantity of material in a laboratory experiment applies equally to a large-scale industrial process when the compositions of the equilibrium phases and T and P are identical to those in the laboratory experiment.

In marked contrast, the kinetics of a reaction are dependent on each of the steps involved in the process. Consider, for example, a heterogeneous process, typical of extraction and refining (few reactions in this field are homogeneous), in which a change of composition is achieved by a chemical reaction at an interface between two liquid phases, or between a gas and a liquid or a solid phase. The process involves mass transport of the reactants to the interface, chemical reaction at the interface and mass transport of the products away from the reaction site. Additional steps, such as adsorption of the reactants at, and desorption of products from the interface may also be involved. The overall chemical reaction may comprise a number of discrete and consecutive or overlapping reactions. Each of these steps can be described by a reaction rate equation dependent on a number of variables.

It is not surprising, therefore, to find that the overall rate constant is markedly dependent on how the process is operated. The value of the constant determined in laboratory studies is rarely even closely approached in commercial practice. Indeed, a value obtained by careful analysis on one production unit may not apply to another seemingly identical unit, and it is often necessary to evaluate the kinetic parameters relative to each individual unit and practice.

The application of kinetic theory, however, can be very profitable. Measurement of the rate of the individual steps in a reaction can identify which step proceeds most slowly under the imposed conditions and thus controls the overall reaction rate. It may then be possible to predict how the conditions should be changed to accelerate the reaction, or at least devise a series of tests to ascertain how the rate can be increased.

3.1. CHEMICAL REACTION KINETICS

3.1.1. Kinetic Theory

Suppose that a reaction occurs between two species, A and B, to produce a compound, A_xB_y:

$$xA + yB = A_xB_y \tag{3.1}$$

The rate (r) of the reaction in the forward direction (i.e., from left to right) is given by:

$$r_f = \frac{dN_{A_x B_y}}{dt} = k_f\, [a_A]^x \cdot [a_B]^y \tag{3.2}$$

and the rate in the reverse direction is:

$$r_r = -\left[\frac{dN_{A_xB_y}}{dt}\right] = k_r[a_{A_xB_y}] \tag{3.3}$$

where k_f and k_r are the specific rate constants for the forward and back reactions, respectively. The rate of the forward reaction decreases and the rate of the reverse reaction increases as the reaction progresses. At equilibrium, $r_f = r_r$ and, in terms of the equilibrium constant for the reaction:

$$K = \frac{[a_{A_xB_y}]}{[a_A]^x\, [a_B]^y} = \frac{k_f}{k_r} \tag{3.4}$$

It is common practice in kinetic equations to express the amounts of the species participating in reactions in terms of concentrations and not activities. This approximation is, of course, strictly valid only for solutions that exhibit ideal behavior. It is also customary to refer to the order of a reaction. This is often defined as the number of reactant molecules participating in the reaction. Thus, a reaction of the type:

$$X_{(gas)} = X_{(metal)} \tag{3.5}$$

in which only one molecule of an element, X, appears as reactant would be classed as a first-order reaction. But this is an oversimplification. Many chemical reactions proceed via a number of intermediary steps, each of which may involve a different numbers of molecules. For example, the reduction of a metallic oxide (MO) by carbon:

$$C + MO = M + CO_2 \tag{3.6}$$

occurs by a sequence of steps in which some carbon is first gasified as CO:

$$C + \tfrac{1}{2}O_2 = CO \tag{3.7}$$

and the CO reacts with the oxide:

$$CO + MO = M + CO_2 \tag{3.8}$$

Some of the CO_2 then reacts with carbon to produce more CO:

$$CO_2 + C = 2CO \tag{3.9}$$

the overall summation being equivalent to Equation 3.6. The order for each step is determined by the number of molecules participating in that step, but the order number for the overall reaction is really the best empirical fit of a rate equation to the experimental data.

Evaluation of the order number can provide useful information about the reaction. Consider a simple reaction similar to that described by Equation 3.5 in which a species is transferred across an interface of area A into a volume of metal, V. If the initial concentration in the metal is C_0, the concentration at equilibrium is C_e, and the concentration at time t is C, then the data for a first-order reaction is fitted by an equation of the type:

$$r = \frac{dC}{dt} = k\frac{A}{V}(C_e - C) \tag{3.10}$$

which is consistent with the requirement that the rate decreases as equilibrium is approached and falls to zero at equilibrium. Since $C = C_0$ at $t = 0$, integration of the equation gives:

$$\ln\frac{(C_e - C_0)}{(C_e - C)} = k\frac{A}{V}t \tag{3.11}$$

The time taken to reach the halfway stage in the reaction is equal to $V/(2Ak)$, which rises as the volume of the metal is increased and decreases with increasing interfacial area, but is independent of the concentration.

Similarly, data for a second-order reaction can be fitted to an equation of the form:

$$r = \frac{dC}{dt} = k\frac{A}{V}(C_e^2 - C^2) \tag{3.12}$$

and the time to reach the half-reaction stage is equal to $V/(2AkC_e)$, which shows that the rate is now also inversely proportional to the concentration.

The order of a reaction can be found by plotting data for the rate of the reaction against the right-hand side of equations in the form of Equations 3.10, 3.12, etc. If the order is a whole number, the plot is linear only for the equation that fits the appropriate order of the reaction. It is often possible to determine which one of a sequence of reaction steps is the slowest and, therefore, controls the overall reaction rate, by plotting data for each of the steps in this way. However, it is sometimes found that the apparent order for a reaction changes if, say, the particle size of the reactants is changed; or, the order may change as the reaction proceeds toward the equilibrium condition and a different rate-controlling step becomes dominant near equilibrium.

3.1.2. Reaction Rate Theories

Two principal theories are used to explain the kinetics of chemical reactions. The **collision theory** assumes that the rate is determined by the frequency with which molecules collide with sufficient energy to break the bonds between the atoms and allow new combinations of bonds to form. The **activated complex theory** postulates that an intermediate complex is formed from the reactants, and the complex then decomposes to form the products:

$$AB + C \rightarrow ABC^x \rightarrow A + BC \tag{3.13}$$

where ABC^x is the activated complex. It is assumed that the complex decomposes spontaneously to form the products, and the rate at which the complex is formed determines the rate of the reaction.

Both theories lead to the premise that a certain minimum quantity of energy—the **activation energy** E_A—is required to enable the reactants to transform into products. This is usually visualized as an energy barrier and is illustrated in Figure 3.1 in terms of the activated complex model. The energy, E_A, is required to raise the reactants from the trough to the peak and form the activated complex. No additional energy is required to transform the complex into the products, which is therefore spontaneous. The free energy change for the reaction, equal to the difference between the

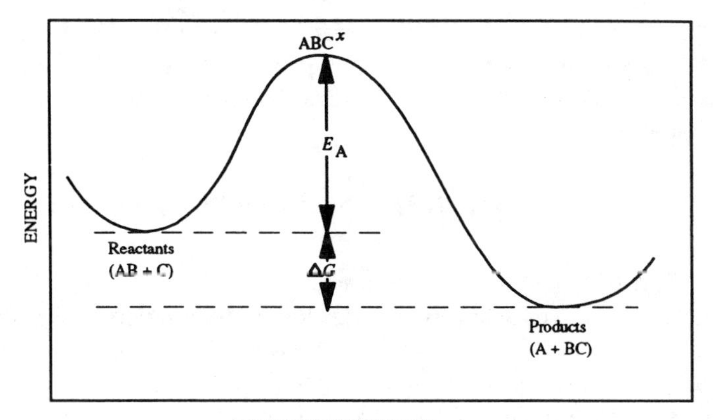

FIGURE 3.1. Schematic representation of the activation energy for a reaction.

energy levels for the products and the reactants, is not affected by the activation energy. In the early stages of the reaction, when ΔG is large, the total energy required to drive the reaction in the forward direction is less than the energy required to transform products into reactants. The energy difference between the forward and reverse directions decreases as equilibrium is approached and falls to zero at equilibrium.

With reference to the collision theory, it is evident that the numerical value of E_A increases with increasing complexity of the molecules participating in the reaction. Thus, a higher activation energy would be expected when polymeric molecules with multiple bonds are involved than for reactions similar to those described, e.g., by Equations 3.7 to 3.9 in which only simple molecules are involved. The activation energy is generally smaller for reaction between molten phases than in the solid state, since both the strength and the number of interatomic bonds decrease when a solid changes to a liquid. There is a marked decrease in the bond strength on transition to a gas and, correspondingly, E_A tends to be low for reactions between gases.

The temperature dependence of the rate of a chemical reaction is determined by the value of E_A. Toward the end of the 19th century, **Arrhenius** found from experimental observations that the logarithm of the specific rate constant varied linearly with $1/T$ (K). This is expressed today in the form:

$$k = A \exp(-E_A/RT) \tag{3.14}$$

or

$$\ln k = \ln A - E_A/RT$$

where A is a constant. Thus, a linear plot is obtained by plotting ln k vs $1/T$, and the slope is equal to E_A/R. The quantity, $\exp(-E_A/RT)$ is called the Boltzmann function, which can be regarded as a measure of the probability that an event will occur.

3.2. TRANSPORT KINETICS

3.2.1. Diffusion

In heterogeneous reactions, the movement of the reactant species up to the site of the reaction (i.e., the interface between the two phases) and the transfer of the products away from the reaction site is frequently controlled by the process of diffusion. This is the only mechanism by which atoms can be transported through a solid substance, so the explanation of the phenomena will be described initially in terms of diffusion in solids.

A characteristic of the solid state is that the atoms are more or less closely packed on a rigid lattice. Hence, an atom on a lattice site can interchange its position with its nearest neighbors only by a number of atoms rotating about a point (ring movement) or, more commonly, by the atom exchanging its position with an adjacent vacant lattice site. The distance an atom is moved through the lattice in a given direction by one such event is referred to as the jump distance (b).

The atoms of some elements (such as carbon, hydrogen and nitrogen) are too small to replace the atoms of the metallic elements on the lattice sites to form substitutional solutions. These small atoms are located instead in the interstices between the metal atoms in a solid lattice and they diffuse by squeezing between the host atoms to move from one interstitial site to another. Since the solubility of these elements in solid metals is usually very low, even on an atom percent basis, the probability that an adjacent interstitial site will be occupied is very low. Consequently, atoms that can migrate by interstitial diffusion can move more rapidly than larger atoms that can only move by substitutional diffusion via the lattice sites.

3.2.1.1. Diffusion Laws

Fick's laws of diffusion describe the rate at which atoms can migrate. Consider the movement of B atoms in a solid solution of A and B species between two planes with concentrations C_1 and C_2, separated by one jump distance, b_x, in the x direction. If the mean frequency with which a B atom moves from an atomic site on plane C_2 is equal to υ and the probability that the jump is in the x direction to plane C_1 is given by p_x, then the number of B atoms leaving C_2 and arriving at C_1 in unit time is equal to $\upsilon p_x C_2 b_x$. The number of atoms making the jump in the reverse direction in the same time interval is equal to $\upsilon p_x C_1 b_x$, and the net gain of B atoms at C_1 is $\upsilon p_x b_x(C_2 - C_1)$. If $(C_2 - C_1)$ is small, then $(C_2 - C_1)/b_x$ is equal to the rate of change of concentration, $(\partial C/\partial x)$. The quantities υ, p_x and b_x are

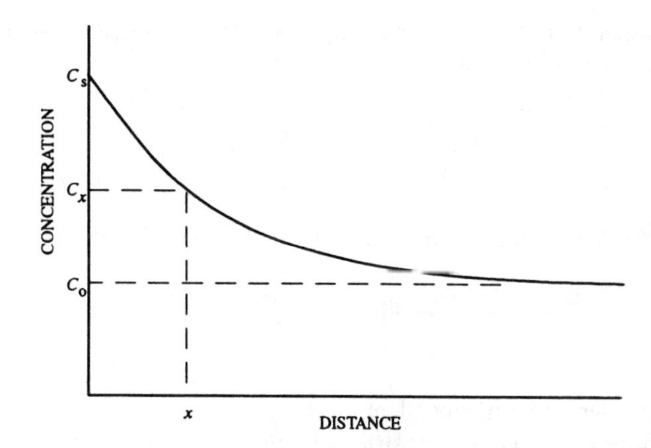

FIGURE 3.2. Variation of concentration with distance from the surface for nonsteady-state diffusion.

dependent only on composition and temperature; hence, when these parameters are constant, all three quantities can be replaced by a constant, D, which is termed the "diffusion coefficient."

The number of B atoms diffusing in unit time across a plane located at any position, x, in the x direction is referred to as the flux, J_x. Hence;

$$J_x = -D_B \frac{\partial C}{\partial x} \tag{3.15}$$

where D_B is the diffusion coefficient of element B. This equation, known as **Fick's First Law of Diffusion,** applies only to a steady-state condition in which the concentration, or the concentration gradient, does not change with time (i.e., $\partial C/\partial t = 0$ for all values of x).

Steady-state conditions frequently do not apply in practice. When an element is transferred from a gas to a solid, the concentration of that element in the surface of the solid quickly approaches equilibrium with the gas. However, transport of the element from the surface into the bulk of the solid is dependent on diffusion, resulting in a nonlinear concentration profile within the solid, Figure 3.2, which changes continuously with elapsed time. Consider, now, a volume of unit cross-sectional area with a very small width, ∂x, in the x direction. The rate of increase of the solute within this volume is equal to the difference between the rate at which it enters (J_x) and the rate at which it leaves ($J_{x+\partial x}$) the volume. But:

$$J_{(x+\partial x)} = J_x + \left(\frac{\partial J}{\partial x}\right)_x \partial x \tag{3.16}$$

for small values of x, so the rate of increase of the solute is:

$$-\left(\frac{\partial J}{\partial x}\right)_x \partial x$$

which is equal to the rate of change of concentration, $\partial C/\partial x$. Equation 3.15 is then modified to the form:

$$-\left(\frac{\partial J}{\partial x}\right)_x = \frac{\partial C}{\partial t} = \frac{\partial}{\partial x}\left(D \cdot \frac{\partial C}{\partial x}\right) \tag{3.17}$$

which is **Fick's Second Law of Diffusion.** This simplifies to:

$$\frac{\partial C}{\partial t} = D\frac{\partial^2 C}{\partial x^2} \tag{3.18}$$

if D is assumed to be independent of composition.

The above equations are not strictly valid. The driving force for diffusion is the gradient in the partial molar free energy of the solute that arises from the concentration gradient. Fick's laws are expressed in terms of concentration units, but the equations should be written in terms of the solute activities to be consistent with the driving force, and activity varies linearly with concentration only in ideal solutions. When the activity of the solute deviates markedly from ideal behavior, the use of concentration units in the diffusion equations can sometimes result in the seemingly absurd situation where a solute diffuses from a region in which it is present at a low concentration into a region where it is already at a higher concentration (uphill diffusion). This phenomenon arises simply because the solute is striving to equalize an activity gradient.

The shape of the concentration profile in Figure 3.2 can be found by integration of Equation 3.17. The boundary conditions must first be prescribed. The solution depends, for example, on the shape of the solid into which diffusion is occurring. Thus, if diffusion takes place across a planar interface of infinite area, assuming that D is independent of concentration, the solute concentration at any distance, x, from the interface is given by:

$$(C_x - C_0) = (C_s - C_0)\left[1 - \text{erf}\left[\frac{x}{(2\sqrt{Dt})}\right]\right] \tag{3.19}$$

where C_x, C_0 and C_s are as shown in Figure 3.2 and erf is an error function, the value of which is obtained from standard tables. It is apparent that if the diffusion time is increased from t_1 to t_2, then:

$$\frac{x_1^2}{t_1} = \frac{x_2^2}{t_2}$$

This implies that the depth below the surface at which the solute concentration is equal to C_x is doubled if the diffusion time is increased fourfold and there is a parabolic relation between the advance of the point, x, and the reaction time.

Solutions for a wide variety of other boundary conditions are available.[45,46]

3.2.1.2. Temperature Dependence of Diffusion Rate

The effect of temperature on the rate of substitutional diffusion can be assessed in terms of the number of lattice vacancies available and into which the atoms can migrate. The equilibrium concentration (N_v) of lattice vacancies per mole of a solid increases exponentially with increasing temperature. From a statistical thermodynamic approach, it can be shown that:

$$N_v = A \exp\left[\frac{-\Delta G_f}{RT}\right] \tag{3.20}$$

where A is a constant with a value of about three for solid metals and ΔG_f is the free energy of formation of the vacancies.

The probability of the movement of a vacancy through the lattice can be expressed similarly as:

$$P = B \exp\left[\frac{-\Delta G_m}{RT}\right] \tag{3.21}$$

where B is another constant and ΔG_m is the free energy of vacancy movement. This must be equal to the probability of the exchange of a vacancy with an atom on a lattice site. Thus, the rate of diffusion of a species is equal to the sum of the number of vacancies and the probability of the vacancy-atom interchange:

$$\begin{aligned} D &= \text{const. exp}\left[\frac{-(\Delta G_f + \Delta G_m)}{RT}\right] \\ &= \text{const. exp}\left[\frac{\Delta S_f + \Delta S_m}{R}\right] \cdot \exp\left[\frac{-(\Delta H_f + \Delta H_m)}{RT}\right] \end{aligned} \tag{3.22}$$

If the entropy terms are regarded as independent of temperature, this can be simplified to:

$$\begin{aligned} D &= D_0 \exp\left[\frac{-(\Delta H_f + \Delta H_m)}{RT}\right] \\ &= D_0 \exp\left[-\left(\frac{Q}{RT}\right)\right] \end{aligned} \tag{3.23}$$

where Q is the activation energy for diffusion and the constant, D_0, is called the frequency factor. As a rough guide, Q is approximately equal to 20 RT_m for solid metals, where T_m is the melting point of the metal on the absolute temperature scale.

The rate of diffusion of a species through a solid is typically of the order of 10^{-12} to 10^{-14} m^2 s^{-1} for substitutional solutes and one or two orders of magnitude higher for those elements such as carbon, hydrogen and nitrogen that occupy and migrate through interstitial sites. The density of the atomic packing is lower in a liquid than in a solid, and the amplitude of the atomic vibrations is markedly greater. The average values for diffusion in molten metals are correspondingly larger, ranging from about 10^{-9} m^2 s^{-1} for substitutional solutes to about 10^{-7} m^2 s^{-1} for interstitials. The rate is lower for large polymeric molecules such as silica in molten slags. Atomic order is completely destroyed in the gaseous state, and D is usually in the range 10^{-4} to 10^{-5} m^2 s^{-1} for gases.

Values of E_A for diffusion are generally smaller than for chemical reactions and, as a result, the rate of transport by diffusion increases less rapidly than the rate of the chemical reactions as the temperature is increased. This often results in a change in the rate-controlling step from chemical control at low temperatures to diffusive transport control at more elevated temperatures.

3.2.2. Heat Transfer

All pyrometallurgical operations involve heating the reactants to an elevated temperature and, since the reaction rates increase with a rise in temperature, the time taken for the reactants to reach the working temperature is an important aspect of kinetics.

If a cold solid is exposed to a hot medium (liquid or gaseous), the surface of the solid is raised almost instantaneously to the temperature of the medium. Heat then flows from the surface into the bulk of the solid by a process analogous to diffusive mass transfer and similar equations apply. Thus, under steady-state conditions, the rate of heat transfer (q) is given by:

$$q = -h\left(\frac{\partial \theta}{\partial x}\right) \tag{3.24}$$

where h is the thermal conductivity of the solid and θ is the amount of heat conducted per unit time in the x direction. This is similar to Equation 3.15 for steady-state diffusion. The corresponding form to Equation 3.17 for heat transfer under nonsteady-state conditions is obtained by substituting h and θ for D and C in that equation. Thus, there is a close analogy between heat and diffusive transfer, and it is sometimes stated that the two mechanisms are coupled, but this should not be taken to imply that the two transfer rates are equal.

3.2.3. Mass Transport

Diffusion is rarely, if ever, rate limiting for mass transfer within the bulk of a liquid or gaseous phase at elevated temperatures. Temperature gradients are invariably present and, even in the absence of artificially induced turbulence, the resultant convective currents cause circulation within the medium that short-circuits the diffusion paths and accelerates the transport of the constituents. Turbulence created, for example, by injection of a gas into a liquid or by mechanical or electromagnetic means further increases the rate of circulation and the rate of mass transport.

At the interface between two liquids or between a liquid and a gas, however, the atoms on either side of the interface adhere to each other and the two media are static or are moving at the same velocity (i.e., zero slip) along the interface. The velocity of movement increases with distance over a short range from the interface until it equals that for movement in the bulk of the medium. The rate of transport changes rapidly over this transition zone and falls to the rate for diffusive transfer at the interface. Hence, for a transport-controlled reaction, the rate of transfer of a substance across the interface and into or out of the medium is controlled by the rate of movement across this boundary layer and not by transport within the bulk phase. Under these conditions, the transport rate, r, is given by equations of the type:

$$r = \frac{\partial n}{\partial t} = k \frac{A}{V} (C_i - C_b) \tag{3.25}$$

where n, k, C_i and C_b are, respectively, the number of moles of the species transported, the mass transfer coefficient, and the concentration of that species at the interface and in the bulk phase. Artificial turbulence may increase the area, A, of the reaction site. Gas bubbling, for example, may eject droplets across the interface, resulting in a marked increase in the effective interfacial area and in the rate of the reaction.

The rate of circulation within the bulk phase by convective or artificial means is dependent on the viscosity of the medium. In general, the viscosity of a melt rises with increasing complexity of its constituent molecules or ions. Thus, the viscosity of slags is usually much higher than for molten metals and rises with increases in the concentration of polymeric anions. The viscosity of a liquid phase decreases with increasing superheating above the liquidus temperature corresponding to the composition of the phase. In contrast, the viscosity of gases well above their critical temperatures is very low and increases only slowly with increasing size of the molecular species. Typical values range from 0.01 to 0.10 cP for gases, through 1.0 to 5.0 cP for molten metals, to several hundred centipoise for acid slags containing complex molecules.

3.3. REACTIONS ACROSS INTERFACES

Reference has been made previously to the way in which a chemical reaction can often be considered as a sequence of steps, any one of which may exert a dominating influence on the rate of the overall reaction. By careful analysis of reaction rate data, it is sometimes possible to identify which step is slowest and is, therefore, rate controlling under a particular set of imposed conditions. Armed with this knowledge, it is then possible to suggest how the imposed conditions should be changed in order to accelerate the process. However, care is required to ensure that all the possible steps constituting the overall reaction and all the factors that may affect the rate of each of these steps have been considered before conclusions are drawn. Some of the relevant factors can be illustrated by considering the kinetics of reactions that occur across different types of interfaces.

3.3.1. Gas-Liquid Reactions

The multiplicity of intermediate steps is well illustrated in a series of papers by Rao[47–49] in which the kinetics of the transfer of nitrogen from a gas into molten iron were examined. The same principles apply to the transfer of any bimolecular gas into any metal. If a species, X, exists as molecules, X_2, in the gas and as separate atoms in the melt, the rates of the individual steps can be expressed by the following series of equations.

Step a: Mass Transport in the Gas Phase

Transfer of X from the gas to the melt results in a concentration gradient in the gas when the gas contains other species in addition to X. The rate of transfer of X down this concentration gradient is given by:

$$r = \frac{k_a A}{RT}(p_{X_{2(b)}} - p_{X_{2(i)}}) \tag{3.26}$$

where the subscripts b and i refer to the partial pressure of the species in the bulk gas and in the gas at the interface, respectively, and k_a is the gas phase mass transfer coefficient. This step does not occur when the gas contains only the X species.

Step b: Adsorption of X at the Interface

For the reaction:

$$X_{2(gas)} = X_{2(gas\ adsorbed)}$$

the rate of reaction is:

$$r = \frac{A}{V} k_b C_s (C^2_{X(eq)} - C^2_X) \tag{3.27}$$

where C_s is the concentration of surface sites available for the adsorption of molecules of X_2, and $C_{X(eq)}$ is the concentration of X in the metal in equilibrium with the gas. Remembering that the concentration of a gas species in the metal is proportional to the partial pressure of that species in the gas phase (Sievert's Law, Equation 2.62), the rate can also be expressed in terms of partial pressures:

$$r = \frac{A}{V} k_b' C_s(p_{X_{2(eq)}} - p_{X_2}) \qquad (3.28)$$

Step c: Dissociation into Atoms

One additional surface site is required to accommodate the two atoms of X released by the dissociation of each X_2 molecule. Hence:

$$r = \frac{A}{V} k_c C_s^2(p_{X_2} - p_{X_{2(eq)}}) \qquad (3.29)$$

Step d: Dissolution of the Atoms in the Metal

The rate equation for this step is:

$$r = \frac{A}{V} k_d C_s(p_{X_2}^{1/2} - p_{X_{2(eq)}}^{1/2}) \qquad (3.30)$$

Step e: Mass Transport in the Liquid Phase

The equation for this step is analogous to Equation 3.25:

$$r = \frac{A}{V} k_e(C_{X(eq)} - C_x) \qquad (3.31)$$

Thus, five steps are involved in this simple process of the transfer of a species from gas to metal. By comparison with Equations 3.10 and 3.12, it is apparent that steps (a), (b) and (c) are second-order reactions, while (d) and (e) are first order. If data for the overall reaction rate are fitted by a first-order plot, then steps (a), (b) and (c) are eliminated as rate controlling and vice versa.

If step (a) is rate controlling, an increase in the flow rate of the gas across the interface will accelerate mass transport in the gas phase and increase the overall reaction rate. However, an increased flow will have no effect if this step does not control the overall rate. Similarly, increased turbulence within the metal will increase the reaction rate only if step (e) is the slowest stage.

Some solute species segregate preferentially to the surface of a liquid and occupy a greater proportion of the surface atom sites than in the bulk of the liquid. The excess surface concentration, τ, of a species, Z, at the interface is given by the Gibbs adsorption isotherm:

$$\tau_Z = -\frac{1}{RT}\left[\frac{\partial\gamma}{\partial \ln a_Z}\right] \tag{3.32}$$

where γ is the surface tension of the melt and a_Z is the activity of the solute. The presence of surface-active solutes may alter the energy required for adsorption of the X solute at the interface and change the kinetics of that step. Thus, the surface adsorption of sulfur in iron retards the adsorption of nitrogen.[47] Likewise, other species in the gas phase that can be adsorbed onto the metal surface reduce the number of sites available for adsorption of the X species and decrease the rates of the interface reactions.

Additional steps are required when the reaction involves the transfer of one element from a gaseous compound. Thus, when carbon is dissolved in the metal from CO gas, extra steps are needed to describe the association of the oxygen atoms released from the CO to form molecules of oxygen, desorption of the oxygen molecules from the surface and mass transport of the oxygen into the gas phase.

Similar equations can be used to describe the kinetics of removal of a volatile species from the metal into the gas. In practice, however, it is often difficult to evaluate the kinetics of the adsorption, dissociation, dissolution, association and desorption stages. Thus, it is arbitrarily assumed that these steps are not rate limiting and they are subsumed into the chemical reaction rate. The kinetic evaluation is then limited to examination of the rates of transport to and from the reaction site and the rate of the chemical reaction at the interface.

3.3.2. Gas-Solid Reactions

A similar series of steps and equations can be used to describe the transfer of a species from a gas to a solid. In this case, however, transport through the condensed phase is entirely diffusion controlled. The concentration profile is changing continuously with time and is described by Fick's Second Law, Equation 3.17.

The time taken to attain equilibrium is dependent on the dimensions of the solid if the overall rate is controlled by mass transport in the condensed phase. Thus, an equilibrium distribution is established more rapidly as the diameter of a spherical particle or the thickness of a tabular particle is decreased. This is the reason why small particle sizes are used wherever possible in extractive metallurgy. Since the ratio of the surface area to the volume of a sphere is equal to $6d$, where d is the diameter, the rate of transfer from gas to solid across the interface is increased as *d* decreases until, eventually, the rate of replenishment of the gas phase at the interface becomes rate controlling. The particle diameter at which the change in rate control occurs can be lowered by increasing the flow rate of the gas over the reaction surface to increase the gas phase mass transport. However, in any reactor, the maximum gas flow rate is limited by the need to avoid unacceptable loss of particles carried out of the reactor in the exit gas. This places a lower limit on the particle size that can be processed satisfactorily in a flowing gas stream.

Continuous renewal of the gas in contact with the solid is readily achieved when the particles are not in close contact, as in flash smelting or in a fluidized bed where the reaction is virtually instantaneous. It is more difficult to obtain continuous replenishment when the solids are packed together in a bed in a blast furnace, for example. This is why fine particles are often sintered together to form larger aggregates of irregular shape, or are formed into pellets, prior to reaction in a packed bed. The voidage between the solids in the bed is thus increased, facilitating the flow of gas to the interfaces, but the reaction rate is still very low in comparison with flash smelting.

3.3.2.1. Gaseous Reduction

Contact between a gas and a solid often results in a chemical reaction that introduces additional stages into the overall reaction. Consider the reduction of a solid oxide ore by a gaseous reducing agent such as H_2 or CO. The chemical reaction occurs on the surface of the solid when the ore is first exposed to the gas. Oxygen is removed from the surface and the solid is soon covered by a layer of metal. Continued reaction now requires transport within the solid, but the rate of diffusion of oxygen through the solid lattice is relatively slow and cannot account for the markedly higher reaction rates found in practice. Solid-state transport may occur by diffusion of metal atoms inward from the surface into the core and a counterdiffusion of vacant lattice site outward to the surface. However, this mechanism can only continue if the metal atoms condense or cause a transformation of a higher oxide into a lower oxide that results in a continuous supply of lattice vacancies. The small hydrogen atoms can diffuse relatively rapidly through the interstitial sites in the lattice. The CO molecule can dissociate on the surface and the carbon atoms thus released can also diffuse through the interstitial sites, but the H_2O, CO and CO_2 molecules produced by the reduction reaction are much too large to escape by lattice diffusion. If this was the only means of transport, then either the reaction would cease after the metal layer had formed or the product gases would build up a disruptive pressure sufficient to crack or shatter the reduced layer.

In the general case, a different mechanism operates. Oxygen atoms constitute half the total number of atoms in an oxide of the MO type and two thirds of the total in an MO_2 oxide. The removal of the oxygen atoms must, therefore, result in a reduction in the volume of the solid. The atoms are usually more closely packed in the metal than on the oxide lattice, making a further contribution to the volume decrease. In some cases, the volume change accompanying the loss of oxygen and the resultant change in the atomic packing (e.g., from a hexagonal lattice in Fe_2O_3 to a cubic lattice in Fe_3O_4) is sufficiently large to cause the formation of shear cracks in the solid.

These changes are not usually accompanied by a corresponding reduction in the overall dimensions of the solid. Quite often, the particle size is increased. This swelling is evidence, which is readily substantiated by metallographic examination, that the shrinkage within the solid is compensated by the formation of fine pores. When the pores are interconnected by channels that reach the outer surface, the reactant gases can circulate through the channels to reach the unreacted solids in the core of the solid. The product gases can escape via the same route. Macroporosity is already present, before reduction, in properly prepared pellet and sinter aggregates. The shrinkage during reduction enhances the macropores and introduces additional micropores that facilitate gas diffusion.

The reaction sequence then comprises transport of the reactants within the gas to the surface of the solid, followed by diffusion of the reducing gas through the pores to reach the reaction site. Chemical reaction between the reductant and the oxide is followed in turn by counterdiffusion of the reaction products through the pores to the surface and transport from the surface to disperse in the bulk gas. The adsorption, desorption, etc. stages are also involved at the reaction site. Since mass transport within the bulk gas is facilitated by the flow of the gas over the solid surface, whereas the atmosphere within the pores is essentially static, it is evident that gas diffusion within the pores is likely to be the rate-limiting step if the overall rate is transport controlled.

The reaction products cannot accumulate within the pores if the reaction is to continue, so the diffusive flux of the product gas species must equal the flux of the reactants. The diffusivities of the gaseous species are dependent on the diameter of the pores, relative to the diameters of the gas molecules. If the diameter of the pores is much larger than the mean free path of the gas molecules, the rate of transport is directly proportional to the amount of porosity and inversely proportional to the sinuosity of the pores. The latter is a measure of the average distance the gas must travel along a pore to reach the surface from a reaction site. Under these conditions, molecular diffusion prevails and the diffusivity is inversely proportional to the partial pressure of the diffusing gas species. At smaller pore diameters, the gas molecules collide frequently with the pore walls, hindering the movement and a slower, Knudsen diffusion rate applies, with the diffusivity independent of the gas pressure. The average pore diameter tends to increase with increasing temperature until a stage is reached when

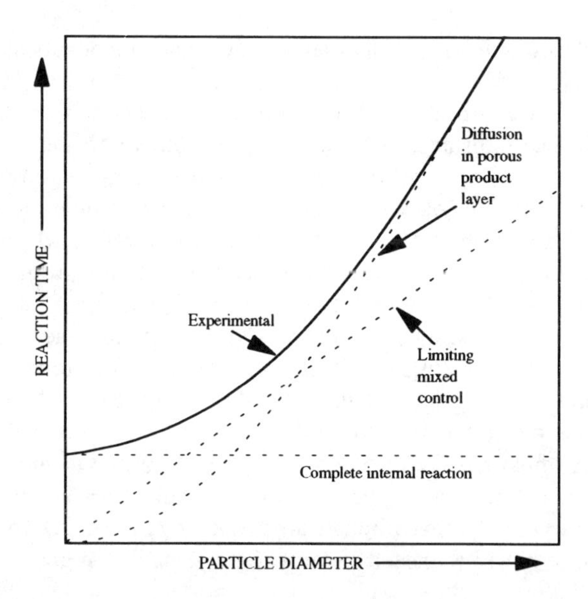

FIGURE 3.3. Schematic representation of time of reaction for porous materials as a function of time for three limiting rate controlling processes. (From Turkdogan, E. T., *Physical Chemistry of High Temperature Technology,* Academic Press, New York, 1980. With permission.)

atomic diffusion within the solid can occur at a sufficiently rapid rate to cause sintering and reduction in the pore volume.

The kinetic equations that apply to this mode of reaction can be quite complex (see, for example, References 50 to 52).

The overall reaction rate is not dependent on the dimensions of the solid particles if the chemical reaction is rate controlling, but is dependent on, e.g., the square of the diameter of a spherical particle if gaseous diffusion through the pores controls the reaction rate. Thus, the controlling step may change as the particle size is increased. The gaseous reduction of solid iron oxides is chemically controlled at very small particle diameters and gas diffusion controlled at larger diameters.[53] A mixed regime, in which both steps proceed at similar rates, applies at intermediate particle sizes, as shown in Figure 3.3. This diagram clearly shows the marked increase in reaction time with increasing particle diameter and emphasizes the importance of producing voidage and linking macropores in the manufacture of sinter. Since the activation energy for chemical reaction is usually larger than for diffusion, it is evident that the transition from one rate control mechanism to the other occurs at smaller particle sizes as the temperature is increased.

Microscopic examination of the cross-section of a partially reduced particle can be a useful aid in the evaluation of the rate-controlling mechanism. A sharply defined interface, separating the reacted and unreacted zones, is usually adduced as evidence that the chemical reaction occurs as rapidly as the reactants arrive at, or the products leave, the reaction site. The overall rate is transport controlled and the reduction is described as ''topochemical.'' A diffuse interface, often with appreciable width indicating an increase in the area over which reaction is occurring, is suggestive of chemical reaction control. This latter type of ''uniform internal reduction'' is often found when the unreacted solid is very porous and provides gas diffusion channels that are available for gas penetration from the start of the reaction.

The kinetics of roasting of sulfide ores to convert the mineral to an oxide depend on similar principles, with oxygen gas diffusing along the pores to the reaction site and the products (SO_2, SO_3, etc.) escaping by counterdiffusion. Since the latter molecules are larger than oxygen molecules, the pore size at which gas diffusion changes from Knudsen to molecular flow is determined by the product species. No inward diffusion is involved when carbonate ores or calcium carbonate are heated to remove CO_2, but a porous layer of oxide must be formed to allow the escape of the gas.

3.3.2.2. Heat Changes in Gas–Solid Reactions

The chemical reactions between a gas and a solid involve heat changes that can alter the temperature distribution within the solid. Hydrogen or CO reduction is exothermic for metal oxides lying above the H_2-H_2O or $CO-CO_2$ line on the oxygen potential diagram and endothermic for oxides lying below the line. Dissociation of carbonates is always endothermic, while the roasting conversion of sulfides can be strongly exothermic. The reaction heat is consumed or released at the reaction site, which is within the solid after the surface layer has reacted. Some of the heat required or released can be compensated by thermal transfer with the reaction gases. The balance can be adjusted only by conductive heat transfer through the solid to the gas-metal interface and then by convective or radiative transfer to the bulk gas. Heat conduction through a solid is relatively slow. It is probable, therefore, that the actual temperature at the reaction site is higher than the measured reactor temperature for strongly exothermic reactions and vice versa. Since the values of the reaction rate constants are temperature dependent, it is evident that a reaction rate predicted in terms of the measured surface temperature of the particles may differ from the rate actually obtained.

3.3.3. Liquid-Liquid Reactions

Reactions occurring across an interface between a molten metal and a molten slag or a matte (i.e., a mixture of molten sulfides) are an important feature of pyrometallurgical operations. As in the preceding case, the kinetics are usually considered in terms of mass transfer to and from the interface and chemical reaction at the interface. The adsorption and desorption steps are assumed to be rapid in comparison with the rates of the other steps and are usually ignored. In the majority of cases, the chemical reaction rate is high and the rate-controlling step is transport to or from the reaction site.

The rate of diffusion of a solute down a concentration gradient in a molten metal is typically about 10^{-3} s mm^{-1}. Diffusive transport in a molten slag may be orders of magnitude slower and decreases as both the viscosity of the slag and the size of the diffusing species are increased. The molten bath may be up to 1 m in depth, so equilibrium would be approached very slowly if diffusion was the only transport mechanism within the metal. The depth of the slag layer is usually much less than the metal but, even so, it is large in terms of diffusion distances. As indicated earlier (Section 3.2.3), the reaction rates achieved in practice are very much faster than would be predicted by assuming that transport throughout the liquid phases is diffusion controlled. The actual rates are lower, however, than would be predicted by assuming only convective or turbulent transport and that diffusion is completely eliminated. Transport kinetics are often considered, therefore, in terms of the boundary layer model or the surface renewal model.

3.3.3.1 The Boundary Layer Model

The model assumes that a static film of liquid is present at the interface. Convective or turbulent flow completely eliminates concentration gradients in the liquid phase beyond this boundary film, but transport across the static layer is controlled by diffusion. It is usually assumed that steady-state conditions exist; thus, Fick's First Law (Equation 3.15), can be used to describe the linear variation of the concentration of the diffusing species across the boundary film. If D is assumed to be independent of composition:

$$\begin{aligned} J &= \frac{\partial n}{\partial t} = -D\left[\frac{\partial C}{\partial x}\right] \\ &= \frac{D}{\delta}(C_b - C_e) \end{aligned} \tag{3.33}$$

where δ is the thickness of the boundary film, C_e is the equilibrium concentration of the species at the interface and C_b is the concentration of that species in the bulk liquid.

If the rate is transport controlled, the reaction proceeds as rapidly as the reactants arrive at, or the products are removed from, the interface. There is no accumulation of reactants or products at the reaction site. Thus, the flux up to the interface in one phase must equal the flux away from the interface in the other phase. For solute transfer from metal to slag, therefore:

$$J_{(m)} = J_{(s)} = \frac{[D(C_b - C_e)]_m}{\delta_m} = \frac{[D(C_e - C_b)]_s}{\delta_s} \tag{3.34}$$

where subscripts m and s refer to the metal and the slag phase, respectively. From this relationship, it can be shown that:

$$J = \frac{\partial n}{\partial t} = \frac{C_{bm} - C_{bs}}{\dfrac{\delta_m}{D_m} + K\dfrac{\delta_s}{D_s}} \tag{3.35}$$

where K is the equilibrium constant for the distribution of the species between the metal and the slag.

It is apparent that when the reacting species has a very low solubility in the metal, compared to the solubility of the product species in the slag, the transport resistance (δ_m/D_m) in the metal is very much larger than the resistance ($K\,\delta_s/D_s$) in the slag and the rate-limiting step is likely to be transport in the metal. The converse applies when the solubility is much greater in the metal than in the slag. Transport in either of the phases may be rate limiting when the solute concentration in both phases is similar.

The thickness of the boundary layer cannot be measured experimentally. The actual concentration profile for a solute shows a smooth change with distance from the interface, as shown by the solid line in Figure 3.4, and not an abrupt change, indicated by the broken line, when the solute concentration reaches that of the bulk liquid, as implied by Equation 3.33. Thicknesses estimated by fitting kinetic data to the boundary layer equations are usually in the range from 0.01 to 0.1 mm for the metal layer and up to five times greater for the slag layer.

When more than one solute is being transferred simultaneously from metal to slag, or vice versa, the estimated thickness of the boundary layer is not constant but varies from solute to solute. This is not surprising when it is remembered that no allowance is made for the variation of the diffusion coefficient with composition, the masses of the reactants and products are expressed in units of concentration and not activities, and the actual concentration profile across the layer is not linear.

Comparing Equations 3.26 and 3.31 for transport in a gas and in a metal with Equation 3.33 shows that the term D/δ is equal to the mass transfer coefficient, k, for the reaction step.

3.3.3.2. The Surface Renewal Model

This model assumes that small packets of material with the average bulk composition are brought to the interface by turbulence within the melt. Solute is transferred across the packet to the interface during the interval in time while the packet resides at the surface before it returns to the bulk phase. The kinetics are then evaluated in terms of the average dwell time for a packet at the surface, the fraction of the surface covered by the packets,

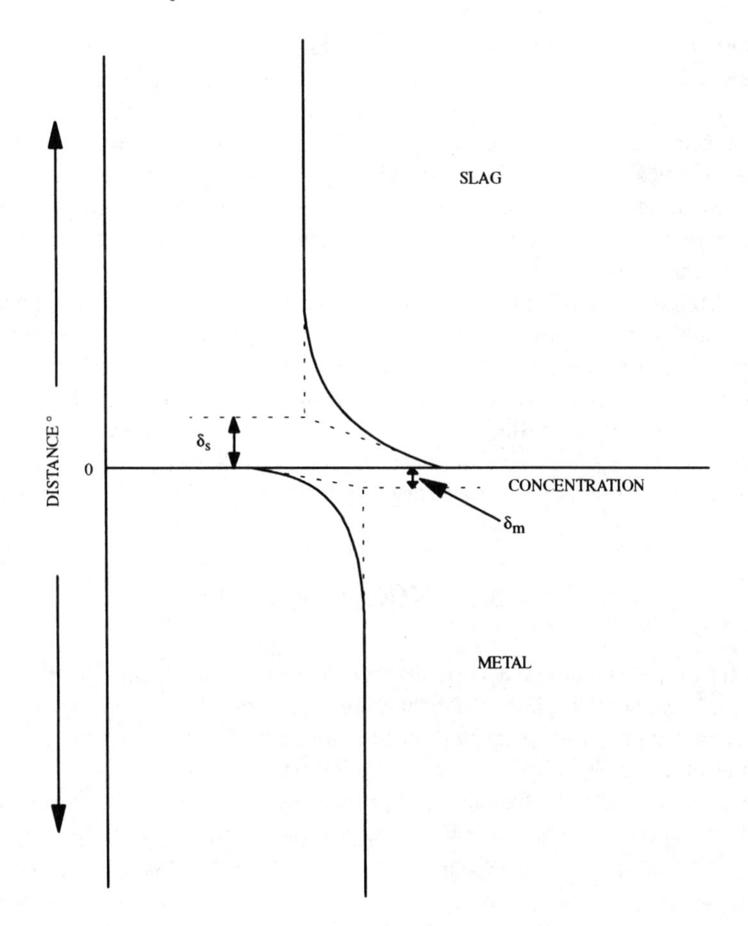

FIGURE 3.4. Concentration profiles across boundary layers.

the diffusion coefficient of the solute within the packet and the concentrations of the solute in the bulk phase and at the interface. The resultant equations reduce again to the form of Equations 3.26 and 3.31 when the surface renewal and diffusion terms are replaced by the mass transfer coefficient.

3.3.3.3. The Overall Reaction Rate

This is determined by the sum of the reaction rates for the individual steps. Thus, for a first-order reaction between metal and slag, the rate is given by:

$$r = \frac{\partial C}{\partial t} = \frac{A}{V} \cdot \frac{(C_m - KC_s)}{\left[\frac{1}{k_m} + K \cdot \frac{1}{k_s} + \frac{1}{k_c}\right]} \tag{3.36}$$

where k_m and k_s are the specific rate constants for the transport steps in the metal and in the slag, respectively, k_c is the rate constant for the chemical reaction at the interface and K is the equilibrium constant for the reaction. This shows that one of the steps controls the overall rate only when k for that step is significantly larger than the rate constants (or the product K/k_s) for the other steps. When two of these terms have similar high values, compared with the other terms, the reaction is under mixed control, in similar manner to that shown in Figure 3.3.

Activation energies for diffusion-controlled reactions in molten phases are usually in the range 5 to 20 kJ mol^{-1}, whereas the values for chemical control are often greater than 50 kJ mol^{-1}. Thus, if the activation energy for the overall reaction is low, it is reasonable to assume that the overall rate is transport controlled, and increased agitation will increase the rate of the reaction. Conversely, increased agitation will have no accelerating effect when the activation energy is high and the rate is chemically controlled.

3.4. NUCLEATION

Extractive metallurgical operations often involve the formation of a new phase. Examples include the formation of deoxidation products within a homogeneous phase, precipitation of a solute as a solid when the temperature of the melt is lowered (e.g., separation of solid Cu from molten Pb by liquation), and the formation of gas bubbles within a single-phase melt. The production of the new phase requires the supply of additional energy to create an interface between the new and the parent phase. This can be regarded as a barrier that must be overcome before the reaction can proceed.

3.4.1. Nucleation of Precipitates

Consider the separation of copper from molten lead containing, say, 3.0 wt% Cu by cooling the melt to precipitate the solid copper. The Cu-Pb phase diagram is shown in Figure 3.5. There is extensive liquid immiscibility, which terminates at 6.5 wt% Cu in Pb-rich solutions. Solid Cu is in equilibrium with molten Pb at lower temperatures and the solubility of Cu in Pb decreases as the temperature falls, until it is eventually equal to the Cu content of the melt (i.e., 3%). The solution becomes supersaturated with further decreases in temperature and the supersaturation provides the driving force for separation. Since the temperature is now well below the melting point of Cu, the excess solute attempts to precipitate as a solid phase. This requires segregation of the solute atoms into clusters in which the atoms are arranged in the face-centered cubic configuration of the solid Cu lattice. Such chance fluctuations can form at any temperature, but they can only become stable below the saturation temperature.

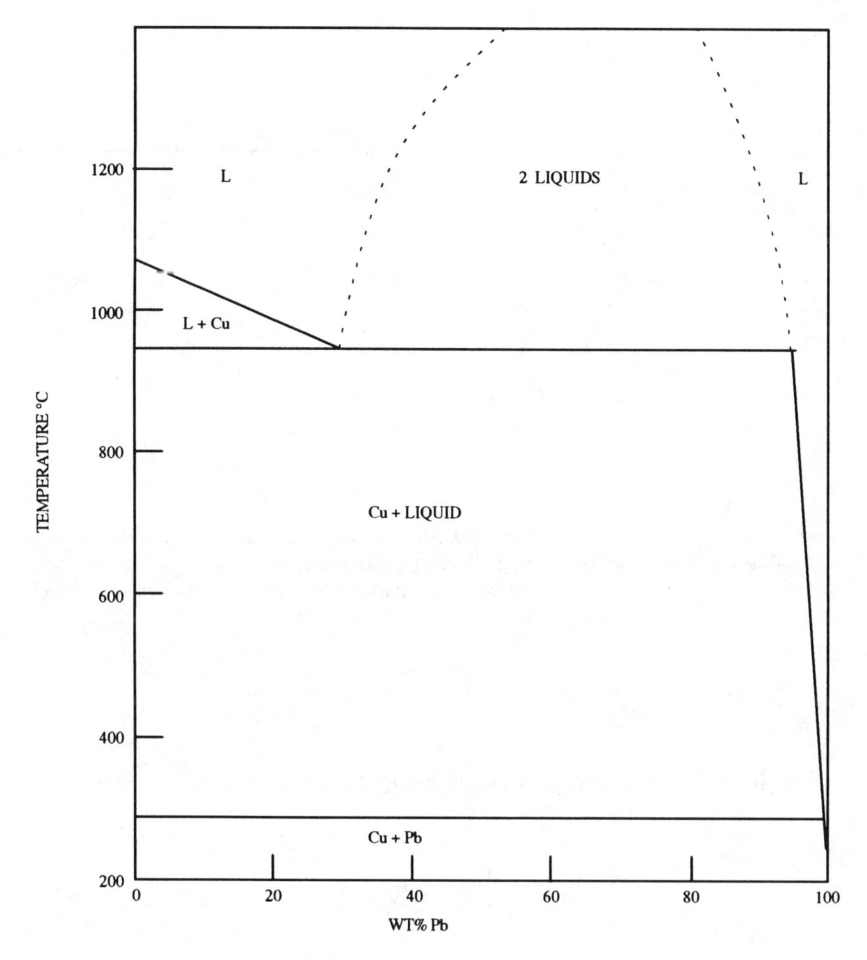

FIGURE 3.5. The copper-lead phase diagram.

An interface between the atomic cluster and the rest of the solution must be formed before it can be recognized as a separate phase. This requires the supply of surface energy. Except when the latter is dependent on crystallographic orientation, the cluster will adopt a spherical form to minimize the amount of surface energy. The surface energy required is then dependent on the surface area of the sphere and is equal to $4\pi r^2\gamma$, where r is the radius of the cluster and γ is the surface energy between the liquid and the solid; this term is always positive.

The driving force for precipitation is dependent on the product of the free energy difference between the liquid and the solid and the volume of the cluster, and is equal to $\frac{4}{3}\pi r^3\ \Delta G_V$. ΔG_V is a negative quantity that increases in value as the temperature falls below the saturation point. Hence,

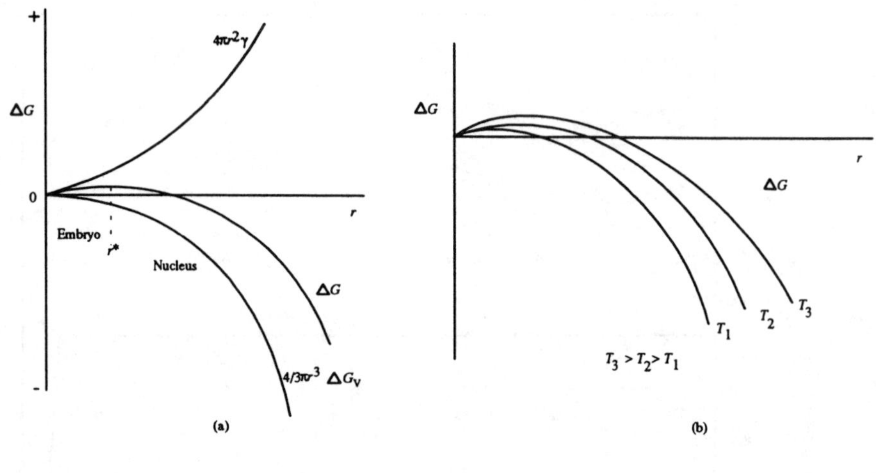

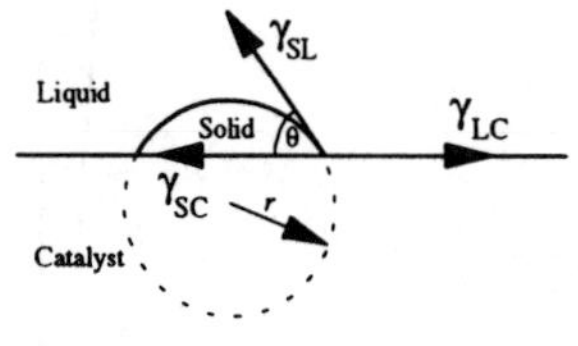

FIGURE 3.6. Energy required for the nucleation of a precipitate: (a) promotion from embryo to nucleus; (b) effect of temperature; and (c) nucleation on a solid substrate.

the total free energy change accompanying the formation of the solid phase is:

$$\Delta G = \frac{4}{3}\pi r^3 \, \Delta G_V + 4\pi r^2 \gamma \tag{3.37}$$

The surface area per unit volume is large for small values of r, and the positive surface energy term is dominant. As the radius increases, a point is reached where the negative volume free energy term becomes predominant. Up to this stage, any growth in the size of the cluster increases r and the positive energy required, so the cluster becomes less stable. Beyond the point, when the volume free energy is dominant, the total energy decreases as r increases. This situation is illustrated in Figure 3.6a. Up to the saddle point, which is marked as r^* on the curve, the cluster is unstable and is called an **embryo.** The cluster is stable at $r > r^*$ and is transformed into a **nucleus,** which can grow continuously until the supersaturation is exhausted.

The critical nucleus size is found by differentiating Equation 3.37 with respect to r:

$$r^* = \frac{2\gamma}{\Delta G_V} \tag{3.38}$$

The surface energy is relatively insensitive to temperature, but ΔG_V becomes more negative as the temperature falls and the supersaturation is increased. As a result, r^* decreases with decreasing temperature, as shown in Figure 3.6b, and a smaller embryo can be promoted into a nucleus. Substituting r^* into Equation 3.37 gives the energy required, ΔG^*, to create the nucleus:

$$\Delta G^* = \frac{16}{3} \cdot \frac{\pi\gamma^3}{\Delta G_V^2} \tag{3.39}$$

This term is also temperature dependent and can be regarded as the activation energy required for the formation of the nucleus. The formation of precipitates in this way, without artificial stimulation, is described as **homogeneous nucleation.**

These concepts can be applied to the consideration of the precipitation of Cu from molten Pb. The segregation of copper atoms in the molten phase to create a cluster of sufficient size to form a nucleus is dependent on diffusion of Cu atoms in the melt, which requires time. The minimum number of atoms required in the cluster decreases as the temperature is lowered. However, the rate of diffusion of Cu atoms through the molten Pb also decreases with falling temperature, reducing the rate at which the clusters can grow. If the temperature is lowered too far before precipitation commences, there is a risk that the Pb will also solidify and the Cu precipitates will then be trapped in the solid. Slow cooling is therefore essential to allow time for the Cu nuclei to form and to float out of the melt.

The need for slow cooling can be avoided by mixing finely divided copper particles into the melt as soon as the temperature falls below the saturation line. If the particles are larger than the critical nucleus size, they will remain stable and the Cu atoms in solution can attach to the solid surfaces as they come into contact. A uniform dispersion of fine particles is required in the melt to reduce the distance over which the solute atoms have to diffuse to reach a solid surface.

Artificial or **heterogeneous nucleation** of a precipitate is not limited to seeding the melt with the same solid. A solute atom in supersaturated solution can attach itself to any solid substrate that has the same crystal structure and a similar lattice spacing to that which the solid solute would normally adopt.

It is possible that solid substrates that do not satisfy these requirements can act as catalysts for nucleation by lowering the magnitude of the surface energy term in Equation 2.27. This is illustrated in Figure 3.6c, where the precipitating phase forms as a hemispherical cap, with an effective radius, *r,* on the surface of the catalyst. Assuming that the interfacial energies are isotropic, the balance of the surface tension forces is given by the appropriate form of Equation 1.4, i.e.,

$$\gamma_{lc} = \gamma_{sc} + \gamma_{sl} \cos\theta$$

where the subscripts, l, c and s, refer to liquid, catalyst and solid, respectively, and θ is the angle subtended by the tangent to the surface of the solid at the point of contact with the catalyst. The volume of the hemispherical solid is equal to $\frac{1}{3}\pi r^3(2 - 3\cos\theta + \cos^3\theta)$ and hence, by substitution in Equation 3.39 and rearranging terms:

$$\Delta G^* = \frac{16}{3} \cdot \frac{\pi\gamma^3_{sl}(2 - 3\cos\theta + \cos^3\theta)}{4\Delta G_V^2} \qquad (3.40)$$

An illustration of the use of this equation to calculate the effectiveness of a solid as a catalyst for nucleation is given in Worked Example 5.

In comparison with Equation 3.39, it is apparent that there is no barrier to nucleation when the precipitating solid completely wets the surface of the catalyst and $\theta = 0°$. Conversely, ΔG^* is given by Equation 3.39 when the catalyst is not wetted and $\theta = 180°$. There is always some degree of wetting but, when θ is less than 40°, the value of ΔG^* is reduced to less than one tenth the value for homogeneous nucleation.

A melt can be deoxidized by adding elements that form oxides which are more stable than the oxides of the host metal. The reaction can usually be regarded as isothermal in the initial stages. The driving force for nucleation of the deoxidation products is then provided by the localized supersaturation with the dissolved oxygen, which is created as the solid pieces of the deoxidant dissolve in the melt. The deoxidation products have a lower density than the metal being refined and, given time, they rise through the melt and are absorbed into the slag. Since the solubilities of both the dissolved oxygen and the added element usually decrease with decreasing temperature, however, further precipitates are nucleated as the molten metal is cooled prior to solidification. This may result in the entrapment of some of the precipitates as inclusions in the solid metal, which are detrimental to the mechanical properties.

3.4.2. Nucleation of Gas Bubbles

The formation of a gas bubble within a molten metal is dependent on similar principles. The total pressure on a bubble within the melt is:

$$p = p_a + p_f + \frac{\gamma}{d} \qquad (3.41)$$

where p_a is the atmospheric pressure above the melt, p_f is the ferrostatic pressure of the melt above the bubble and γ is the surface tension between the gas and the melt. The total pressure increases with decreasing bubble diameter, d, and is very high for bubbles of extremely small diameter. For example, γ is about 1.5 J m^{-2} at 1600° C for liquid steel containing 0.5% C in contact with a bubble of CO. Thus, the total pressure on a bubble only 0.001 mm in diameter just below the surface of the melt is about 60 atm. This is roughly 20 times the value of p_{CO} in equilibrium with the carbon and oxygen dissolved in the iron. The pressure on the bubble rises with increasing depth below the surface of the melt, due to the increase in the ferrostatic pressure.

A very large number of C and O atoms would have to cluster together, excluding Fe atoms from the volume, in order to nucleate a gas bubble of even this minute size. As the diameter of the initial bubble decreases to dimensions more appropriate to the size that might be achieved by chance clustering of atoms, the pressure within the bubble increases to several thousand atmospheres. Clearly, homogeneous nucleation of a gas bubble within a melt is a rather unlikely occurrence and some form of artificial nucleation, similar to the effect of a catalytic surface described above, is necessary.

The mechanism of gas release from a liquid can be observed if a carbonated beverage (e.g., soda water) is poured into a glass. The bubbles form on the sides and bottom of the glass and not within the liquid. The surface of the glass may appear to be smooth but, on a microscopic scale, it contains numerous minute cavities and scratches. Many of these are filled with the liquid, but surface tension prevents the liquid entering the finer cavities. The minimum pore radius, r, that the liquid can penetrate is:

$$r = -\frac{2\gamma \cos \theta}{p_a + p_f} \tag{3.42}$$

where θ is the contact angle between the gas in the cavity and the surface of the glass. When the glass is filled, air remains trapped in cavities smaller than this critical size and the CO_2 dissolved in the liquid can diffuse into the air pocket. The size of the hemispherical gas cap thus increases until it eventually exceeds 90°. A bubble then detaches and rises to the surface, leaving a residual pocket of gas in the crevice in which the process of bubble growth and detachment can be repeated. A similar mechanism operates for gas release from a molten metal, crevices in the furnace lining providing the residual gas pockets from which the bubbles are released.

It is evident from Equation 3.41 that the pressure within the bubble decreases both with decreasing ferrostatic pressure (i.e., as the bubble approaches the melt surface) and with increasing size. As the bubble rises through the melt, it absorbs more of the atoms that form the gas and its size increases, thus approaching closer to equilibrium with the gas atoms that remain in the melt. Usually, however, the depth of the melt is restricted in order to achieve a large interfacial area per unit volume with the slag. Consequently, the transit time of the bubble through the melt is short and there is insufficient time for equilibrium to be fully attained. Turbulence in the bath may cause some recirculation of the bubbles, but the increased average residence time in the melt is still inadequate for full attainment of equilibrium. Thus, the amount of carbon and oxygen dissolved in molten steel is greater than the quantity in equilibrium with $p_{CO} = 1$ atm when this is the only mechanism of gas evolution, as in the top-blown BOS steelmaking process.

Two methods are commonly used to remove more of the dissolved gas atoms from the melt. If an inert flushing, or purging, gas such as argon is bubbled through the melt, the gas atoms in the metal dissolve in the bubbles in an attempt to achieve the partial pressure in the bubbles in equilibrium with the activity of the dissolved species, in accordance with Sievert's Law, Equation 2.62. This scavenging action relies upon the dissolved species coming into contact with the bubble surface, but there is usually insufficient time for full attainment of the equilibrium partial pressure before the bubbles escape from the melt. Consequently, the amount of the dissolved species removed per Nm^3 of the flushing gas increases as the bubble size decreases. In practice, the inert gas is usually introduced through some type of disperser or diffuser to produce a large number of bubbles that are only a few millimeters in diameter.

Alternatively, the melt may be exposed to a partial vacuum. This lowers the atmospheric pressure in Equation 3.41 and aids bubble nucleation. It also reduces the partial pressure of the gaseous species above the bath. This shifts the equilibrium in the formation reaction to lower contents of the dissolved elements and decreases the rate of the reverse reaction, relative to the rate of the forward reaction, according to Equations 3.2 and 3.3.

FURTHER READING

Belton, G. R. and Worrell, W. L., Eds., *Heterogeneous Kinetics at Elevated Temperatures,* Plenum Press, New York, 1970.

Geiger, G. H. and Poirer, D. R., *Transport Phenomena in Metallurgy,* Addison Wesley, New York, 1973.

Guthrie, R. I. L., *Engineering in Process Metallurgy,* Clarendon Press, Oxford, 1989.

Schlichting, H., *Boundary Layer Theory,* McGraw-Hill, New York, 1979.

Sohn, H. Y. and Wadsworth, M. E., *Rate Processes in Extractive Metallurgy,* Plenum Press, New York, 1979.

Szekely, J. and Themelis, N. J., *Rate Phenomena in Process Metallurgy,* Wiley, New York, 1971.

Szekely, J., Evans, J. W. and Sohn, H. Y., *Gas Solid Reactions,* Academic Press, New York, 1976.

Chapter 4

PYROMETALLURGICAL EXTRACTION

Four major factors influence the choice of an extraction process:

1. The cost of production of the metal from the available feedstock (ore, concentrate, flue dust, etc.) by each of the possible extraction methods, relative to the selling price of the product, as discussed in Chapter 1.
2. The feasibility of producing a metal with a composition suitable for the next stage of processing. This involves consideration of the extent to which the possible reactions can proceed and is determined by thermodynamic constraints.
3. The rate at which the composition changes can be achieved, which is controlled by the kinetics of the reactions.
4. Environmental and safety legislation, which is progressively becoming more restrictive and limiting the freedom of choice.

These requirements are often in conflict and the problem is to select the most profitable compromise.

One of the major requirements in the production of metals is the separation of the metallic minerals from the gangue that remains after the ore has been subjected to the mineral dressing and beneficiation treatments. This can be accomplished by extracting the metal from the ore as a soluble compound, leaving the gangue as a solid residue, or by dissolving the impurities to leave a pure metallic mineral. Further treatment is then necessary to convert the soluble compound or the mineral to metallic form. The principles of this mode of hydrometallurgical production are considered in Chapter 6. The major route, however, in terms of the annual tonnage of metals produced, incorporates the separation of the mineral from the gangue and the reduction of the mineral to the metallic form in one or more sequential steps. This is the pyrometallurgical extraction route. Efficient separation of metal and gangue is very difficult to achieve in the solid state, so the ore is usually heated until the reduced metal, or both the metal and the gangue, are molten. The difference in the density of the two constituents then results in separation of the melt into two discrete layers.

4.1. REDUCING AGENTS

Metallic minerals most commonly occur as oxide, sulfide, carbonate and hydroxide compounds. On heating, CO_2 and water vapor are readily expelled from the carbonate and hydroxide ores, leaving an oxide or sulfide residue, thus, the extraction stage is concerned mainly with the reduction of oxide and sulfide minerals.

The relative stabilities of the oxides are illustrated in the oxygen potential diagram (Figure 2.8). The corresponding standard free energy-temperature plot for the formation of some of the metallic sulfides is shown in Figure 4.1. When these two diagrams are compared, the most obvious difference is the bunching together of most of the sulfide stability lines at small negative free energy values, relative to the spread of the oxide stability lines. Closer inspection shows that Ag_2S is more stable than Ag_2O. This explains why silver artifacts tarnish more rapidly when traces of sulfur species are present in the atmosphere. Hg, Pd and Pt behave similarly. Cuprous oxide, Cu_2O, is slightly more stable than Cu_2S at low temperatures, but the sulfide becomes the more stable compound as the temperature is increased. The oxides of most of the other common engineering metals are markedly more stable than their sulfides at all temperatures.

A compound may dissociate into its constituent elements at any temperature wherein its standard free energy of formation is positive. Au_2O is unstable at room temperature, Ag_2O becomes unstable above 190 °C and the oxide and sulfide of mercury are unstable above about 600 °C when, in each case, the pure compound is in equilibrium with pure oxygen gas at a pressure of 1 atm. These metals can be reduced from their ores simply by heating. The dissociation temperatures are lowered when air instead of oxygen gas at 1 atm pressure is in contact with the ore. The temperature can be lowered still further by heating the ore in a vacuum but, within the range of vacuum pressures that can be readily attained in commercial practice, a high temperature is still required for the complete dissociation of most of the other naturally occurring metallic compounds. A higher operating temperature than that required for equilibrium dissociation would also be required to give sufficient driving force for the reaction. Consequently, the more stable oxides and sulfides cannot be reduced economically by this means.

When thermal dissociation is not feasible, some other technique must be used to reduce the mineral compounds to the metallic form. The obvious way is to mix the ore intimately with some other substance that forms an oxide or a sulfide which is more stable than the metallic compound to be reduced. Inspection of the oxygen potential diagram shows that the Si-SiO_2 line lies roughly across the middle of the plot. This means that metallic silicon could act as a reducing agent for the oxides of Cu, Fe, Ni, Pb, Sn and all the other metals that lie above the silicon line on the diagram. Conversely, SiO_2 is oxidizing to all the metals that oxidize at lower (i.e., more negative) oxygen potentials. As explained in Chapter 2, however, the lines on the free energy-temperature plots are shifted upward to less negative values when the activity of the metal is less than the activity of the oxide, and vice versa. So, it is only safe to conclude that one metal can

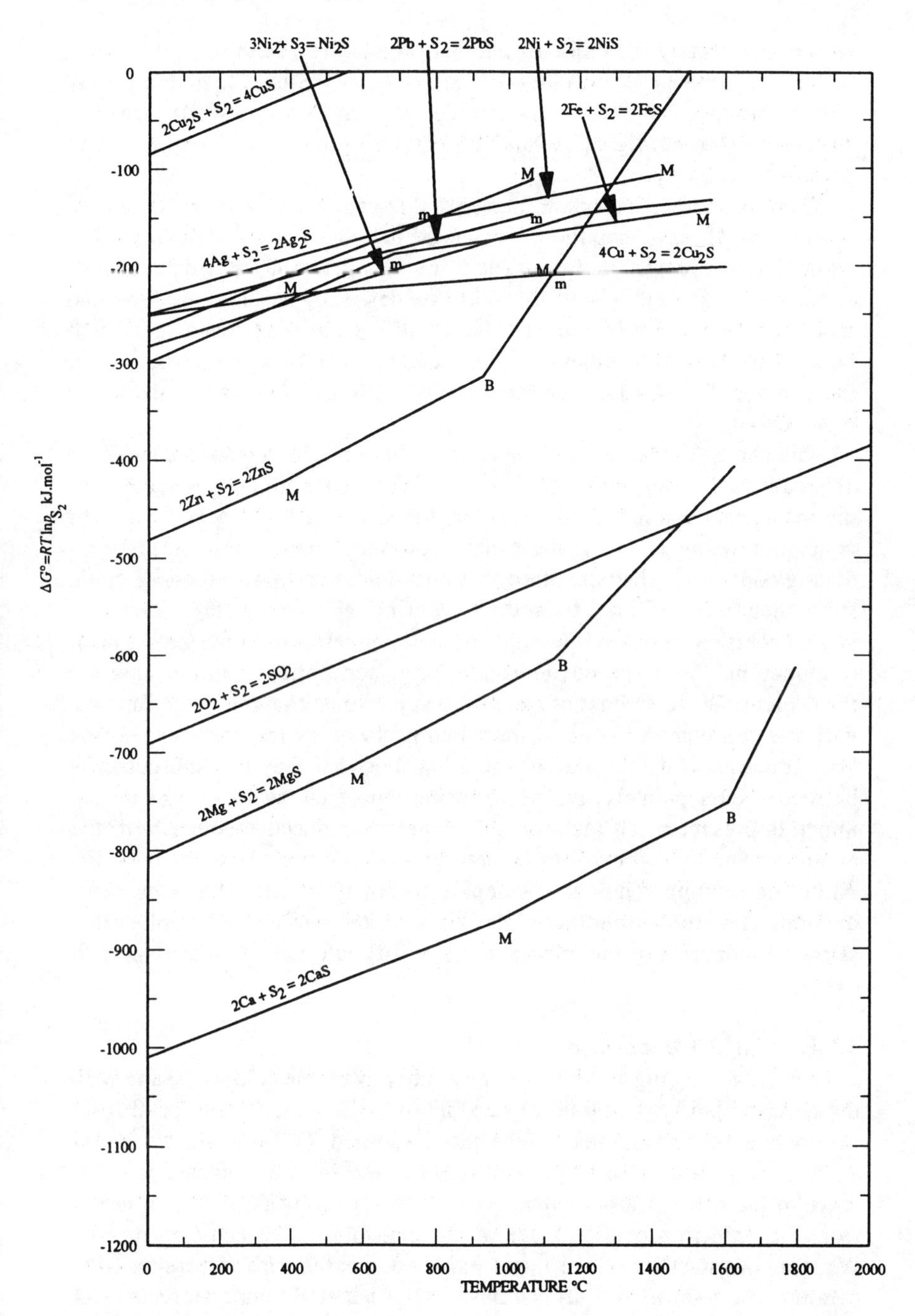

FIGURE 4.1. Temperature dependence of the free energy of formation of some metal sulfides.

reduce completely the oxide of another metal when there is a fairly large difference between the stabilities of the two oxides. Since the lines are more closely grouped together on the sulfide diagram, it follows that there is a more restricted choice of metals that can be used to reduce sulfide compounds in this way.

There is another important inference that can be drawn from these considerations. If, say, metallic Si was used to reduce SnO_2, then any other oxide that was present in the ore and less stable than SiO_2 would be reduced simultaneously. Only the more stable oxides, such as alumina, lime and magnesia, would not be reduced. Hence, the probability that a metal will be contaminated with other metallic elements increases as the stability of the compound formed by the reducing agent (e.g., silica in this example) is increased.

This can be expressed in another way. Only mildly reducing conditions are required for the reduction of Cu_2O. When such mildly reducing conditions are maintained in the reactor, the Cu is reduced selectively and contains few impurities. If the metal is produced under more strongly reducing conditions, then the other elements that are now also reduced could subsequently be oxidized from the Cu during the refining stage. Hence, it is not necessary to remove the other metallic minerals from the Cu ore prior to reduction. The major advantages of prior concentration in this case are the decrease in the amount of material that has to be charged to the furnace, with the consequent saving in the amount of energy required for melting and reduction, and the value of the other metals if they can subsequently be recovered separately. At the opposite end of the scale, almost all the minerals present in an alumina ore would be reduced together with the aluminum, and the impurities cannot be oxidized preferentially from the Al during refining. Thus, it is essential to purify the alumina prior to reduction. The close grouping of the lines on the sulfide plot implies that selective reduction of the mineral is more difficult with sulfides than with oxides.

4.1.1. Carbon Reduction

Ideally, a reducing agent should be readily available at low cost and with the required purity. It should be capable of fully reducing the metal from the ore at a temperature that can be easily attained. The reducing agent and its reaction product should be readily separated from and should not dissolve in the metal. Carbon comes nearest to satisfying these requirements from the reduction of the oxides of the majority of the common metals. Whereas the stability of the metal oxides decreases with increasing temperature, the stability of CO_2 is almost independent of temperature and CO becomes more stable as the temperature is raised. Hence, carbon can reduce the oxide of any metal if a high enough temperature is attained.

Inspection of the oxygen potential diagram reveals that the precious metals and Cu, Ni, Pb and Sn can be reduced with almost complete conversion of the carbon to CO_2. Oxides with stabilities equal to or greater than ZnO can be reduced with carbon only if the product gas is primarily CO; but the minimum reaction temperature rises rapidly and becomes more difficult to attain as the oxide stability increases. Alumina can be reduced with carbon, but only at a very high temperature and the process is not commercially viable.

Coal is the cheapest and most readily available form of carbon, but even the purest coals contain significant amounts of inorganic material (ash), hydrocarbons and moisture. The latter two can be removed by heating the coal in the absence of oxygen to produce coke, which can then be used as the reductant. The inorganics are intimately mixed with the carbon and are only released as the carbon is consumed by the reducing reaction. They mix with the gangue to increase the volume of unwanted material.

Carbon can be partially combusted in a separate rector to produce a gas rich in CO, which is then brought into contact with the ore. The coke ash does not enter the reaction vessel, and both the carbon reactant and the carbon product species are present as gases, which are readily separated from the metal, in the reduction stage. However, more carbon is consumed per ton of metal produced since the carbon is already oxidized to CO before the start of the ore reduction reactions, thereby increasing both the total energy consumption and the environmental pollution.

The CO-CO_2 line on the oxygen potential diagram represents the conditions where $p_{CO} = p_{CO_2}$. The equilibrium CO/CO_2 ratio decreases progressively below unity as the line is rotated upward about its origin at 0 K and, conversely, the ratio increases as the line is rotated downward on the diagram. Hence, the maximum extent to which CO can be reacted with the ore and form CO_2 increases with increasing displacement above the CO-CO_2 line of the M-MO line for the metal being reduced. For example, the ratio CO-CO_2 at any temperature in equilibrium with Sn and SnO_2 is lower than the ratio in equilibrium with Pb and PbO at the same temperature. Conversely, the maximum achievable efficiency of CO utilization decreases as the M-MO line moves further below the CO-CO_2 line. Thus, when CO gas is used as the reductant, the total carbon (and hence energy) consumption increases rapidly as the stability of the metal oxide is increased. The energy consumption can be lowered by removing the CO_2 from the reacted gas and recycling the CO.

The metal produced tends to be saturated with carbon when carbon is soluble in the metal. Sometimes, this can be beneficial. Thus, carbon is quite soluble in molten iron and lowers the melting point from 1537 °C for the pure metal to 1147 °C at the eutectic point in the Fe-C system with 4.3% C in solution. This permits the blast furnace process for the reduction of iron ores to be operated at a much lower temperature than would be feasible if the product was pure iron. But carbon cannot be used for the reduction of oxides of elements such as Mo, Nb, Ta, Ti and Zr, which form metal carbides that are more stable than the metal. Carbon reduction would then result in the direct production of the metal carbide:

$$MO + 2C = MC + CO \tag{4.1}$$

where M represents the metal.

Solid or gaseous carbon cannot be used to reduce a metal from a sulfide ore. The compounds CS, CS_2 and COS, formed between carbon and sulfur, are less stable than most of the metal sulfides and the metal would be converted to the sulfide if exposed to these gases. Consequently, in most cases, the sulfide mineral must be oxidized to expel SO_2 and SO_3 prior to reduction with carbon.

4.1.2. Hydrogen Reduction

As is apparent from the oxygen potential diagram, oxide minerals capable of reduction with CO gas can also be reduced with hydrogen. Similar limitations apply to the maximum possible efficiency of H_2 conversion of H_2O in terms of the displacement of the M-MO line above or below the H_2-H_2O line on the oxygen potential diagram. The hydrogen reactant and water vapor product are again readily removed from the reaction vessel and there is no ash contamination. The water vapor is easily condensed and there is no environmental hazard. However, hydrogen is slightly soluble in some metals and must be subsequently removed from solution in the metal in order to avoid embrittlement and loss of ductility. Some metals form hydrides that are more stable than the metal and hydrogen reduction is then not acceptable. H_2S is also less stable than most of the metal sulfides and sulfide minerals must first be oxidized before reduction with hydrogen.

In comparison with carbon and CO gas, hydrogen is expensive to produce and, correspondingly, hydrogen reduction is a higher cost process. One method for the production of hydrogen gas is the partial oxidation (reforming) of a hydrocarbon:

$$2C_xH_y + xO_2 = yH_2 + 2xCO \tag{4.2}$$

followed by removal of the CO; but the CO is also a reducing agent and the reformed gas can be used for reduction without further treatment. Ore reduction can then be regarded as proceeding via two separate reactions:

$$MO + CO = M + CO_2\text{: } K_a = \frac{a_M \cdot p_{CO_2}}{a_{MO} \cdot p_{CO}} \tag{4.3}$$

$$MO + H_2 = M + H_2O\text{: } K_b = \frac{a_M \cdot p_{H_2O}}{a_{MO} \cdot p_{H_2}} \tag{4.4}$$

These reactions are linked through a third (water gas) reaction:

$$H_2O + CO = H_2 + CO_2: K_c = \frac{p_{H_2} \cdot p_{CO_2}}{p_{H_2O} \cdot p_{CO}} \tag{4.5}$$

When equilibrium is established at constant temperature and at the same activities for the condensed species, the ratio p_{CO_2}/p_{CO} in Equation 4.3 is equal to $K_c \times p_{H_2O}/p_{H_2}$ in Equation 4.4. The oxygen potential diagram shows that the lines for CO-CO_2 and H_2-H_2O cross at about 800 °C when all the gaseous species are present in their standard states. At the intersection temperature, all four gas species are equally stable, their partial pressures are equal at equilibrium, and K_c is equal to unity. At lower temperatures, CO_2 is more stable than H_2O, Equation 4.5 moves to the right, and K_c is greater than unity, the value increasing with decreasing temperature. The converse applies at higher temperatures and K_c becomes progressively less than unity as the temperature is raised. Thus, with regard solely to equilibrium conditions, H_2 is a more efficient reductant than CO at high temperatures, while CO is more efficient at low temperatures. Kinetic factors may change the relative efficiencies of the gaseous reductants when conditions deviate from equilibrium.

4.1.3. Metallothermic Reduction

As explained earlier, any metal can be used to reduce oxides that are significantly less stable than the oxide of that metal. Al, Mg and Si are commonly used for this purpose, but they are expensive to produce from the ore, and their use can be justified only for those metals that command a much higher selling price than Al or Si and that cannot be readily reduced by a less costly route. They can be used, for example, to reduce ores that can only be reacted with carbon at very high temperatures, or which form carbides and hydrides when in contact with hydrogen or carbon.

The oxidation of Al is strongly exothermic, as evidenced by the heat released when it is ground to a fine particle size and then combusted spontaneously in an incendiary bomb, or when metal powder is used similarly to generate heat in the thermite cutting and welding of metals. When the finely divided metal is mixed with a less stable compound, such as chromium oxide, the heat released is sufficient to melt the constituents, facilitating separation of the metal from the reaction products by density difference. The reaction is very rapid, however, and it is difficult to ensure that the reaction just goes to completion. As a result, there is a risk that either some Al remains unreacted and dissolves in the reduced metal, or some of the more valuable mineral is not reduced and is lost in the molten alumina. An alternative treatment, in which the valuable metals are extracted from the ore as volatile halides that are subsequently reduced to the metal, is described later in this chapter.

4.2 ROASTING

Since, with few exceptions, sulfide minerals cannot be reduced directly to the metal, the first stage in processing these ores is the conversion of the mineral from the sulfide to a more amenable form. This is accomplished by roasting, which is a solid-state process. The composition of the mineral can be changed in three different ways to suit the feed requirements for the next stage of metal extraction. These are:

1. Dead roasting, wherein the sulfide is converted to an oxide according to reactions of the type:

$$2MS + 3O_2 = 2MO + 2SO_2 \quad (4.6)$$

$$MS + 2O_2 = MO + SO_3 \quad (4.7)$$

2. Sulfating roasting, where the mineral is converted into a soluble sulfate:

$$MS + 2O_2 = MSO_4 \quad (4.8)$$

3. Chloridizing roasting, in which chlorine (usually in the form of NaCl) is added to the mix to convert the minerals of elements such as Ag, Cu and Pb into water-soluble chlorides:

$$2MS + 4NaCl + 3O_2 = 2MCl_2 + 2SO_2 + 2Na_2O \quad (4.9)$$

This technique, using Cl_2 or HCl gas, is used also to convert oxides into soluble chlorides in the presence of carbon:

$$MO + C + Cl_2 = MCl_2 + CO \quad (4.10)$$

Only the first two methods will be described, but similar principles apply to all three.

The ore feed to the roaster is usually in the form of a finely divided concentrate from a flotation plant. It is desirable that a small particle size is retained in the product if a soluble salt is produced that is subsequently extracted by leaching, or if the oxide is reduced in a process that can handle finely divided solids, in order to increase the reaction rate in the next processing stage. This places an upper limit of 800 to 900 °C for the roasting operation, depending on the composition and particle size of the concentrate, to avoid the risk of the particles agglomerating together by partial fusion (see Section 4.3.1.). The rate of chemical reactions decreases rapidly with decreasing temperature, and this fixes a lower working temperature of about 500 to 600 °C, again depending on the composition of the concentrate, in order to obtain an acceptable rate of reaction.

The overall reaction for the sulfating roast can be considered a series of consecutive steps, such as:

$$2MS = 2M + S_2 \tag{4.11}$$

$$2M + O_2 = 2MO \tag{4.12}$$

$$S_2 + 2O_2 = 2SO_2 \tag{4.13}$$

$$2SO_2 + O_2 = 2SO_3 \tag{4.14}$$

$$MO + SO_3 = MSO_4 \tag{4.15}$$

the sum of which is equal to Equation 4.8. The reaction described by Equation 4.6 can be expressed as a series of steps in a similar way. Thus, four gaseous species—S_2, O_2, SO_2 and SO_3—are involved in the reactions. The concentrations of all four can be determined through the equilibrium constants for Equations 4.13 and 4.14 if, say, the partial pressures of O_2 and SO_2 in the gas phase are known at the roasting temperature. These are the two gas species plotted along the axes of the predominance area diagram (PAD) shown in Figure 2.9. Hence, from consideration of the PAD, it is evident that dead roasting is favored by high p_{O_2} and low p_{SO_2}, whereas a sulfating roast requires a high partial pressure for both species. This is an alternative way of expressing the fact that the stability of SO_2 increases with respect to the stability of SO_3 as the temperature is raised. Since the boundary between the oxide and the sulfate phase moves downward on the diagram to lower p_{SO_2} as the temperature is lowered, the range of gas compositions suitable for a sulfating roast is extended as the temperature is depressed toward the lower operating limit. Conversely, the range of gas compositions suitable for a dead roast is extended as the temperature is raised toward the upper limit.

By superimposing PADs for Cu and Fe, for example, it is possible to select roasting conditions that will produce soluble $CuSO_4$ while the iron remains as the insoluble oxide. Likewise, conditions can be selected to ensure that the Cu remains as the sulfide, Cu_2S, while the iron is oxidized to FeO or Fe_3O_4. It must be remembered, however, that the lines drawn on the PAD do not represent phase boundaries, but merely the limits of the range over which a particular species is the predominant one in a mixture of other species, all of which are at unit activity. The metallic elements in the ore may not be present as separate minerals, but as solid solutions in which the activities of the species are less than unity. The boundaries for any phase are expanded as its activity is decreased, relative to the activities of the adjoining phases. Hence, a PAD is useful for preliminary selection of suitable operating conditions, but it may be necessary to solve the appropriate equilibrium constants in order to determine the optimum conditions.

The rate of reaction of the concentrate particles is determined by the kinetics of gas-solid reactions, as described in Section 3.3.2. The surface of each particle becomes covered with a relatively dense oxide layer as sulfur is removed during dead roasting. Continued reaction then requires counterdiffusion of oxygen and SO_2 through the oxide layer, between the outer surface and the reaction site. However, the volume change is negligible when compared to the shrinkage that occurs when a compound is reduced to the metal. Consequently, the porosity of the oxide layer is low and gas transport to and from the reaction site is hindered. Topochemical reaction results, with a clearly defined reaction interface moving toward the core of the particle at a fairly uniform rate. The volume of the reaction product is greater than that of the initial material when a sulfating roast is performed. The particle dimensions are then increased, increasing the length of the diffusion path, and the porosity is decreased as the reaction proceeds. This volume expansion, coupled with the effect of the low operating temperatures required for a sulfating roast, results in a low reaction rate.

4.2.1. Dead Roasting

Traditionally, roasting is performed in a multiple-hearth furnace, wherein the ore is fed onto the periphery of a circular plate to form a bed a few centimeters thick. Slowly rotating arms equipped with blades (rabbles) rake the ore toward the center of the disk, where it falls onto a second disk on which the direction of movement is reversed to fall from the disk at the periphery. This sequence is repeated several times as the ore descends through the furnace. Gas-solid contact is poor within the bed on each plate, thus, reaction occurs mainly on the upper surface of each bed and during free fall between the plates.

Dead roasting today is commonly conducted in a fluidized bed, where the ore particles are suspended by buoyancy in an upward flow of gas. This facilitates rapid replenishment of the gas at the particle surfaces and ensures a rapid rate of reaction, with a marked increase in throughput per hour, but requires that all the particles are contained within a very small size range. The buoyancy is affected by the density and the size (e.g., diameter) of the particles in a manner similar to the settling of the ore in gravity mineral dressing. Oversized particles tend to settle out rapidly in a layer at the bottom of the fluidized bed, through which gas contact is restricted. Undersized particles are entrained in the gas stream and are carried out of the reactor in the exhaust gas. The gas velocity must be sufficiently high to fluidize the bed, and only limited time is available in which to establish gas-solid equilibrium. The minimum gas flow rate required to cause fluidization decreases as the average particle size is lowered but, even with very finely divided concentrates, it is probable that thermal and chemical equilibrium between the gas and the solid is not fully attained.

Even less time is available for the attainment of equilibrium in a flash roaster, in which the particles fall through a tall cylinder countercurrent to a slowly upward-flowing stream of gas. If the particle size is too large, the reaction may not be completed in the time available during free fall, leaving a core of unreacted sulfide in each particle that will release sulfur during the subsequent reduction stage. Very fine particles may again be lost in the exhaust gas.

The heat of reaction for Equation 4.6 is given by:

$$\Delta H^{\ominus} = 2H^{\ominus}_{MO} + 2H^{\ominus}_{SO_2} - 2H^{\ominus}_{MS} - 3H^{\ominus}_{O_2} \tag{4.16}$$

Since the reactions occur in the solid state, the heats of formation at 298 K can be used to assess the overall heat change at the roasting temperature without serious error. In general, metal oxides are more stable than the corresponding sulfides and, correspondingly, the enthalpies of formation of the oxides are more negative than the values for the sulfides. The standard enthalpy of SO_2 is -298.6 kJ mol^{-1} at 298 K. The sum effect is a large release of heat when the reaction occurs. That is, the reaction is strongly exothermic. Typical values per mole of SO_2 released for some sulfide minerals are:

Mineral	**$\Delta H^{\ominus}$ kJ mol^{-1}**
CuS	395
FeS	464
Ni_3S_2	447
PbS	420
ZnS	442

In practice, the concentrates are not pure, but contain some residual gangue and absorbed moisture, so less heat is released per unit weight of concentrate than these figures suggest. Nevertheless, the heat released is more than sufficient to heat the concentrate to the reaction temperature. The reaction is usually self-sustaining (autogenous) when the ore has been heated to about 400 °C by the ascending gas and the reaction is continuous without any other source of heat. Indeed, if the heat is released too rapidly, as in the fluidized bed or flash smelter, steps must be taken to remove the surplus heat and prevent the ore or the concentrate from being heated into the temperature range where the particles can sinter together.

Air is normally used to supply the oxygen required for the reaction. Thus, in addition to the energy required to heat the oxygen, additional heat is consumed to bring the nitrogen in the air to the reaction temperature. However, the total quantity of energy needed to heat the concentrate and the stoichiometric quantity of air to the required temperature is usually less than the quantity of heat released by the reaction. The surplus heat can be consumed either by admitting excess amounts of air or by injecting water vapor into the reactor.

4.2.2. Environmental Constraints

The sulfur remains fixed in the ore with a sulfating roast, but is removed in the exhaust gas during dead roasting. Originally, the contaminated gas was discharged to the atmosphere and the location of a dead roaster was clearly revealed by the blighted landscape, devoid of all vegetation, on the downwind side of the plant. Eventually, it was realized that a valuable by-product could be obtained if the gas was used to produce sulfuric acid, large amounts of which are consumed in hydrometallurgical extraction and in electrorefining, or the sulfur could be recovered in elemental form. More recently, environmental legislation is placing increasingly stringent limits on the amount of sulfur-bearing species in the gas that can be discharged to the atmosphere and more attention is now being directed to control of the gas composition produced by the roaster.

If the stoichiometric amount of air is used for dead roasting and it is assumed that all the sulfur is converted to SO_2, then the maximum concentration of SO_2 in the exhaust gas is equal to the oxygen content of the air (20.9 vol%). However, some SO_3 is always present, even at the highest roasting temperatures and gas-solid equilibrium is rarely, if ever, attained in the roaster; thus, some unreacted oxygen remains in the exhaust gas. As a result, the sulfur-bearing species rarely exceed 18 vol% when air unenriched with additional oxygen is used, and the concentration falls as the quantity of excess air or steam is increased to control the maximum temperature within the roaster. It is possible to precipitate elemental sulfur from a gas containing as little as 4 vol% SO_2 and to manufacture sulfuric acid with as little as 2 vol% SO_2, but the cost of gas treatment escalates rapidly as the concentration of the sulfur species is lowered by dilution in an increasing volume of gas. It is normal practice, therefore, to operate a fluidized bed roaster to give an SO_2 content in the exhaust gas in the range 10 to 15 vol%. Only about half this concentration is achieved with the much slower rates of reaction that occur in a multiple-hearth roaster, and this is one of the principal reasons why the latter are rapidly being replaced by dead roasting.

In view of the environmental constraints, there is renewed interest in the possibility of fixing the sulfur as the sulfide of a more stable species within the reactor. Calcium forms one of the most stable sulfides and is readily available in the form of lime (CaO). But CaO is more stable than CaS and the standard free energy for the reaction:

$$2CaO + S_2 = 2CaS + O_2 \qquad (4.17)$$

is positive at all temperatures. The standard free energy for the conversion of metal sulfides to oxides:

$$2MS + O_2 = 2MO + S_2 \tag{4.18}$$

is, with few exceptions, a negative quantity, but the overall free energy change for the sum of these two reactions:

$$CaO + MS = CaS + MO \tag{4.19}$$

is still positive and the reaction goes to the left if the reactants and products are present in their standard states at unit activity. The reaction can be driven in the required direction, however, if carbon is present to give:

$$CaO + MS + C = CaS + M + CO \tag{4.20}$$

The free energy change for this reaction is positive at low temperatures but changes to a negative sign as the temperature is raised, Figure 4.2. This is due to the entropy contribution from the formation of CO gas in the reaction. Thus, it is possible to reduce the sulfide directly to the metal in this way, without the need for intermediate conversion of the mineral to an oxide. However, the CaS remains as a solid and the metal product is also solid if the reaction is conducted below the melting temperature of the metal, so the two solids remain mixed in the product.

4.3. AGGLOMERATION

The importance of intimate gas-solid contact and frequent renewal of the gas at the interface, with respect to the kinetics of transport-controlled reactions, is evident from the low reaction rates obtained in multiple-hearth roasting relative to the rates prevailing in a flash roaster. The ore bed is only a few centimeters deep in a multiple-hearth roaster. The depth of the bed above the fusion zone in a blast furnace is measured, not in centimeters but in tens of meters. The ore and coke occupy the entire cross-section of the furnace and the ascending reducing gas must percolate through the bed. Thus, the permeability of the bed is critically important in the kinetics of the reactions in the upper part of a blast furnace. Furthermore, the material is charged at the top of the furnace. Devices are used to ensure that the charge is spread as evenly as possible over the top surface (as indicated by the rotating trough in the upper part of the furnace in Figure 1.7b), but some unevenness is unavoidable. If a hump is formed at the top surface, the larger pieces of ore roll down the sides of the hump, leaving smaller pieces concentrated at the peak. This results in variations in the density of packing

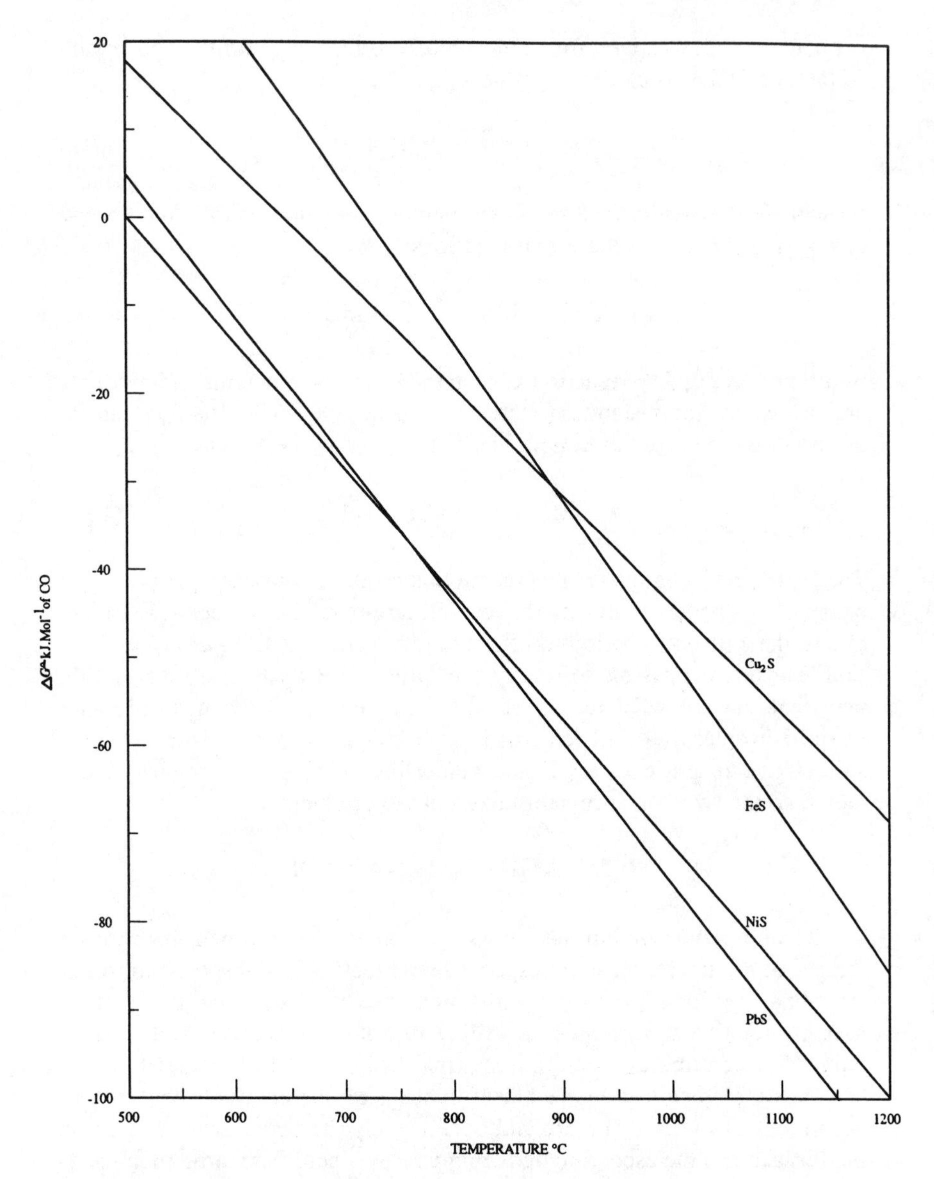

FIGURE 4.2. Temperature dependence of the free energy change for the carbothermic reduction of metal sulfides by the reaction:

$$MS + CaO + S = M + CaS + CO$$

where MS refers to the metal sulfide as indicated on the diagram.

across the furnace and creates short circuits for the passage of the gas through the bed.

In order to achieve uniform gas flow over the whole cross-section, it is essential that the size range of the particles fed into the furnace is restricted within narrow limits. The average particle size is also very important. The pressure resistance to the passage of the gas through the bed is increased as the particle size is decreased. With decreasing particle size, a stage is eventually reached where the pressure required to force the gas through the bed is so great that the velocity of the gas emerging from the top of the bed is sufficient to entrain the finer particles and remove them from the furnace. Conversely, the time required to attain chemical and thermal equilibrium in a gas-solid reaction that is transport controlled increases with increasing particle dimensions.

A compromise is reached between these two requirements by sizing the charge usually in the range from 6 to 30 mm. The furnace operation becomes more uniform as the mean particle size is decreased within this range. But the product from a flotation plant is often much finer than 6 mm. If the ore is merely crushed to −30 mm, a significant fraction will be reduced to much finer size. Sintering and pelletizing are therefore used to agglomerate the fine ore particles to a suitable size for the blast furnace feed. Practical details of these processes are described in Section 1.4.

4.3.1. Sintering

Agglomeration is achieved in sintering by heating the ore to a temperature at which the particles can weld together at the points of contact, followed by atomic transport to the necks thus formed to reduce the radius of curvature and decrease the total surface energy. A higher temperature than used for roasting is required to enable welding to occur. However, atomic transport to the weld neck is by a solid-state diffusion mechanism, which is a slow process even at elevated temperatures. Thus, bonding is accelerated by heating to a temperature at which small amounts of liquid are formed by the gangue and any added fluxes. The liquid is drawn by capillary action into the interstices between the particles and, on cooling, solidifies as a cement that holds the grains together and increases the strength.

The sinter must be produced with a microstructure that enhances its reducibility in the blast furnace. The ore does not begin to melt rapidly in an Fe or Zn blast furnace until it has descended roughly to the base of the upper conical section (the stack), which constitutes the major part of the overall height of the furnace. The upper surface of the charge is near the top of the furnace and the crushing load on the solids increases as they descend through the stack. The sinter that is charged into the blast furnace must have adequate strength to prevent deformation or disintegration under this superimposed load at the elevated temperatures that are encountered before the ore starts to melt. The strength of the sinter is critically dependent on the amount, composition and distribution of the various phases and compounds present in it but, in general, the strength rises as the contact area between the particles is increased. Since the height of a modern iron blast furnace is markedly greater than those used for nonferrous metal extraction, the strength requirements are greatest for iron ore sinters.

Prior to the onset of melting, the ore is reduced by gas-solid reaction as it descends through the blast furnace stack. Since the external surface area per unit volume is less in the sintered aggregate than in the original fine particles, it is important that the sinter supplied to the furnace has a large volume of coarse pores to allow ready access of the reducing gas to the reaction site. Hence, the reducibility increases as the volume of the interparticle bridges is lowered. However, the glassy cement of solidified melt, which bonds the particles during sintering, also coats the surface of the particles and obstructs the contact of the furnace reducing gas with the mineral surface. The conditions required for optimum reducibility and optimum strength are thus diametrically opposed. Furthermore, at the high temperatures reached in sintering, the gangue materials or the added fluxes may react with the metallic mineral to produce compounds, e.g.,

$$2FeO + SiO_2 = 2FeO \cdot SiO_2 \tag{4.21}$$

in which the activity of the metal oxide is less than unity and that further lower the reducibility. This can be avoided, in the example given, by mixing finely divided lime into the sinter blend. The silica is then converted into a calcium silicate compound and the reducibility is enhanced.

As the basicity (i.e., CaO/SiO_2) of iron ore sinter is increased, some of the lime combines with the iron to form calcium ferrite compounds, which appear as an interlocking acicular structure and increase the strength of the sinter when the heat input is low. The reducibility decreases as the CaO/Fe_2O_3 ratio in the ferrite compound is increased from 0.5 to 2.0 (i.e., $2CaO \cdot Fe_2O_3$). Unreacted particles of lime remain if the basicity is too high, with a consequent deterioration in the strength, so the basicity is not usually allowed to exceed about 2.2.

Some types of ore undergo very small volume changes during reduction, so the porosity is poorly developed in the reduced metal layer and gas access to the reaction site is impeded. This applies particularly to the reducibility of magnetite (Fe_3O_4). The ore is exposed to strongly oxidizing conditions during sintering after the maximum temperature has been reached and the aggregate is being cooled by the air drawn through the bed. Conditions are sufficiently oxidizing to convert the magnetite to hematite (Fe_2O_3), which undergoes a much larger volume change on reduction. Although more oxygen now has to be removed during smelting, the improved reducibility, resulting from the increased porosity accompanying the volume change, causes a marked increase in the overall rate of reduction.

With careful preparation, the reducibility of sinter can be markedly superior to the reducibility of lump ore in the same size range. It is now common practice to crush all of an ore to a fine size, blend with other ores to reduce the flux requirements, add the required fluxes and then sinter the mixture. The improved blast furnace productivity more than compensates for the additional costs of crushing, grinding, blending and sintering.

The strength and reducibility of the sinter depends on the temperature profile as the combustion zone moves progressively through the bed during the sintering operation. Ideally, the temperature rises rapidly to a peak, usually in the range 1100 to 1300 °C depending on the composition of the ore, remains constant for only sufficient time to form the required volume of liquid bond and then falls rapidly before the liquid has time to spread widely over the solid surfaces. When oxide ores are sintered, the ore is mixed with coke fines (coke breeze) and water. The top surface of the sinter bed is ignited with burners and, thereafter, the heat is supplied by combustion of the coke. The temperature is controlled both by varying the amount of gaseous fuel used to ignite the coke at the start of the reaction and by the amount of moisture added to the sinter mix. The moisture addition also helps to ball up the particles into aggregates that are sintered more readily and to reduce dust losses. After discharge, the sinter is crushed to the required maximum size range and the undersized material is returned to the sinter mix. Wood charcoal is used as the fuel in some countries (e.g., Argentina, Brazil) where suitable coking coals are not readily available. This is highly reactive, resulting in a wide combustion zone and the risk of either excessive fusion or poor strength in the sinter.

All the heat required is provided by sulfur oxidation in the sintering of sulfide ores. In fact, too much heat is released when high-grade sulfide concentrates are sintered, and it is necessary to recirculate large amounts of sinter to act as a heat sink and limit the maximum temperature attained. Alternatively, limestone ($CaCO_3$) may be added to absorb heat by the endothermic decomposition of the carbonate. Care must also be taken, particularly with Pb ores, to ensure that the SO_2 released is diluted sufficiently with air and nitrogen (remaining from the air) to avoid the risk of the formation of basic sulfates.

The hot gases from the combustion zone are drawn through the unreacted ore and heat is transferred from the gas to the solid as the gas passes through the bed. When carbon is used as the heat source for sintering oxide ores, the equilibrium gas mixture leaving the combustion zone should be rich in CO since CO_2 is less stable than CO when in contact with carbon at the sintering temperature. Equilibrium is not attained, however, and the gas is much richer in CO_2. However, it is still sufficiently reducing to react with any less stable oxides present in the sinter mix. If the metals produced are volatile and swept away from the reaction site by the combustion gases, the minimum temperature for the reduction of those oxides is lowered and the volatility is enhanced. Arsenic and antimony, which are harmful impurities in most metals, form oxides with relatively low stabilities and the vapor pressures of these metals or their suboxides are sufficiently high at, say, 1000 °C for significant amounts of these species to be removed by volatilization during sintering. This is beneficial, so far as refining is concerned, but care must be taken to avoid contamination of the working environment and the release of the toxic gases into the atmosphere. Pb and Zn may also be partially removed from the ore during sintering. Again, this is beneficial when these elements are classed as impurities that have to be removed during subsequent refining, but it is undesirable when Pb or Zn ores are being sintered. In the latter case, it is necessary to ensure that a more marked excess of air flows through the bed to reduce the risk of reduction of the oxides to the volatile metallic form.

4.3.2. Pelletizing

Pelletizing, in which the ore fines are mixed with a suitable binder and coagulated into small balls at ambient temperature, is more suitable for the agglomeration of very finely divided concentrates. There is then less risk of loss of mineral particles through the apertures between the grate bars of the firing pallets during the subsequent firing (induration). Coke breeze may be intimately mixed with the ore when pellets of oxide ores are formed, but no fuel is required with sulfide ores. A more uniform size and shape of the aggregated material is obtained, but the two-stage treatment of pelletizing and induration incurs a higher cost than for sintering.

It is important to ensure that the pellets contain adequate porosity to allow gas access to the reaction sites and adequate strength to withstand the superimposed load at elevated temperature during the subsequent reduction stage. Since heat transfer to the center of a pellet requires a longer time than when the individual particles are heated, the heating time for bonding is longer than with sintering and care is required to avoid excessive fusion of the outer layers, but well-prepared pellets have similar reducibility to a good sinter. Volatile impurities in the ore are evolved during firing and lower oxides are oxidized to a higher state of oxidation in a similar manner to the reactions occurring during sintering.

The volatile elements evolved during both sintering and pelletizing must be removed before the exhaust gas is vented into the atmosphere. Removal of sulfur-bearing gases can be a major problem when sulfide ores are agglomerated, since it is difficult to prevent dilution of the gas leaving the bed with large amounts of air drawn from around the heated mass and into the extractor hood.

4.4. CARBOTHERMIC REDUCTION

4.4.1. The Blast Furnace Process

A reaction temperature of over 1500 °C can be attained in the blast furnace. This is adequate for the reduction of iron ores by carbon, and the process has been the principal route for the production of iron for several centuries. If iron oxides can be reduced in this way, it follows that any metal having a melting point no higher than iron and forming oxides that exist only at less negative oxygen potentials than ferrous oxide on the oxygen potential diagram (Figure 2.8) can also be reduced in the furnace. Today, the process is used primarily for the reduction of the oxides of Fe, Pb and Zn, but it has also been used for the reduction of Cu and Sn ores.

The lines for the formation of the oxides of Cr, Mn and Si intersect the C-CO line on the oxygen potential diagram at temperatures within the range that can just be achieved in the blast furnace, but it is very difficult to produce these metals by this route. If the activity of the metal product is lowered by solution in another metal, however, the reduction temperature is lowered and can be attained more readily. Thus, the blast furnace can be used to produce alloys of iron (ferro alloys) containing quite large amounts of these elements. The metals thus produced can be utilized as a source of the elements (master alloys) in the production of alloy steels or as deoxidizers for the removal of the residual oxygen in the final stages of steel refining.

4.4.1.1. The Heat Balance

A convenient starting point for consideration of the blast furnace process is an examination of a heat balance for a process in which heat is generated by the combustion of carbon. The balance is an assessment of the heat input and output for the process. Under steady-state conditions, the input must equal the output.

The heat input arises from:

1. Sensible heat of the input materials (solids and gases)
2. Heat released by the combustion of carbon
3. Heat released by the exothermic reactions

while the heat output comprises:

1. Heat absorbed by the endothermic reactions
2. Heat lost as sensible heat in the output materials
3. Heat lost by conduction and convection to the surroundings

If the inlet and outlet temperatures for each substance are known, the sensible heat for each substance can be calculated at the appropriate temperature, *T*, using Equation 2.3, i.e.,

$$\Delta H = H_T - H_{298\,\mathrm{K}} = \int_{298}^{\mathrm{T}} C_\mathrm{p}\mathrm{d}t \tag{4.22}$$

on the assumption that the reference temperature of 298 K for standard enthalpies corresponds approximately to the ambient temperature. Latent heats of transformation must also be incorporated, as in Equation 2.6, when a solid substance undergoes a change of state (phase change, melting, etc.) within the temperature range considered. The enthalpy change accompanying each reaction at the reaction temperature can also be determined in terms of the heat capacities:

$$\Delta H_T - \Delta H_{298\,\mathrm{K}} = \int_{298}^{\mathrm{T}} \Delta C_\mathrm{p}\mathrm{d}t \tag{4.23}$$

The calculation of heat balances is illustrated in Worked Examples 1 to 3.

Construction of a full heat balance also requires preparation of a mass balance to quantify all material inputs and outputs. Equations 4.22 and 4.23 can then be evaluated in terms of the actual amounts of each of the reactants and products involved in the reaction. The heat balance is usually calculated in terms of the heat requirements for the production of 1 t metal product. This requires a detailed knowledge of the process, as the material and heat balances vary with the operating practice. But an indication of the limitations can be obtained with a more simple approach.

Consider a closed, insulated box in which an oxide ore is reduced by carbon. It is necessary to raise the temperature sufficiently to melt the metal and fuse the gangue and the flux materials to form a slag with sufficient fluidity to allow the slag to separate by density difference and float on top of the molten metal. In blast furnace practice, this requirement is satisfied when molten lead can be tapped from the furnace at about 1050 °C and molten iron is tapped at about 1450 °C.

If the ore and the carbon are introduced into the box at ambient temperature (say 298 K), then, according to Hess' Law, the reaction can be considered to occur at 298 K and the products are heated up to the final reaction temperature by the heat released in the reaction.

The heat of formation of CO_2 is 393.5 kJ at 298 K, so this amount of energy is released when 1 mol carbon is oxidized to CO_2 at 298 K. But, in the presence of carbon, CO_2 is not stable at the high temperatures required for separation of the metal and the slag. Hence, the carbon is oxidized to CO with the release of only 110.5 kJ of energy per mole of carbon reacted.

This is less than the heat of formation of the metal oxides at 298 K, so the reduction reaction is endothermic. Thus, for the reduction of hematite:

$$Fe_2O_3 + 3C = 2Fe + 3CO: \Delta H^{\circ}_{298\ K} = +489.7 \text{ kJ} \qquad (4.24)$$

and at least this amount of energy must be supplied to drive the reaction to the right. Since the atomic weight of iron is 55.85 g, the energy required per ton iron is:

$$(489{,}700/2 \times 55.85) \times 10^6 = +4.38 \text{ GJ t}^{-1}$$

Inserting the appropriate heat capacity data for Fe into Equation 4.22, the energy required to raise the temperature of the iron to 1450 °C can be evaluated as approximately 1.0 GJ t^{-1}. An extra 0.35 GJ is required to melt the Fe and more energy is consumed in dissolving about 4 wt% C in the Fe. Forty-six kilojoules of energy are consumed in heating each mole of CO produced in the reaction to 1450 °C. This is equal to 138 kJ mol^{-1} Fe_2O_3 or 1.2 GJ t^- Fe if only the stoichiometric amount of carbon is consumed.

A small amount of energy is released by the formation of inorganic compounds in the slag, but this is not sufficient to heat the slag up to the reaction temperature. The sensible heat contained in the slag depends on its mass and composition, but is usually in the range 0.2 to 0.5 GJ t^{-1} metal. Thus, at least 7.2 GJ of energy must be supplied for each ton of metallic iron produced. PbO has a smaller heat of formation than Fe_2O_3 and the reaction temperature is lower, so less energy is required for the production of Pb; ZnO has a larger heat of formation but a lower melting point than Fe and a similar quantity of energy is required for Zn and Fe production.

If excess carbon is charged into the box and the stoichiometric amount of air is admitted to burn the excess, then 110.5 kJ of energy is released for each mol of carbon burnt to CO. Each mole of carbon combines with 0.5 mol O_2. Since 1 mol of air contains 0.21 mol O_2 and 0.79 mol N_2, the gas produced by combustion comprises 1 mol CO and [(0.5/0.21) × 0.79] = 1.88 mol N_2 per mole of carbon consumed. The heat released by combustion is insufficient to heat the product gas to the reaction temperature of 1450 °C. An approximate indication of the temperature to which the gases can be heated may be obtained by inserting in Equation 4.22 a value of 33 J mol^{-1} K^{-1} as the average value of the heat capacities of CO and N_2 and equating the heat supplied by the reaction with the sensible heat in the gas:

$$110{,}500 = H^{\circ}_T - H^{\circ}_{298} = 2.88 \times 33(T - 298) \qquad (4.25)$$

Solving this equation gives the gas temperature, T as 1435 K. Thus, combustion of carbon to CO does not supply sufficient heat to raise even the gas to the reaction temperature. It cannot supply the energy required for reduction of an oxide ore when all the materials (ore, coke, air, etc.) are admitted to the furnace at ambient temperature and the carbon is oxidized only to CO.

The blast furnace has evolved slowly over a longer period of time as a means of altering the heat balance and making feasible the reduction of oxides using carbon as the medium both for heating and reduction.

Only a small part of the potential energy available from carbon is released within the furnace if the carbon is only oxidized to CO. Some of the remaining energy can be recovered if the gas is fully combusted to CO_2 outside the furnace (in the absence of solid carbon) and the heat released is transferred partly to preheat the incoming air blast. This is achieved in Cowper stoves located adjacent to the blast furnace that function as heat exchangers. In this way, the incoming air can be preheated to above 1000 °C, transferring sensible heat into the furnace and raising the adiabatic flame temperature from the combustion of the carbon in the furnace to as high as 1900 °C. Even higher temperatures are possible, but a limit is imposed by the thermal stability of the refractory materials used to build the furnace. It is a very expensive task to empty a blast furnace and lining repairs cannot be effected while the furnace is in use, so the lining is intended to last for several years of continuous operation. Iron blast furnace campaigns of up to 12 years with the production of over 40 million tonnes of iron between relinings are now being achieved. This justifies the use of more expensive refractories, such as carbon blocks or silicon carbide, for construction of the highest temperature zone of the furnace and allows flame temperatures of over 1900 °C to be used safely if the flame does not impinge directly onto the refractory. Some of the sensible heat in the combusted gas is also used in the Zn blast furnace for preheating the coke to about 800 °C prior to charging into the furnace.

4.4.1.2. Reactions in the Upper Furnace Zones

If coke containing 90% C (10% ash) is burned with the stoichiometric amount of air and the maximum flame temperature is 1900 °C (2173 K), then the sensible heat in the combustion gases is approximately:

$$[2.88 \times 33(2173 - 298) \times (0.9 \times 10^6)/12\text{J}]$$

which is equal to 13.4 GJ t^{-1} coke. If the gas is cooled to, say, 1450 °C by heat transfer to the molten metal and slag, only 2.8 GJ is transferred to the charge and 10.6 GJ sensible energy remains in the gas. The high stack of the iron blast furnace allows time for very effective heat transfer from the ascending hot gas to the descending ore and coke charge. The gas leaves the furnace at 200 to 300 °C, so most of the sensible heat is transferred to the solid charge and reduces the amount of energy that must be supplied in the highest temperature zone. A shorter stack is used in the lead blast furnace, so heat transfer is less efficient and the gas is discharged at 550 to 600 °C.

Before the ore reaches the temperature zone in which it begins to melt, the reaction between the ore and the coke is limited to the points of contact between the two solids. Such reactions are invariably slow and very little reduction with solid carbon is achieved until the ore melts. The stability of CO_2 increases with respect to CO, however, as the temperature of the gas falls during ascent through the stack and the ore may then be reduced also by gaseous reactions. For example, hematite can be reduced in a series of steps:

$$3Fe_2O_3 + CO = 2Fe_3O_4 + CO_2 \tag{4.26}$$

$$Fe_3O_4 + CO = 3FeO + CO_2 \tag{4.27}$$

$$FeO + CO = Fe + CO_2 \tag{4.28}$$

Some of the potential heat remaining in the gas is recovered within the furnace by these reactions, but sufficient potential energy remains in the gas, leaving the furnace to preheat the air blast to the required temperature. The reactions are often referred to as indirect reduction steps since carbon is not directly consumed. However, this can be confusing, for carbon is simultaneously consumed by the linking **Boudouard reaction:**

$$C + CO_2 = 2CO \tag{4.29}$$

in an attempt to restore equilibrium between carbon and the gas.

The extent to which an ore can be reduced by gaseous reactions depends on a number of factors, including the composition of the mineral phases in the ore, the volume of gas ascending the stack per tonne of ore charged, the temperature profile in the stack, the variation with temperature of the equilibrium constants for the reactions and the reaction kinetics. The equilibrium constant for a generalized gas reduction reaction can be written in the form:

$$MO + CO = M + CO_2\text{: } K = \frac{a_M \cdot p_{CO_2}}{a_{MO} \cdot p_{CO}} \tag{4.30}$$

which is equal to p_{CO_2}/p_{CO} if the condensed substances are present at unit activity. Hence, the limiting conditions for gaseous reduction can be represented on a plot of the equilibrium gas composition versus temperature. Conventionally, the gas composition is represented by the ratio CO/(CO + CO_2) for $p_{CO} + p_{CO_2} = 1$. The appropriate diagram for Fe and Zn oxides

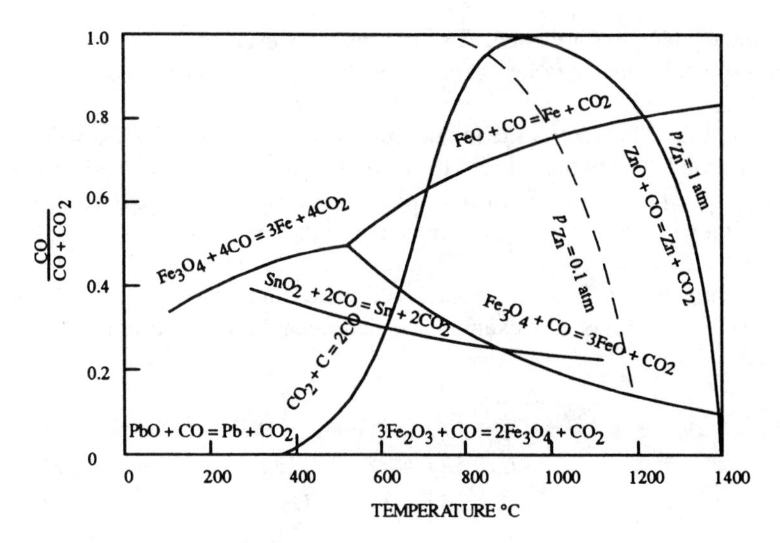

FIGURE 4.3 The stability of iron, lead and zinc oxides in CO-CO_2 atmospheres.

is shown in Figure 4.3. This is a different form of a predominance area diagram.

Before examining this diagram, it is useful to first consider the thermodynamic significance of metal-oxygen phase diagrams. Figure 4.4 shows the iron-oxygen phase diagram for the composition range in which the three oxides—wustite, magnetite and hematite—are formed. The wustite phase exists over a range of compositions that do not include the stoichiometric Fe/O ratio and is only stable above 570 °C. Magnetite is the oxide in equilibrium with Fe at lower temperatures. Oxygen is slightly soluble in Fe. The solubility is 0.16 wt% in liquid Fe at the melting point, but decreases to 0.0086 wt% in solid Fe at 1500 °C[54] and continues to decrease with decreasing temperature. It is too small to show on the diagram.

It is shown in Chapter 2 that the partial molar free energy of a species in a binary solution is given by the intercept of the tangent to the free energy-composition curve at the ordinate for the pure species (Figure 2.1). The activity of the species in the solution is related to its partial molar free energy through Equation 2.30. It is evident from Figure 2.6 that the partial molar free energy and hence the activity of a solute species in a binary solution increases with increasing solute concentration in a single-phase field, but remains constant as the solute content is increased across a two-phase field.

Consider, now, the activity of oxygen in the iron-oxygen system at, say, 1000 °C. Starting from zero activity in oxygen-free pure iron, the activity of oxygen increases with increasing amounts of oxygen dissolved in iron up to the solubility limit at the temperature considered. As the oxygen

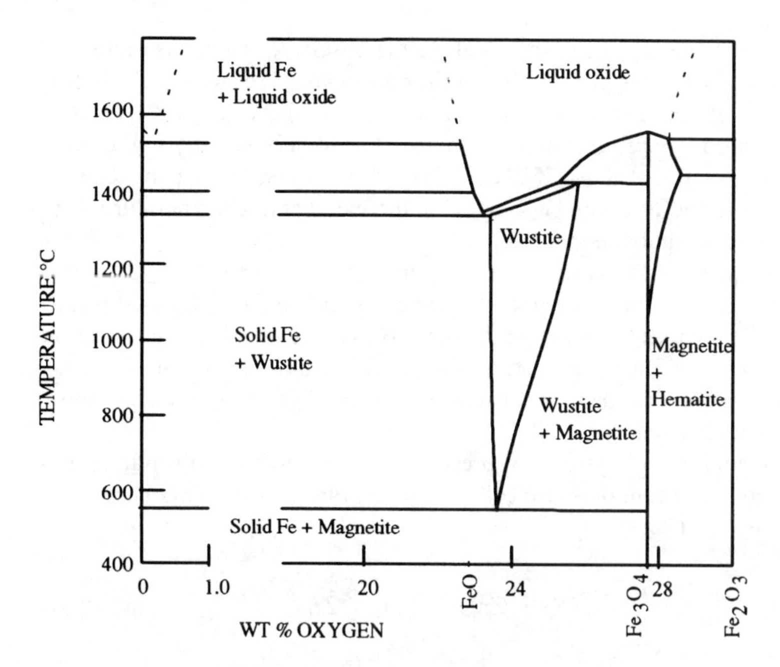

FIGURE 4.4. The iron-oxygen phase diagram.

content continues to increase; the iron-plus-wustite phase mixture is entered, but the activity of oxygen remains constant across the phase mixture until the boundary defining wustite in equilibrium with iron is intersected. The oxygen activity increases again as the single-phase wustite field is crossed and then remains constant across the wustite-magnetite phase mixture. Similar changes occur as the oxygen content is increased further, up to the composition at which the hematite phase appears.

At any chosen composition, the CO/CO_2 ratio in equilibrium with the condensed phases can be calculated from the equilibrium constant for the relevant reaction (i.e., Reaction 4.26, 4.27 or 4.28). Thus, the line designated by Equation 4.26 in Figure 4.3 identifies the change in the gas composition with temperature for any solid composition at the iron-plus-wustite phase boundaries. Similarly, the line described by Equation 4.27 indicates the temperature dependence of the gas mixture in equilibrium with a mixture of wustite and magnetite. The space between these two lines at any chosen temperature corresponds to the change in the equilibrium gas composition as the wustite phase field is crossed. The two lines meet at 570 °C, below which wustite is no longer stable. Metallic iron is stable only in the upper part of the diagram, the oxygen in solution in the iron decreasing progressively below the solid solubility limit at the Fe_3O_4-Fe or the FeO-Fe line as the gas ratio is increased toward unity.

The Fe_2O_3-Fe_3O_4 line is shown along the abscissa. The concentration of CO in a CO-CO_2 gas mixture in equilibrium with these two oxides is less than 10^{-10} at 100 °C and only increases to about 10^{-4} at 1300 °C. The CO concentration in equilibrium with Pb and PbO is only slightly higher than this, rising to 10^{-4} at about 750 °C. This means that hematite can be reduced to magnetite and PbO can be reduced to the metal at any temperature in an atmosphere of almost pure CO_2.

Conversely, Zn is oxidized even in pure CO below about 950 °C, but the metal can exist in the presence of increasing amounts of CO_2 as the temperature is increased above that point. However, metallic zinc boils at 907 °C, and at higher temperatures the product from the reduction of ZnO is a gas and not a condensed substance if the total gas pressure is not more than a few atmospheres.

The solid line in Figure 4.3 representing the ZnO-Zn equilibrium relates to Zn vapor at 1 atm pressure (i.e., at unit activity). If the ZnO is reduced completely by CO:

$$CO + ZnO = CO_2 + Zn \tag{4.31}$$

then, p_{Zn} cannot exceed p_{CO_2}. When the CO is produced by combustion of carbon with the stoichiometric quantity of air, the gas contains 1.88 mol N_2 for each mole carbon created, and the gas comprises 25.7 vol% Zn, 25.7 vol% CO_2 and 48.4 vol% N_2 if all the CO is oxidized to CO_2. In practice, the CO conversion is far from complete. The reduction of ZnO is strongly endothermic and extra carbon must be burned to produce the necessary heat, generating more CO in the furnace than is required for the reduction of all the ZnO in the charge. Most of the ZnO is actually reduced by solid carbon in the lower zones of the furnace. Consequently, p_{Zn} is usually about 0.05 to 0.07. The broken line in Figure 4.3 shows the equilibrium gas compositions for $p_{Zn} = 0.1$ atm. It indicates that metallic Zn remains stable in the presence of higher concentrations of CO_2 when p_{Zn} is lowered.

It is essential that the gas composition in the Zn blast furnace remains to the right of the Zn-ZnO line appropriate to the actual value of p_{Zn} until the gas leaves the furnace; otherwise, the Zn would be reoxidized. (Recovery of the Zn from the outlet gas is considered later in this chapter.) In practice, the gas is released at about 1050 °C. This is partially achieved by shortening the height of the stack, in comparison with the iron blast furnace, and using the sensible heat remaining after the Zn has been stripped from the gas to preheat the incoming charge to an average temperature of 400 to 500 °C. In contrast, Fe and Pb are discharged from the bottom of the furnace and exposed to atmospheres that become more reducing as the charge descends through the furnace. Since the rate of chemical reactions decreases rapidly at low temperatures, the temperature and composition of the gas at the top of the furnace are then of major concern only with respect to the energy balance.

One other line in Figure 4.3 describes the equilibrium condition for the Boudouard reaction (Equation 4.29). For any composition to the right of this curve, CO_2 is less stable than CO and reacts with coke in an attempt to restore the equilibrium. However, the kinetics of this reaction play a very significant role in the blast furnace process.

When the ore and coke are present as solids, both the gaseous reduction reactions and the Boudouard reaction to consume carbon are dependent on gas-solid contact. The rate of transport of CO and CO_2 in the bulk gas affects the kinetics of the two reactions to a similar extent. Thus, the relative rates of the two reactions depend on the transport of the gas to and from the reaction site on or within the solids and the rates of the chemical reaction steps. The importance of porosity in the ore, sinter or pellets to allow ready access of the gas to the reaction sites has been emphasized previously. Coke is very porous and the Boudouard reaction can occur on the surface of the coke particle, since there is no solid reaction product to obstruct the gas contact. But the activation energy for the chemical reaction of coke gasification is much larger than for the gaseous reduction of the metal oxide. This causes the rate of the Boudouard reaction to decrease more rapidly with falling temperature than the gaseous reduction reactions and carbon gasification has practically ceased by the stage where the gas has cooled to about 900 °C. As a result, the CO/CO_2 ratio of the gas in the hotter regions of the Fe blast furnace stack is much lower than would be expected if equilibrium conditions prevailed and the energy efficiency of the process is increased.

Conversely, as the gas temperature continues to fall, the gas enters the regime to the left of the Boudouard line in Figure 4.3. The equilibrium condition for Reaction 4.29 then moves to the left and carbon should be deposited as small graphite flakes that would block the pores in the ore and also increase the dust losses in the exit gas. Fortunately, this reaction is also slow in the absence of a suitable catalyst and very little carbon is precipitated before the gas leaves the furnace. If equilibrium was established, the gas would leave the furnace with a CO/CO_2 ratio of almost zero but, in practice, the ratio is usually between 1.5 and 2.2, leaving sufficient potential energy in the gas for preheating the air blast.

4.4.1.3. Reactions in the Lower Regions of the Furnace

Direct reaction between solid carbon and the metal oxide only occurs at a significant rate when the molten oxide begins to drain down over the surface of the coke. Even then, Reactions 4.26 to 4.29 can account for a large part of the overall reduction. As the temperature increases, a stage is reached when the gangue in the ore and the ash released by combustion of the coke begin to react with the fluxes in the charge to form a molten slag, which also drains over the surface of the remaining coke to accumulate as a slag layer in the furnace hearth. Owing to their higher density, molten Fe and Pb fall as droplets through the slag and form a separate layer below the slag. The final composition of the metal is determined primarily by the partitioning of the solutes between the metal and the slag as the metal droplets pass through the slag layer.

The residual amounts of volatile species, such as As, Bi, Cd and Sb, are volatilized at the high temperatures encountered in the blast furnace. Some of this vapor recondenses as the gas cools during ascent of the stack and recirculates in the burden or reacts with the refractory lining. The remainder is discharged with the outlet gas. Since Zn is also discharged as a gas, the volatile impurities may contaminate the metal. Only very small amounts of these species are left to dissolve in the metals recovered from the hearth of the furnace. Alkalies, which are often present in the coke ash, are also volatile and may recirculate by condensation on the descending burden, but are removed mainly by solution in the slag phase.

Precious metals are rarely present to any appreciable extent in iron ores, but are often a valuable by-product from nonferrous ores. They are readily reduced to the metallic form and are then virtually insoluble in the slag, but can dissolve in the metal. Almost complete recovery of the precious metals is obtained as solutes in the metal during lead smelting, but any of these elements contained in Zn ores are lost if a blast furnace is used solely for the production of Zn. This is one reason why ores containing both Pb and Zn are commonly smelted together in the blast furnace. The Zn is recovered from the outlet gas and the Pb containing all the bullion from both ores is recovered from the hearth.

Relatively strongly reducing conditions are required for the reduction of iron oxides in the blast furnace. However, iron oxides are often present in Pb and Zn ores and, if similar reducing conditions are applied in the reduction of these ores, the Fe is also reduced to the metallic form. When the latter ores are reduced, therefore, the metal/carbon ratio in the charge is restricted to limit the amount of carbon available in the hottest zones of the furnace. Carbon can then be partially oxidized to CO_2 in this zone to hold the CO content below the Fe-FeO line but above the Zn-ZnO line in Figure 4.3. The iron then remains as FeO, which can form low melting point ferrous silicate or aluminosilicate components and thus produce a fluid slag at lower temperatures than would be feasible in the absence of FeO. This allows the Pb and Zn blast furnace processes to be operated at lower temperatures than for iron production. Magnetite has a high melting point, however, and does not dissolve in a silicate slag, so sufficient carbon must be present in the hot zone to ensure that the gas composition remains above the Fe_3O_4-FeO line in Figure 4.3. Otherwise, the magnetite accumulates as a solid in the bottom (hearth) of the furnace and reduces the capacity of the furnace. Any metallic iron formed by more strongly reducing conditions also remains as a solid at the lower operating temperature and accumulates in the hearth.

4.4.1.4. The Production of Iron

Carbon is soluble in molten iron and more than 4 wt% can dissolve in the molten metal. The metal is rapidly saturated with carbon and the concentration changes as the composition and temperature of the metal changes to maintain the activity of this element at, or very close to, unity.

Metals that form less stable oxides than Fe (such as Co, Cu, Ni, Pb and Sn) are reduced in the furnace and dissolve in the metal. The oxides of Si and Mn are more stable but, at the high temperatures required for reduction of Fe, these elements are also reduced until their activities in the metal reach the value in equilibrium with both the activities of SiO_2 or MnO in the slag and the gas atmosphere at the reaction temperature.

The reduction of MnO is straightforward. In the inverted conical section of the furnace (the bosh), the temperature is high enough for the reaction:

$$\mathrm{MnO + C = Mn_{(in\ Fe)} + CO} \quad (4.32)$$

to occur, and the partition ratio (i.e., the distribution of Mn between the metal and slag) is given by:

$$\frac{a_{\mathrm{Mn}}}{a_{\mathrm{MnO}}} = \frac{K}{p_{\mathrm{CO}}} \quad (4.33)$$

since a_{C} is unity in the presence of excess carbon. The value of K increases with temperature such that, at constant p_{CO}, the partition ratio is increased fourfold by a 200 °C increase in temperature in the bosh. However, there is strong evidence that the equilibrium partitioning is not attained during the descent of the metal droplets through the bosh to the top surface of the slag.

The reduction of Si also commences in the bosh and appears to occur via the formation of a volatile suboxide, followed by reduction of the volatile species on the surface of the iron droplets:

$$\mathrm{SiO_2 + C = SiO + CO} \quad (4.34)$$

$$\mathrm{SiO + C = Si_{(in\ Fe)} + CO} \quad (4.35)$$

giving the overall reaction:

$$\mathrm{SiO_2 + C = Si_{(in\ Fe)} + 2CO} \quad (4.36)$$

and

$$\frac{a_{Si}}{a_{SiO_2}} = \frac{K}{p_{CO}^2} \tag{4.37}$$

The equilibrium constant for Equation 4.36 is more markedly temperature dependent than for Equation 4.32 for Mn reduction and the silicon partition ratio increases rapidly with increasing temperature. Equation 4.34 is also strongly temperature dependent, and sufficient SiO can be generated by reduction of unfluxed silica in the coke ash and in the ore at the very high temperatures attained in the combustion zone to transfer significant amounts of Si into the Fe.

The activity of the silica from which SiO is volatilized is at or very close to unity. As the burden continues to descend through the furnace, the fluxes and the ash released from the coke react with the gangue materials to form a fluid slag in which the activity of silica is less than unity. The cumulative effect of the change in the activity as the slag is formed, followed by the fall in the temperature of the metal and the slag as they descend below the combustion zone, is to displace the equilibrium for Equation 4.36 to the left. However, the molten metal droplets are isolated from the gas atmosphere when they pass below the surface of the slag. The oxygen required for oxidation of any excess Si in the metal can then be supplied only by continuation of the Mn reduction Equation 4.32 in an attempt to reach equilibrium also for the Mn partitioning at the prevailing temperature with a_{MnO} in the slag.

The sum of Equations 4.32 and 4.36 can be written as:

$$2MnO + Si = 2Mn + SiO_2 \tag{4.38}$$

and

$$\left[\frac{a_{Mn}}{a_{MnO}}\right]^2 = K\left[\frac{a_{Si}}{a_{SiO_2}}\right] \tag{4.39}$$

The Si and Mn in the iron should also be in equilibrium with the carbon dissolved in the iron:

$$MnO + C_{(in\ Fe)} = Mn_{(in\ Fe)} + CO \tag{4.40}$$

$$SiO_2 + 2C_{(in\ Fe)} = Si_{(in\ Fe)} + 2CO \tag{4.41}$$

since, at equilibrium, p_{CO} should have the same value in all of these partition reactions. The equilibrium Si and Mn partition ratios can thus be calculated from knowledge of the activity of carbon in the iron and the temperature in the hearth. Analysis of samples taken from operating blast

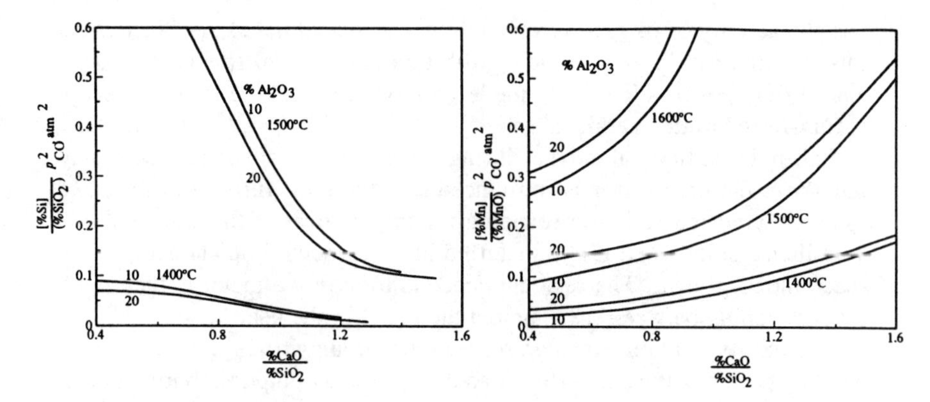

FIGURE 4.5. Equilibrium metal-slag distribution for (a) silicon and (b) manganese as a function of basicity of CaO-Al_2O_3-SiO_2-MnO slags saturated with graphite. (From Turkdogan, E. T., *Met. Trans.*, 9B, 163, 1978. With permission.)

furnaces has demonstrated that the actual partition ratios for these elements are greater than the equilibrium values, but the Mn partition is controlled by the Si partitioning.[55] Within the restricted range of slag compositions that are fluid at the working temperature, a_{MnO} increases[56] and a_{SiO_2} decreases[57] as the basicity (e.g., CaO/SiO_2) is raised. The observed partition ratios show a corresponding change with the basicity, Figure 4.5. The partition ratios also increase with increases in the metal temperature.

The furnace is usually operated to produce molten iron containing less than 1% Si to reduce the load in the subsequent refining operations. This is achieved with a slag basicity of about 1.1 and a hot metal temperature of not greater than 1500 °C. The simple CaO/SiO_2 ratio is not a precise measure of the slag basicity, however, since the slag also contains other compounds. Magnesia and alumina together may total as much as 30% of the slag weight, MgO functions as a base and small amounts of MgO have a similar effect to the CaO. Alumina is amphoteric. It forms aluminate ions and acts as an electron acceptor in basic melts, but it behaves as a base in slags of low basicity, pertinent to blast furnace practice. So, the addition of alumina at constant CaO/SiO_2 ratio lowers the Si content of the metal.

Most iron ores contain some phosphorus as mineral phosphates, which are less stable than FeO and are readily reduced in the blast furnace. Phosphorus can be stabilized in the slag as a calcium phosphate, but the formation of the phosphate depends on the reaction:

$$2P_{(in\ Fe)} + 5O_{(in\ Fe)} = P_2O_5 \tag{4.42}$$

The equilibrium constant for this reaction is given by:

$$K = \frac{a_{P_2O_5}}{a_p^2\, a_O^5} \tag{4.43}$$

and the activity of oxygen is very low in carbon-saturated Fe. Since a_O is raised to the fifth power in the equilibrium constant for this reaction, the equilibrium position lies far to the left and virtually all the P in the charge is transferred to the metal.

Sulfur is partially removed during sintering, but small amounts may remain in the ore. Sulfur is introduced into the blast furnace in the coke ash. The atmosphere is too reducing for the removal of this element as a volatile oxysulfide and it is transferred into the metal droplets before they reach the slag layer. The residual concentration in the metal is then fixed by partitioning between the slag and the metal in the hearth.

The partitioning reactions presented thus far are written in terms of molecular species in the slag. This is an acceptable assumption, but an unsaturated slag is really composed of ions and not molecules. The basic oxides, such as CaO and MgO, decompose into positively charged ions, Ca^{2+} and Mg^{2+}, and negatively charged oxygen anions, O^{2-}. The oxygen ions are associated with the acidic oxides to form complex anions such as SiO_4^{4-}:

$$SiO_2 + 2O^{2-} = SiO_4^{4-} \tag{4.44}$$

The sulfur partitioning reaction is usually written in the ionic form:

$$S_{(in\ Fe)} + O^{2-}_{(in\ slag)} = S^{2-}_{(in\ slag)} + O_{(in\ Fe)} \tag{4.45}$$

$$K = \frac{a_{S^{2-}} \cdot a_O}{a_{O^{2-}} \cdot a_S} \tag{4.46}$$

This shows that the transfer of S from metal to slag is favored by the strongly reducing conditions that exist in the blast furnace hearth. The S species in the slag is a negatively charged ion that draws attention to the need for cations to stabilize them. Since silica forms anion complexes in the slag, but CaO and MgO both form cations, it follows that the transfer of sulfur from metal to slag increases as the basicity of the slag is raised within the fluid composition range. The numerical value of the equilibrium constant for desulfurization increases with a rise in the hot metal temperature so that, at constant basicity, the amount of sulfur remaining in the metal is lowered as the metal temperature is increased. Partition ratios, $S_{(slag)}/S_{(metal)}$, of over 40 can be obtained with the slag compositions used in the iron blast furnace. This is in marked contrast to the much poorer partitions obtained at the lower temperatures and with the lower slag basicities that are used in lead smelting. Fortuitously, S is often added to Pb to remove Cu in the subsequent refining stage, so S contamination in the molten Pb is not a major problem. Removal of the residual S remaining in the iron obtained from the blast furnace is a major task, however, in the refining of steel. Thus, the iron is often treated with calcium carbide, Mg or other strong desulfurizing agents after tapping from the furnace and separation of the blast furnace slag, before charging into the refining furnace. The iron tapped from the blast furnace may contain 4.3 to 4.7% C, 0.5 to 1.0% Mn, 0.2 to 0.8% Si, 0.02 to 0.04% S and 0.06 to 0.12% P. After treatment in the transfer ladle, the sulfur content can be lowered to 0.005–0.010%.

The slag removed from the iron blast furnace contains only a very small amount of Fe, which is insignificant when compared with the Fe content of the as-mined ores, so the slag is discharged to waste. This is again in marked contrast to the treatment of the slag removed from the Pb-Zn blast furnace. The slag here may contain about 1% Pb and a higher concentration of Zn. The volume of slag is also much greater than that formed from the rich iron ores used at present. The as-mined Pb and Zn ores contain only low concentrations of these elements and recovery of the residual amounts from the slag by volatilization (fuming) at about 1200 °C or by other treatment is often justified on economical grounds.

4.4.1.5. The Lead-Zinc Blast Furnace Process

Less strongly reducing conditions are used for the reduction of ores containing Pb and Zn. Consequently, only Pb, which is stable in CO_2 at elevated temperatures, is reduced in the stack. ZnO is reduced in the hottest zone of the furnace and the metal vapor produced ascends with the combustion gas. The vapor must be cooled to condense the Zn before it is reoxidized. The boiling point of Zn in equilibrium with its vapor at 1 atm pressure is 907 °C, but condensation from a gas containing p_{Zn} at 0.05 to 0.07 atm begins only when the gas has cooled to below 670 °C. The vapor is transported in a mixture of CO, CO_2 and N_2, the CO content of which is limited to below the Fe-FeO line in Figure 4.3 to prevent the formation of metallic Fe in the hearth of the furnace. Any CO/CO_2 ratio in the gas phase that satisfies the latter requirement would start to oxidize the vapor before the temperature falls sufficiently to cause condensation.

An ingenious solution is applied in the Imperial Smelting Process.[58] The gas leaving the furnace is conveyed in an insulated pipe to a chamber in which the gas is shock-cooled by sprays of molten Pb. It is essential that the gas remains above the temperature at which Zn reverts to the oxide until it reaches the spray chamber. This is achieved by admitting small amounts of air to the upper stack of the blast furnace to burn part of the CO to CO_2 and raise the gas temperature at the top of the stack to at least 1000 °C. The gas leaving the furnace then contains about 20.5 vol% CO and 11.0% CO_2; so, the ratio, $CO/(CO + CO_2)$, is about 0.65. Thus, the increase in the Zn reversion temperature accompanying the resultant change in the CO/CO_2 ratio in the gas is less than the increase of the gas temperature, so the risk of reversion to ZnO is reduced.

Zinc dissolves in molten Pb at temperatures greater than 417.8 °C, but separates from the Pb as a solid phase at lower temperatures. Allowing for errors in temperature measurement and control, this places a lower limit of about 450 °C for the Pb quenchant if the Zn is to be recovered as a liquid phase. Sensible heat is transferred from the gas to the Pb during the quench, raising the temperature of the Pb. The partial pressure of Zn in equilibrium with the pure metal is 0.005 atm at 550 °C. If the temperature of the Pb is not allowed to rise above this level and thermal equilibrium is attained between the Pb and the gas, then 90 to 95% of the Zn can be dissolved in the Pb, depending on p_{Zn} in the outlet gas from the furnace. The remainder of the vapor is recovered as an oxide dust during subsequent cooling of the gas and is sold for other uses, such as paint manufacture.

The gas must be passed through a large quantity of spray in order to limit the increase in the temperature of the Pb. If the total pressure of the gas is 1 atm and p_{Zn} in the gas is 0.06 atm, then 1 mol Zn is accommodated in 16.7 mol furnace gas. Assuming similar heat capacities for the gas and for molten Pb, that 90% of the Zn is condensed and that the gas is cooled from, say, 1000 to 550 °C while the Pb is heated from 450 to 550 °C, then a minimum of

$$[16.7 \times (450/100) \times (100/90) = 83.5 \text{ mol Pb}$$

must be sprayed through the gas for each mole Zn recovered. However, this is not a sufficient quantity to recover all of the Zn. The solubility of Zn in Pb is 2.5 wt% at 550 °C, but falls to 2.2% at 450 °C. So, the Zn in excess of 2.2% can be separated by recirculating the Pb to an external tank and cooling it to 450 °C. Zn has a lower density and floats on top of the Pb, from where it is recovered. The atomic weights of Zn and Pb are 65.4 and 207.2, respectively. If only 0.003 g Zn is recovered from each gram of Pb cycled through the spray chamber, the actual quantity of Pb required to recover the Zn is

$$[(65.4/0.003) \times (1/207.2) \times (100/90) = 117 \text{ mol}$$

In practice, the temperature of the gas leaving the top of the furnace may be as high as 1100 °C, to ensure that the Zn vapor is not reoxidized before it is condensed. This may require the recirculation of still larger quantities of Pb at the highest gas temperatures to absorb the additional sensible heat.

When ores containing both Zn and Pb are reduced in the blast furnace, the exothermic gaseous reduction of the PbO in the stack generates heat where it is required to sustain the temperature of the Zn-bearing gas. The Pb recovered from the furnace hearth supplies the metal required for the Zn vapor quench.

Zinc is extracted primarily by the hydrometallurgy and electrolysis route, but any Pb present in the ore must then be recovered separately and must not be allowed to contaminate the Zn electrolyte solution. Consequently, the Imperial Smelting Process is the major route for the extraction of Zn from ores containing significant amounts of Pb.

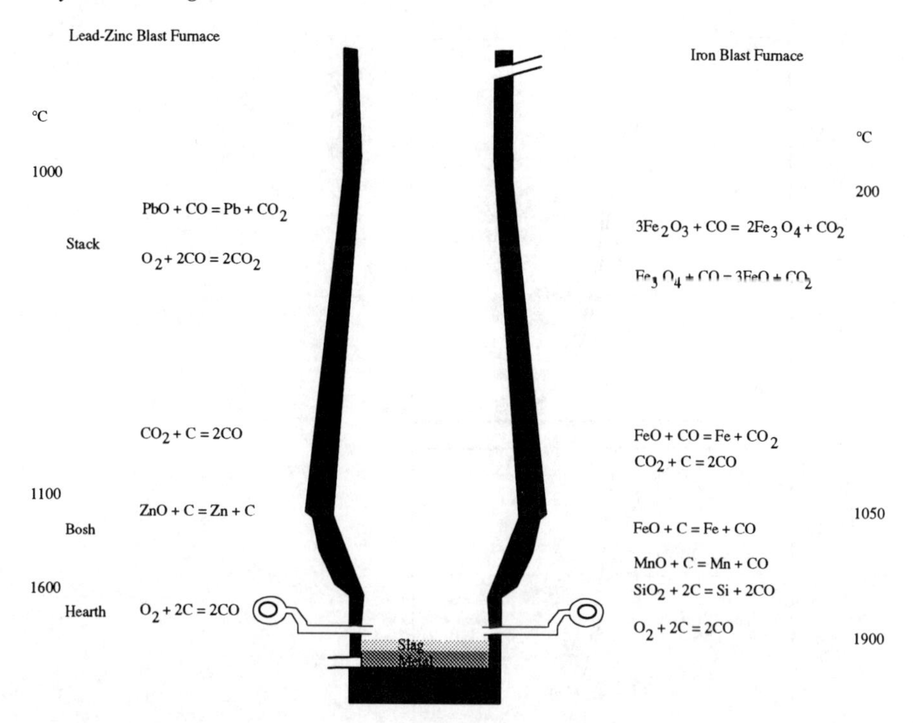

FIGURE 4.6. The principal reactions occurring in the zones of the iron and the lead-zinc blast furnace.

The principal reactions and the zones in which they occur in the Fe and in the Pb-Zn blast furnace processes are summarized in Figure 4.6. The furnaces are not equal in height and the temperature gradients are very different, but the diagram is presented in this way to compare and contrast the reactions.

4.4.1.6. The Rist Diagram

The temperature distribution along the vertical axis of an iron blast furnace above the gas inlet or tuyere level is illustrated in Figure 4.7a. The gas leaves the combustion zone at about 1900 °C and cools as it ascends through the stack. The charge enters the furnace at ambient temperature and is heated during descent by transfer of sensible heat from the gas. Additional heat is generated in the charge by the gaseous exothermic reduction of the higher oxides of iron to FeO, which more than compensates for the heat consumed by the endothermic dissociation of any $CaCO_3$ and other carbonates and hydrates present in the charge. As a result, the gas and solid temperatures converge as the charge descends through the stack. The endothermic direct reduction reactions and the heat consumed in the

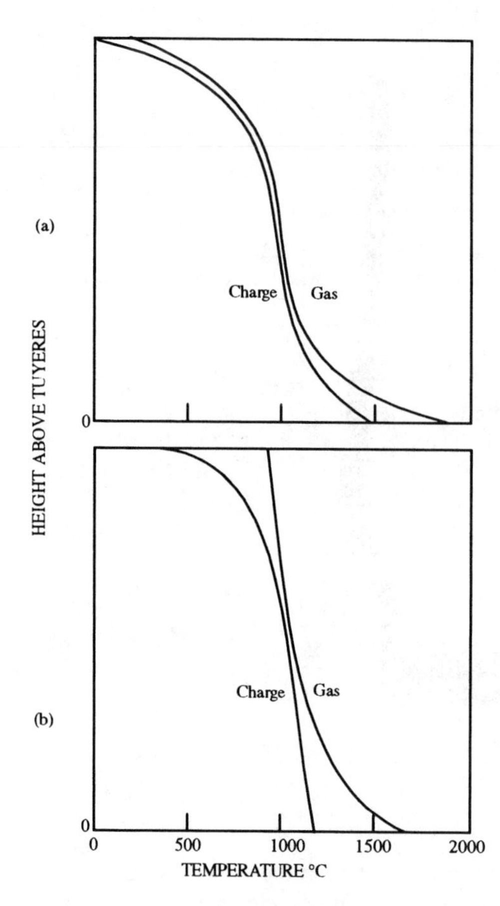

FIGURE 4.7. Temperature distribution along the vertical axis of (a) the iron blast furnace and (b) the lead-zinc blast furnace.

fusion of the metal and the slag results in a divergence of the two temperature profiles in the lower regions of the furnace.

The gas and solid temperatures are very similar in the region between these two zones. This is the region in which most of the FeO is reduced by CO gas and it is called the "thermal reserve zone." The zone usually extends over a temperature range from about 900 to 1000 °C and varies in height up to about 4 m. In effect, it separates the lower zone of the furnace, which serves as a fusion reactor, from the upper zone, which functions as a countercurrent gas-solid reactor.

The temperature distribution in the Pb-Zn blast furnace differs in a number of ways. The temperature of the gas leaving the combustion zone is lower, due to the higher ore/carbon ratio compared to the Fe blast furnace process. Initially, the gas temperature falls more rapidly as heat is consumed by the endothermic reduction of all of the ZnO in the lower region of the furnace. Conversely, the preheating of the charge, the exothermic reduction of PbO and the partial combustion of some of the CO within the furnace by a secondary air supply results in a more rapid temperature rise in the solids in the stack (Figure 4.7b). The net result is a small increase in the temperature of the thermal reserve zone to about 1050 °C.

The importance of the thermal reserve zone is apparent when it is recognized that, effectively, the enthalpy released as the gas is cooled to the zone temperature is equal to the enthalpy absorbed by the burden above the zone. Thus, if a furnace is operating smoothly with a thermal reserve zone, then the enthalpy available from the gas could be increased, for example, by increasing the temperature of the hot air blast and the amount of coke charged could be reduced to restore the thermal balance. However, the extent to which the coke consumption can be reduced in this way is limited because the amount of CO available for gaseous reduction is lowered as the quantity of coke combusted is decreased. More carbon is then consumed by the direct reduction reactions and the thermal efficiency of the process is lowered.

A useful way of studying the performance of the process is the diagrammatic representation, based on the mass balances for the metal, oxygen and carbon, that was first proposed by Rist and Meysson.[59] The mass balances, expressed as O/Fe and O/C, are illustrated in Figure 4.8 for the 1000 °C isotherm in an iron blast furnace reducing a hematite ore. The points W, M and H define the O/C ratio (i.e., the gas composition) in equilibrium at that temperature with wustite, magnetite and hematite. The sloping line between W and W′ indicates how this ratio changes as the Fe/O ratio changes across the wustite phase (shown in the lower diagram). Positive intercepts along the ordinate represent the oxygen content of the iron oxides, while negative intercepts indicate the amount of oxygen introduced in the air blast (C-G) and from the reduction of other oxides such as Mn and Si (O-G). The abscissa to the left of the unit intercept relates to gas generation by direct reduction, while the portion between 1.0 (corresponding to CO) and 2.0 (representing CO_2) indicates the utilization of the gas for gaseous reduction in the stack.

The line, ABC, is called the operating line of the furnace and the portion, AB, represents the change in the gas composition accompanying the gaseous reduction of the ore. This must be a straight line, since there is no change in the number of gas moles accompanying the gaseous reactions (cf Equations 4.26 to 4.28). So B-E is a measure of the amount of oxygen removed from the ore by gaseous reduction, while D-B is the amount removed by direct reaction with carbon. Since O/C = 1 corresponds to pure CO and O/C = 2 represents pure CO_2, the intercept, A, on the top boundary of the diagram indicates the ratio CO/CO_2 in the furnace top gas.

The operating line must pass through, or to the left, of the point W; otherwise, the ore would be reoxidized in the zone where the operating line overlapped the shaded area. A layer of wustite exists at the level in the stack corresponding to the temperature for which the diagram is constructed

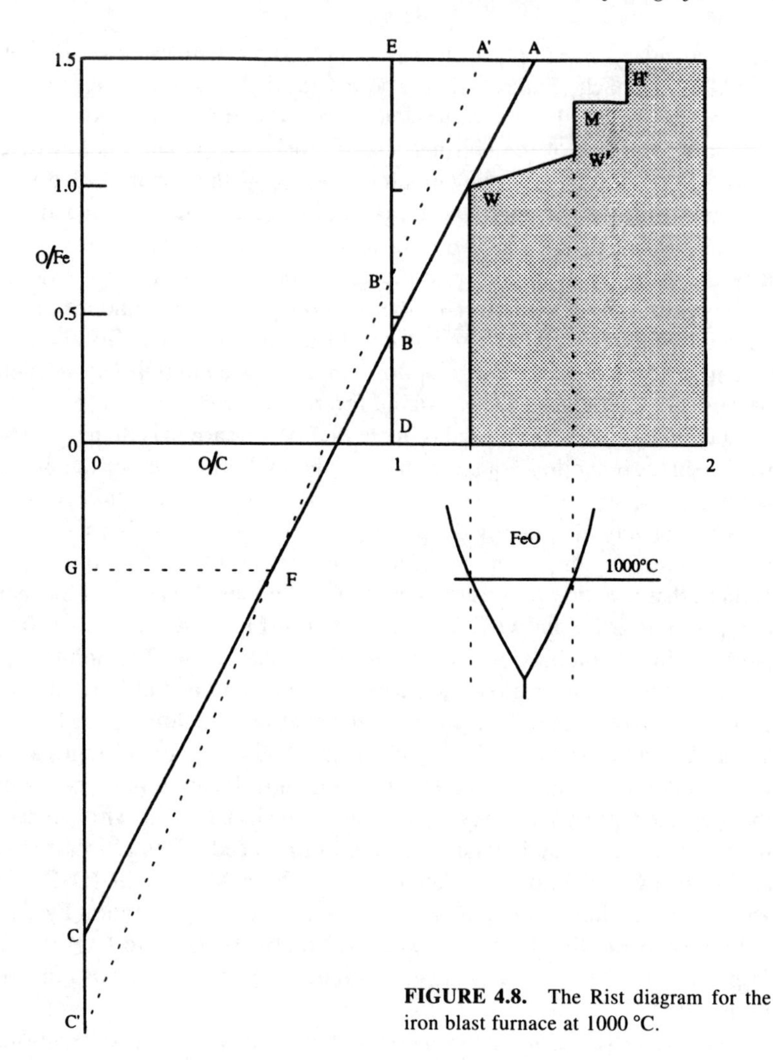

FIGURE 4.8. The Rist diagram for the iron blast furnace at 1000 °C.

if the line just passes through the point W. The intersection implies that the gas and the ore are in chemical equilibrium at that point. It is then called a "chemical reserve zone," a zone that separates the upper zone where the higher oxides are reduced to FeO from the lower zone in which FeO is reduced to the metal. The existence of a thermal reserve zone is usually a prerequisite for the occurrence of a chemical reserve zone.

The reducibility of the ore is lowered if the iron is present as compounds such as ferrous silicates, if the size of the ore lumps is increased or if the porosity of the ore is decreased. The operating line may then pass to the left of W, but the position of the line is constrained by the requirement that a certain quantity of heat must be generated below the thermal reserve zone to satisfy the endothermic and sensible heat requirements. This quantity can be evaluated[59] from a heat balance and identified by some position F, on the operating line. A decrease in the reducibility of the ore, therefore, causes a rotation of the operating line through the point F to A′B′C′, resulting in an increase in the amount of direct reduction (D-B′) and a decrease in the amount of gaseous reduction (B′-E). The more negative intercept at C′ indicates that more carbon must then be supplied, while the position of A′ shows that the CO/CO_2 ratio of the top gas is lowered.

The slope of the operating line is equal to the atom ratio, C/Fe, which identifies the quantity of carbon required to satisfy the heating and reduction requirements. This is more commonly expressed as the coke rate in kg/C per tonne Fe, which is equal to:

$$[1000 \times (12/55.8) \times (\text{slope C/Fe})]$$

to which must be added the quantity of carbon dissolved in the molten Fe. Since the slope of the operating line is increased, it is evident that the coke rate is increased by a decrease in the reducibility of the ore. The position of point F is fixed by the thermal requirements in the lower regions of the furnace. Thus, F is moved downward on the diagram and the coke rate is increased by increasing the gangue content in the ore (and hence the volume of slag that has to be melted) by increasing the metal temperature or by reducing larger amounts of elements such as Si into the Fe. Conversely, the hinge point is moved upward by an increase in the temperature of the hot blast, decreasing both the coke rate and the CO/CO_2 ratio in the top gas.

4.4.1.7. Decreasing the Coke Requirement

The enthalpy of formation of Fe_2O_3 is 821.3 kJ at 25 °C and 1 mol oxide contains 111.6 g of Fe. Hence, the energy required to dissociate the oxide at 25 °C is

$$[(821{,}300/111.6) \times 10^6] = 7.4 \text{ GJ t}^{-1} \text{ Fe}$$

Adding 1.34 GJ for the energy required to melt the iron and raise the temperature to 1450 °C gives a total heat requirement for the reduction of the oxide of 8.74 GJ t^{-1}. The energy actually consumed in a modern, high-productivity blast furnace smelting a high-grade iron ore when the operating line passes through the point W is about 11.45 GJ t^{-1}.[60] A large part of the additional energy is consumed in heating the gangue, coal ash and fluxes to form a molten slag, in the reduction of other elements to form solutes in the metal and in the solution of carbon in the iron. Heat losses through the furnace lining and as sensible heat in the waste gas constitute only about 10% of the total energy requirement. Thus, the blast furnace process has evolved into an energy-efficient operation; but the total energy required for the manufacture of iron by this route must include the energy consumed in sintering and in coke manufacture. Sintering absorbs about 1.9 GJ t^{-1} Fe, but the energy consumption in the blast furnace would be increased if unsintered fine ore was included in the burden. Coke production accounts for an additional 2.4 GJ t^{-1} Fe.[60] Attempts are being made, therefore, to reduce the dependence of the process on coke.

There is a further reason for diminishing the coke requirement. Although vast quantities of coal are still available in the ground, only very limited stocks remain that are suitable for processing into the high-strength, low-reactivity coke required for the blast furnace. In recent years, major advances have been made in blending together different coals and in the coking process to produce coke with the required properties. However, coke ovens are expensive to construct and operate and there are serious environmental problems associated with coke production, so major efforts are being made to restrict the amount of coke required in the blast furnace charge. Improvements have been achieved through a better understanding of the overall thermal balance and the thermal loads in different zones of the furnace, as portrayed in the Rist diagram.

The coke consumption can be reduced if part of the energy required for heating is met by other means. This has been achieved by injecting gaseous hydrocarbons or granulated coal with the hot blast through the tuyeres. These additives, however, alter the heat requirements in the lower regions of the furnace because:

1. They are not preheated before injection into the hottest zone of the furnace
2. The decomposition of the hydrocarbons (which includes the volatiles in the coal) is endothermic
3. The enthalpy of reduction of iron oxides by H_2 is lower than for reduction by CO

The hot blast temperature cannot be raised above about 1300 °C in conventional hot blast stoves, so the additional thermal load is compensated by increasing the oxygen content of the hot blast to, say, 28 vol%, thus decreasing the volume of nitrogen in the air that has to be heated from the air inlet temperature to the combustion temperature. The efficiency of the coke combustion is thereby increased and the velocity of the gas through the stack is lowered, decreasing the dust losses. The volume of ore that can be charged is increased as the coke consumption is lowered and hence the furnace throughput is raised. The H_2 released by the dissociation of the hydrocarbons contributes to the gaseous reduction of the oxides. The abscissa in Figure 4.8 must then be changed to $(O + H_2)/(C + H_2)$ when a significant amount of hydrogen is present in the gas.[61]

Thirty years ago, a typical blast furnace consumed at least 750 kg of coke per tonne of iron. Coke rates of under 300 kg have now been achieved[62] and further reductions are possible. This change has been brought about by fuel injection through the tuyeres (up to 250 kg coal/ton Fe), smelting of higher grade ores with consequent reduction in the slag volume, improved burden preparation by sintering or pelletizing all of the ore, increase in the top gas pressure in the furnace (up to 2 atm) to ensure more uniform gas flow over the furnace cross-section and reduced heat losses per tonne of metal as a result of increased furnace size. With maximum fuel injection at the tuyeres, the gross energy consumption has been reduced to 14.5 to 14.8 GJ t^{-1} Fe. These changes have also resulted in marked increases in furnace productivity and outputs of over 3 tonnes per day per cubic meter of furnace volume have been achieved, with consequent reduction in the real operating costs.

4.4.2. Carbothermic Reduction Without Coke

4.4.2.1. Reverberatory Smelting

The capital cost of a blast furnace plant with the associated coke oven battery and gas by-product equipment is very high. The plant must be operated continuously, 7 days a week for many years at full production to amortize the cost. It is not surprising, therefore, to find that Sn and Cu ores, which are only mined in relatively small quantities, are no longer smelted in the blast furnace. The reverberatory furnace, so named because the heat is "reverberated" to the charge from the furnace roof, is heated by the product gases from the combustion of coal, gas or oil. It is thermally inefficient, since the gas leaves the furnace at a higher temperature than the metal bath and the sensible heat can only be utilized if it is captured in heat exchangers for use in downstream processes. However, the capital cost is very much lower and small capacity units can be commercially viable, so it is more suited to the small-scale production that is typical for Cu and Sn smelting.

The cost of sintering or pelletizing can be avoided if the ore or concentrate is injected into the surface of the molten bath in the furnace in order to avoid dust losses. The air/fuel ratio can be adjusted readily to ensure that the CO/CO_2 ratio in the gas leaving the furnace is only slightly in excess of the ratio in equilibrium with the metal and its oxide, thus ensuring that most of the potential energy in the fuel is available within the furnace. Cu and Sn ores often contain large amounts of iron oxides, and this ability to control the reducing potential aids the separation of the metals from the iron oxides.

Metallic Cu is stable in an atmosphere of almost pure CO_2, so the enthalpy available from carbon combustion can be fully released within the furnace when copper oxide ores are reduced. The atmosphere is then oxidizing with respect to iron and, if silica is present in the charge, the iron is removed as ferrous silicate in the slag. The gas composition must not fall below the Fe_3O_4-FeO in Figure 4.3; otherwise, magnetite is formed. The operating temperature is below the melting temperature of magnetite, so it accumulates below the molten Cu in the hearth of the furnace and is difficult to remove from the furnace.

Tin oxide, SnO_2, has roughly similar stability to magnetite and separation of metallic Sn from Fe is more complex when Fe is a major impurity in the ore. This is where the flexibility of the reverberatory process is a major advantage. By controlling the air/fuel ratio, the atmosphere above the bath is maintained at a level where most of the Sn can be reduced from the ore at about 1350 °C, using coal as the reductant. The oxide is reduced via the formation of the volatile suboxide, SnO, that condenses on the surfaces of the carbon reductant where it is reduced to the metallic form:

$$SnO_2 + CO = SnO + CO_2 \tag{4.47}$$

$$SnO + C = Sn + CO \tag{4.48}$$

$$C + CO_2 = 2CO \tag{4.49}$$

Some SnO is lost in the exhaust gas if the C/Sn ratio in the charge is less than about 3/1.[63] Some Fe is also reduced during this stage, the amount increasing with decreasing rate of reduction, which allows more time for the metallic Fe to dissolve in the Sn. The remainder of the SnO_2, together with the iron oxides and the gangue and the fluxes, forms a slag containing 20 to 25% Sn and about 20% Fe.

After separation from the liquid Sn, the slag is heated to about 1300 °C, lime and coal are added and somewhat less reducing conditions are maintained. Both a_{FeO} and a_{SnO_2} in the slag are lowered by the lime additions. The activity of FeO is lowered to a greater extent, which aids the preferential reduction of the Sn, but some Fe is also reduced and forms an Sn-Fe alloy containing 3 to 5% Fe. This is called "hardhead," since a hard, brittle, intermetallic compound, $FeSn_2$, is formed in equilibrium with Sn in the solid state. This alloy is recycled to the first smelting stage, or is remelted with more SnO_2 concentrate and a silica flux to oxidize the Fe into a silicate slag:

$$2Fe + SnO_2 + 2SiO_2 = Sn + 2FeO \cdot SiO_2 \tag{4.50}$$

The slag remaining after resmelting contains only about 1% Sn. When the selling price of Sn is high, this is sufficient to justify further treatment of the slag, for example, by adding a sulfur-bearing compound such as gypsum to volatilize SnS, which is recovered and recycled.

4.4.2.2. Electric Smelting

Reverberatory smelting is an environmentally undesirable process in view of the large amounts of CO_2 (and possibly CO) per tonne of metal produced that are released into the atmosphere from the combustion gases. Electric smelting in electric arc furnaces is much more attractive in this respect, particularly in countries where cheap hydroelectric power is available. The environmental benefit is less significant when the electric power required for smelting is generated by combustion of fossil fuels, but is still attractive. Only the gases generated by the reduction reactions plus nitrogen from the air supply are released from the furnace. It is much easier to remove dust, fume and noxious volatiles from the smaller volume of gas before it is vented to the atmosphere. The reactions that occur are similar to those in the reverberatory furnace, but the oxygen potential within the furnace is more readily controlled by adjusting the carbon and oxygen supply rates.

The higher temperatures that can be achieved with electric arc heating facilitate the full or partial reduction of oxides that are too stable for reduction in the reverberatory furnace. The temperatures required for the reduction of relatively very stable oxides of elements such as chromium and silicon can be lowered by reducing the metal to form an alloy with another metal in which its activity is less than unity. High-grade ferrochromium and ferrosilicon are produced in this way. A variety of other alloys and compounds are also produced in this manner.

4.4.2.3. Direct Reduced Iron (DRI)

A modern blast furnace can produce 2 to 3 million tonnes of iron per annum. A furnace of this size requires a large associated plant to convert the molten iron into steel and a market for the steel products produced. The process efficiency of the blast furnace decreases if productivity has to be curtailed and it lacks flexibility to adjust rapidly to changes in the market demand. Although, as described earlier, the dependence of the process on coke supplies has been eased somewhat in recent years, coke manufacture is an increasingly complex production and environmental problem. An alternative method of iron production that can be operated economically on a smaller scale and does not require a supply of coke is, therefore, attractive.

The direct reduction process can satisfy these requirements. It depends on the gaseous reduction of iron oxides below the melting temperature, in a way similar to the gaseous reduction that occurs in the shaft of the blast furnace. A wide variety of techniques has been proposed for the production of DRI,[64,65] but two processes—Midrex and Hyl—dominate in the present manufacture of about 20 million tons per annum of this product. They both employ a furnace similar in shape, but much smaller in size than the iron blast furnace. The reducing gas is introduced through tuyeres at the base of a conical stack in which the ore is reduced. The inverted cone below the tuyeres serves as a cooling zone to decrease the risk of reoxidation when the DRI is discharged into the atmosphere. It can also be used to carburize the DRI. Fluidized beds and rotary kilns are also used for DRI production.

Oil or natural gas is most commonly used as the reductant. This makes the process particularly attractive in countries where excess gas is available from oil wells, and the major locations of DRI plants are currently found in those countries. The higher hydrocarbons are easily dissociated to lower H/C ratios on heating, with the release of hydrogen, but the basic molecule, methane, does not readily react with the oxygen in the ore to form CO and H_2 at temperatures below about 850 °C. The gaseous reductants are usually reformed, therefore, with an oxidant in the presence of a catalyst in a separate unit, i.e.,

$$CH_4 + CO_2 = 2H_2 + 2CO \tag{4.51}$$

$$CH_4 + H_2O = 3H_2 + CO \tag{4.52}$$

The reformed gas is cooled, if necessary, and then fed to the reduction furnace.

The composition of the reformed gas can be adjusted within wide limits. The ratio $(p_{CO} + p_{H_2O})/(p_{CO_2} + p_{H_2})$ is fixed by the water gas reaction (Equation 4.5). The equilibrium constant for this reaction is temperature dependent (Figure 4.9). Below 880 °C, the reaction goes to the right and the water vapor is reduced by CO. This can be utilized to further increase the H_2 content of the gas by oxidizing most of the CO and absorbing the CO_2 into a hot carbonate solution. When CO and H_2 are present in the correct ratio, the reduction process is autogenous and heat is required only to bring the ore up to the reaction temperature (see Worked Example 3).

The gas leaving the reduction unit contains more potential energy (i.e., $CO + H_2$) than the exit gas from the blast furnace. Part of this energy is recovered by combustion of some of the gas in a heat exchanger to supply the energy needed for the endothermic reforming reactions and to preheat the reformed gas and thus supply the heating requirements for the ore. The remainder of the gas is recirculated with added hydrocarbons to convert the CO_2 and H_2O formed by the ore reduction back into CO and H_2. The recirculated gas can be reformed in an external unit, or it can be reformed within the reducer by using the freshly reduced Fe surfaces as the catalyst for the reforming reactions if the temperature is high enough to activate the methane dissociation.[66] Since reforming with added hydrocarbons increases the H_2/CO ratio, the gas is first cooled rapidly to condense some of the H_2O and preserve the ratio when autogenous operation is required (see Worked Example 7).

Coal can also be used as the reductant, but the efficiency of the operation is lower than with a hydrocarbon reductant because coal is not readily gasified at the relatively low reduction temperatures.

The metal and the gangue are not separated by melting during direct reduction, so it is important that the ore contains not more than about 5 wt% gangue. Otherwise, the thermal load on the subsequent melting and

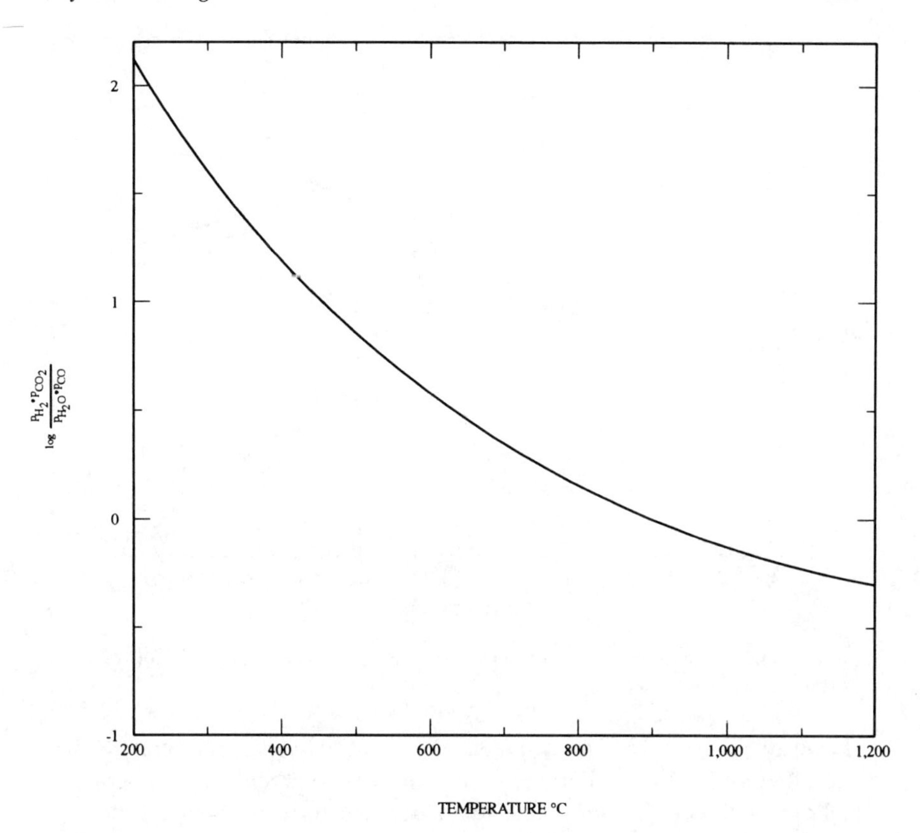

FIGURE 4.9. Temperature dependence of the equilibrium constant for the water gas reaction.

refining unit becomes excessive. The ore is heated more rapidly than in the blast furnace, resulting in the imposition of an increased thermal stress on the volume expansion accompanying the reduction from Fe_2O_3 to Fe_3O_4. This can result in the production of a large volume of fine particles of ore that fill up the interstices between the lumps and restrict access of the gas to the reaction sites. Finely divided Fe is also a hazard in the material discharged from the furnace because it is pyrophoric. Accordingly, ores that are very friable or swell excessively during reduction are crushed and pelletized prior to feeding to the furnace.

The maximum reduction temperature is limited by the temperature at which the ore particles begin to sinter together, again restricting access of the gas to the reaction sites. This maximum temperature is dependent on both the type of ore and the particle size and is usually in the range 800 to 950 °C.

If equilibrium is attained in the reduction reactions, H_2 is a more efficient reducing agent for iron ores than CO at temperatures above 840 °C, as

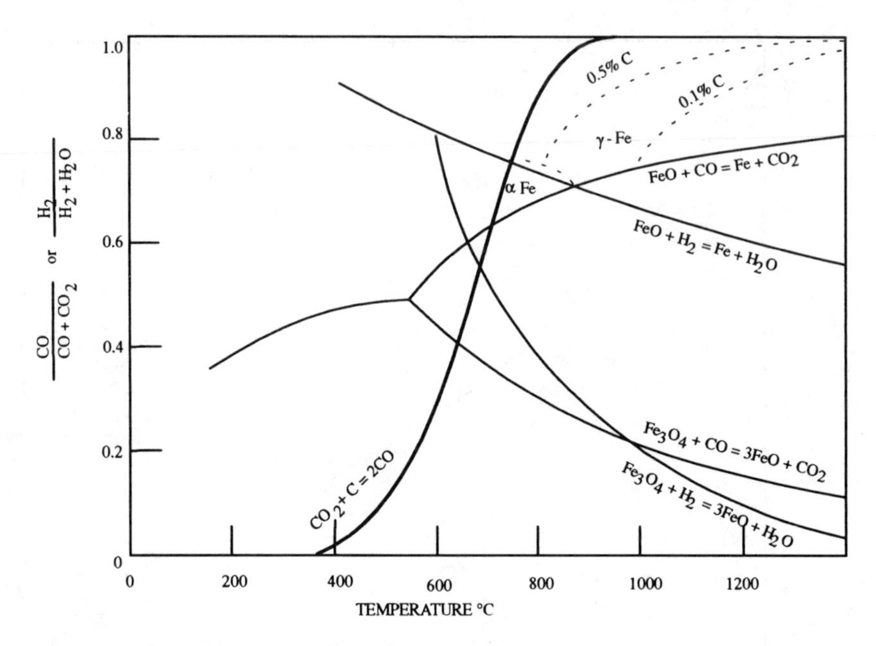

FIGURE 4.10. Equilibrium relations in the systems iron-carbon-oxygen, iron-hydrogen-oxygen and carbon-oxygen.

shown in Figure 4.10. The converse applies at lower temperatures. However, the ore is charged into the reduction furnace at ambient temperature and chemical equilibrium is not attained in the time of about 3 h during which the ore is progressively heated and reduced during passage down the stack. Kinetic factors then result in H_2 becoming a more efficient reducer than CO at the lower temperatures. The small size of the H_2 molecule allows molecular diffusion of the gas along narrow pores through which CO can only pass by the markedly slower Knudsen diffusion. As the ore temperature rises, the cross-section of the pores increases and CO begins to play an increasing role in the reduction mechanism. But it is probable that H_2 continues as the more reactive species at the reaction site and CO is involved primarily in the reduction of the H_2O product via the water gas reaction, which is catalyzed by the freshly reduced Fe.[67] As the maximum temperature is approached and the pore size is further increased, CO plays an increasingly important role, probably because the exothermic CO reduction reaction raises the ore temperature more rapidly and thus improves the reaction kinetics.[68]

The rate of reduction of the higher oxides of Fe is rapid when compared to the rate of reduction of FeO and this is the slowest reaction step. As with any reaction, the rate decreases as 100% reduction is approached and a long time at temperature is required to obtain complete removal of FeO. The energy required for subsequent melting and refining of the DRI is increased by about 41 MJ for each 1% Fe not metallized in the reducer.[69] Consequently, the degree of reduction aimed for is a compromise between the cost of additional time in the reducer to increase the reduction and the energy cost in subsequent refining. DRI is usually marketed in the range 92 to 95% reduced.

Carbon can dissolve in the Fe when either CO or a hydrocarbon is present in the gas phase and the equilibrium carbon concentrations are indicated by the broken lines in Figure 4.10. Theoretically, the Fe should be saturated with graphite if the final reducing conditions lie to the left of the line for the Boudouard reaction, but graphite is not readily nucleated in solid Fe and metastable Fe_3C is formed instead. It is evident, therefore, that in the temperature range normally used for DRI production, the metal contains very much less carbon than blast furnace iron. If the ore is treated at 900 °C to give about 90% reduction, the DRI contains about 0.1 to 0.2% C.[66] A C/O ratio of 0.75 is required in the DRI to provide sufficient carbon to react with and remove the residual oxygen when the DRI is subsequently melted for conversion into steel. Up to 2.0% C can be dissolved in the DRI by circulating a gas containing a small amount of methane through the cooling zone of the reducer. This is sufficient carbon to react with the residual FeO when the DRI is melted, thereby increasing the metal yield and reducing the slag bulk.

The rate of DRI carburization is diffusion controlled and, for a given time of exposure, the carbon content increases rapidly as the particle size of the ore is decreased. However, the risk of particle agglomeration by sintering is raised as the particle size is reduced and the maximum reduction temperature is lowered correspondingly. It is possible to produce almost complete conversion to Fe_3C by reducing very finely divided ore in a fluidized bed at temperatures no higher than about 700 °C in a gas mixture that lies to the left of the Boudouard line in Figure 4.3.[70]

Elements in the gangue such as Mn, Si and P are not reduced into the Fe at the comparatively low temperatures that prevail in the direct reduction processes. The gangue is usually mixed intimately with the reduced Fe and physical separation by magnetic or other means requires first grinding the product to a very fine size, with consequent increase in cost and risk of surface reoxidation or even a pyrophoric reaction. So, the gangue is usually left *in situ* until the DRI is melted for refining. The oxygen potential is higher in the refining stage, so there is less risk of these elements being reduced into the iron before they are stabilized in a molten slag in which their activities are too low for reduction to occur. The volume of the refining slag is increased, however, by the gangue and the additional fluxes required to form a slag of the required composition. The heat requirements for slag formation are increased correspondingly.

DRI processes only reached the stage of commercial operation to produce a feedstock for steelmaking about 30 years ago. The forecasts of growth made in the early years of exploitation have not yet been realized, however, and the quantity of iron produced via this route remains only a very small fraction of the total iron production in the blast furnace. There are several factors that have constrained more rapid development.

The energy consumed in the gas reforming and the reduction process is 10.5 to 11.0 GJ t^{-1} DRI (11.7 to 12.5 GJ t^{-1} Fe). When the DRI is cooled to ambient temperature before transfer to the steelmaking furnace, approximately 1.8 GJ t^{-1} Fe is consumed in heating the DRI and the additional fluxes required to react with the gangue to about 1600 °C for conversion into steel. Thus, the total energy consumption is only slightly less than that for hot metal from the blast furnace. Against this, the cost of the hydrocarbon fuel is greater than the cost of the coal mixtures used for coke manufacture, and renewed attempts are being made to improve the efficiency of direct reduction when coal is used as the reductant. The cost of transportation of the DRI from a remote production site to a steelmaking plant is also increased by the precautions that must be taken to minimize the risk of surface oxidation and prevent spontaneous combustion. So, DRI is used mainly today as a means of diluting the concentrations of impurities, such as Cu and Sn, which are often present in steel scrap and cannot be easily removed during refining of the scrap.

4.4.2.4. Smelting Reduction

A new concept, which will probably replace the blast furnace for the reduction of iron ores in the next century, has provoked renewed interest in the principles of direct reduction. A furnace in which the oxides can be melted, reduced and separated from the gangue is coupled to a gaseous reducer, similar in principle to either the shaft or fluidized bed type of DRI unit. Gaseous, liquid or solid sources of carbon are injected into a molten bath to reduce the oxides. The carbon gases emerging from the bath are partially combusted within the furnace to generate the heat required for smelting and reduction. The hot gases leaving the furnace pass to the gaseous reducer, where thermal energy is transferred from the gas to the descending ore and some of the residual reducing potential of the gas is used to convert the higher oxides at least to FeO, thus conserving energy and reducing the load on the melting furnace. The balance between the amount of gaseous reduction in the upper chamber and reduction by carbon in the molten bath can be adjusted for thermal efficiency by varying the degree of combustion of the gas in the melting furnace. Alternatively, hydrocarbons can be injected into the hot gas before it enters the upper chamber, using the freshly reduced Fe surfaces to catalyze the endothermic reforming reaction, cooling the gas and yielding additional H_2 and CO to increase the reducing potential of the gas.

A variety of names are used to describe this approach, but it is most commonly known as "smelting reduction." Only the COREX process[71] has thus far been fully tested in pilot-scale production. This uses a vessel shaped like a blast furnace in which oxygen and air are injected into a fluidized bed of partially reduced ore and coal. The exhaust gas from the furnace is used for pre-reduction of the ore in a shaft kiln. Several different variants of the process are under development at the present time. The Direct Iron Ore Smelting (DIOS) process undergoing development in Japan uses a vessel similar in shape to the basic oxygen steelmaking (BOS) converter (Figure 1.4c) coupled to a fluidized bed pre-reducer. A horizontal smelting-reducer is coupled to a fluidized bed pre-reducer in the High Intensity Smelting (HISMELT) process that is being developed in Australia. All of these processes can use iron ore fines, obviating the need for sintering.

Most of these processes aim to produce hot metal with a similar carbon content to the blast furnace product, but there is no reason in principle to prevent the direct production of metal with a lower carbon content in the range required for steel. This is the aim of the process being developed by the American Iron and Steel Institute (AISI process). Here, ore is fed into one end of a cylindrical vessel. Coal is injected at the opposite end and oxygen jets combust the reaction gases above the bath. Molten low carbon steel is discharged continuously from the ore feed end of the vessel, while a slag with a low FeO content is drained from the opposite end. Fluxes charged with the ore combine with the sulfur in the coal to give continuous desulfurization.

An estimate of the energy requirements indicated a possible saving of at least 1 GJ t^{-1} Fe, compared to blast furnace-BOS practice.[72] Coke is not required, removing the need for coke ovens and ancillary plant. A smelting-reduction unit with an annual production of only 0.5 M tonnes per annum, which is within the range of metal requirements for a mini-steelworks (e.g., a small-scale operation, often serving a single local market), could produce iron at a lower cost per tonne than a large conventional blast furnace.[72]

4.5 REDUCTION OF SULFIDE ORES

Although sulfide ores are normally roasted prior to pyrometallurgical reduction, some sulfides can be reduced directly to the metal. Mercury is found in the earth as the mineral cinnabar, HgS, that can be converted to the liquid or gaseous metal, depending on the reduction temperature, by oxidation of the sulfur:

$$HgS + O_2 = Hg + SO_2 \qquad (4.53)$$

since SO_2 is more stable (i.e., a more negative free energy of formation) than HgO. Copper and silver sulfides can be reduced similarly.

The reduction of Cu sulfides is an interesting example of the way in which thermodynamics can be used to manipulate the extraction operations. Copper ore is frequently found in association with Fe in the form of the iron sulfide, pyrrhotite (Fe_7S_8), and in minerals such as chalcopyrite, $CuFeS_2$. Most of the pyrrhotite can be separated in the flotation process, but the Fe is chemically combined in the chalcopyrite and the flotation concentrate may contain roughly equal amounts of Cu and Fe. It is possible to reduce the Cu to the metallic form in the molten state, while the Fe is converted into an oxide that is absorbed into a slag.

The nickel mineral pentlandite, $(NiFe)_9S_8$, is often found associated with Cu sulfides. Flotation of these ores produces a Cu-rich concentrate and a second, Ni-rich concentrate that contains sufficient Cu to justify pyrometallurgical treatment to recover both metals. However, it is not feasible to reduce the Ni mineral directly to the metal. The reason why can be seen from consideration of the standard free energies for the two reactions:

$$MS + O_2 = M + SO_2 \tag{4.54}$$

$$\tfrac{2}{3}MS + O_2 = \tfrac{2}{3}MO + \tfrac{2}{3}SO_2 \tag{4.55}$$

The relevant data are plotted in Figure 4.11, which shows a markedly more negative free energy for Reaction 4.55 than for 4.54 when M represents Fe or Ni. The converse applies when M signifies Cu and the lines for both Cu reactions lie at less negative free energies than those for the formation of FeO and NiO. An alternative approach is by comparison of the relevant predominance area diagrams.

4.5.1. Treatment of Copper Sulphides

4.5.1.1. The Traditional Reverberatory—Converter Route

The traditional route for the smelting of copper sulfides involves partial roasting of low-grade (i.e., high Fe content) ores, to remove some of the sulfur, prior to melting in a reverberatory (or electric arc) furnace in which air is blown onto the surface of the charge. Part of the Fe is oxidized into the slag to produce, typically, a liquid sulfide matte containing 40% Cu, 30% Fe, 27% S and a slag comprising 45% ($FeO + Fe_3O_4$), 43% SiO_2, 10% CaO and 0.6% Cu. Any Cu_2O present in the ore or formed during melting is altered to the sulfide form by the reaction:

$$Cu_2O + FeS = FeO + Cu_2S \tag{4.56}$$

for which the standard free energy is negative, as shown in Figure 4.11.

The matte is then converted to metallic Cu at 1200 to 1300 °C either in a cylindrical, horizontal Pierce Smith converter or in an inclined, Bessemer-type converter. Air is blown into or onto the matte during converting to

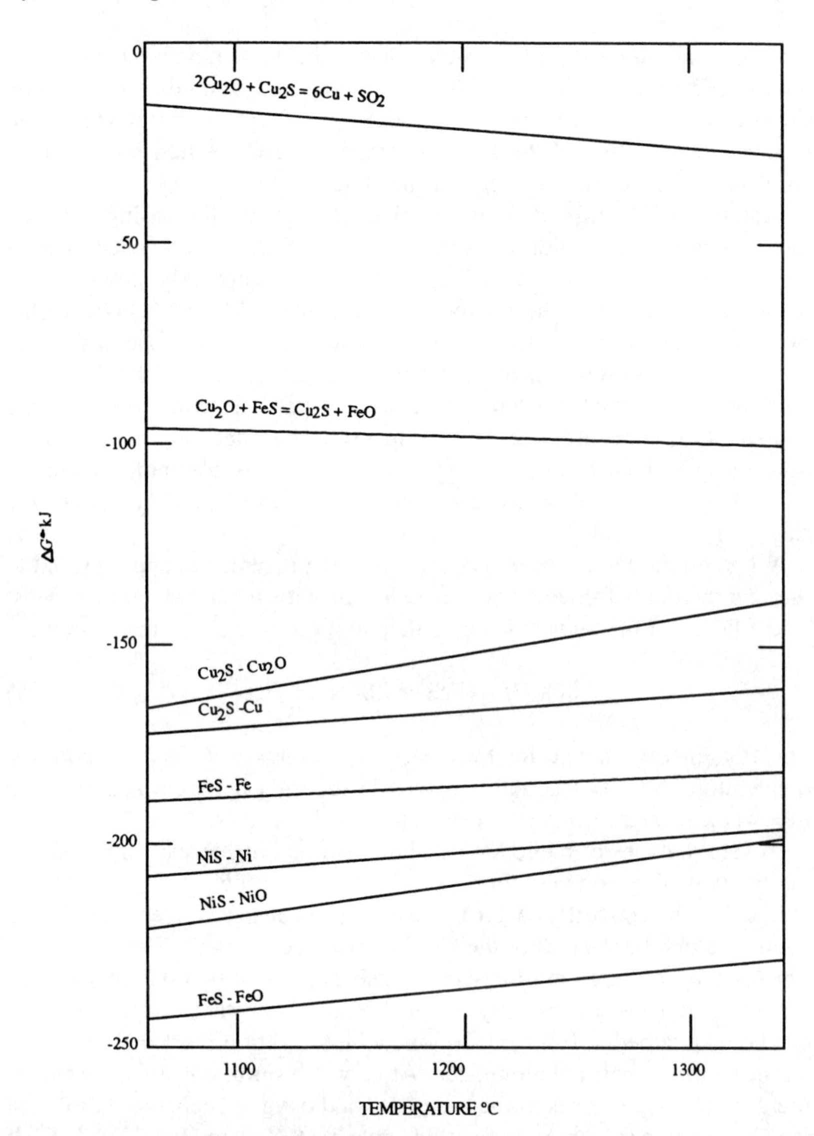

FIGURE 4.11. Comparison of the standard free energies of formation of some metal oxides and sulfides and for reactions between the compounds.

give a partial pressure of oxygen in the bath of about 10^{-7} atm, which is sufficient to oxidize the remaining Fe. The FeO produced is highly soluble in the matte, but its activity is reduced markedly by adding silica to form an immiscible ferrous silicate slag. Cu_2S is slightly soluble in the slag, so small amounts of lime and alumina are added to lower the solubility and to increase the fluidity of the slag. The density of the slag (about

3.5 Mg m^{-3}) is only slightly less than that of the matte (about 4.0 Mg m^{-3}), but the difference is sufficient to cause the slag to float on top of the matte. Toward the end of the blow, however, as the FeS concentration is reduced to a low level, some of the Fe is oxidized to Fe_3O_4, which remains as a solid and increases the viscosity of the slag.

About 0.6% Cu dissolves in the slag, the amount increasing with the melt temperature. In addition, small droplets of Cu_2S are carried upward into the slag by the ascending SO_2 bubbles and return only slowly to the matte layer because of the relatively small density difference between the two layers. The rate of return is slowed further by the increase in the slag viscosity when Fe_3O_4 is formed and the total Cu loss to the slag increases with increasing Fe_3O_4 content. The loss also increases with increasing Cu content of the concentrate. This is compensated by a decrease in the volume of the slag formed from a rich concentrate, but it is common practice to mix a high-grade with a low-grade concentrate to restrict the loss to the slag.

When all the Fe has been oxidized, the slag is removed and recycled to the reverberatory furnace where it is mixed with fresh concentrate. Most of the Fe_3O_4 in the recycled slag is then reduced to FeO by the reaction:

$$3Fe_3O_4 + FeS = 10FeO + SO_2 \tag{4.57}$$

The free energy change for this reaction is almost zero at the operating temperature, but the reaction is driven to the right by again adding silica to lower a_{FeO} and form a silica-rich slag.

Several batches of matte are usually converted in succession to remove the Fe, with slag removal after each stage, until sufficient Cu_2S has accumulated in the converter. After the last slag has been removed, the blow is continued in order to reduce the Cu_2S to metal according to Equation 4.54. The Cu-S phase diagram shows an extensive region of liquid immiscibility, Figure 4.12. Thus, the activity of Cu_2S remains constant and the air blow can be maintained at full force to give a high reaction rate until the sulfur has been almost entirely eliminated. At very low sulfur contents, the partial pressure of oxygen rises in the gas phase and oxygen begins to dissolve in the Cu, the amount increasing with decreasing %S up to about 2 wt% O. If blowing is continued thereafter, Cu_2O begins to appear on the surface of the melt. If the blow is interrupted before all the Cu_2S has been eliminated, however, the Cu_2O can be reduced by the residual sulfide (Figure 4.11):

$$2Cu_2O + Cu_2S = 6Cu + SO_2 \tag{4.58}$$

A calculation of the equilibrium conditions in this process is shown in Worked Example 13.

Any precious metals present in the ore are insoluble in the slag and are recovered in the metal. Since the concentrates are not sintered prior to

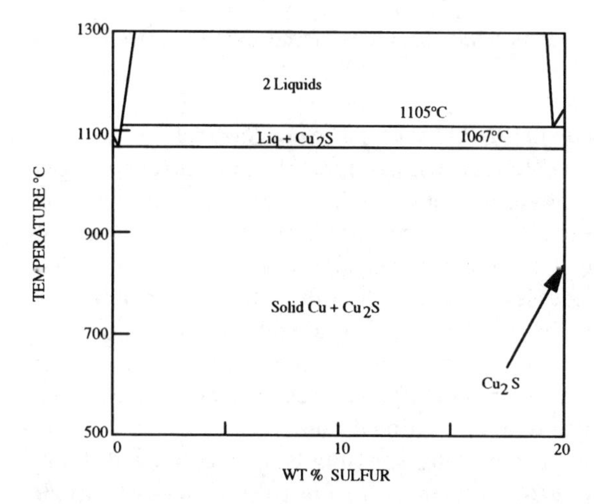

FIGURE 4.12. The copper-sulphur phase diagram.

smelting, the volatile impurities are not removed before the ore is charged into the reverberatory furnace. Elements such as As, Pb, Sb and Zn are then partially volatilized during melting and converting, and the dust recovered from the exit gas is sometimes sufficiently rich to justify recovery of these elements. However, the Cu produced in the converter still contains residual amounts of these impurities, together with some Fe, S and oxygen, and is usually in the range 98.5 to 99.3% Cu. This is traded under the name of "blister copper."

The heat released by the exothermic oxidation reactions is not sufficient to satisfy the heat requirements in the reverberatory furnace when a wet concentrate is blown with cold air. Sufficient heat is released if the charge is dried, the air preheated to about 500 °C and the oxygen content of the air increased to about 30%. Heat is generated more rapidly by the high rate of reaction in the converting stage and it is often necessary, particularly with high-grade mattes, to add cold concentrates or scrap copper metal to the converter to prevent an excessive temperature rise in the early part of the blow. As the sulfur content falls, toward the end of the blow, the heat supply changes from surplus to deficit and the bath temperature begins to fall.

The air is blown into the top half of the matte in the Pierce Smith converter. This serves two purposes. The metallic Cu has a higher density than the matte, so it accumulates in the bottom of the converter where it is protected from oxidation by the matte layer. The air is immediately heated by the exothermic chemical reaction as it enters the matte. But if the air is blown through the bottom of the converter, as it is in the bottom-blown steelmaking processes, it extracts heat from the metallic Cu with the resultant risk that the metal may solidify. When an alternative type of top-blown Bessemer type of converter is used, the air is blown downward through a lance with sufficient force to penetrate the slag and impact on the matte surface. The metal is again protected from oxidation by the intervening matte layer.

4.5.1.2. Modern Methods for Reduction of Copper Sulfides

The traditional route is a batch process involving the three discrete stages of roasting, smelting and converting. In general, capital and operating costs rise as the number of process steps is increased, and there is a financial incentive to combine the steps into more simple processes. There is also increasing interest in processes in which the ore is fed in a continuous stream into the reactor and the metal and the slag are removed continuously, since this type of process is more amenable to automation and computer control. A number of processes that satisfy these requirements, in whole or in part, are gradually replacing the traditional route.

The first two stages of roasting and smelting can be combined if the temperature attained is sufficiently high to melt the charge. In **flash smelting,** the sulfide concentrate is injected together with silica and fluxes either downward in a flow of preheated and oxygen-enriched air (Outokumpu) or horizontally in a blast of pure oxygen (INCO). Oxygen enrichment of the gas reduces or eliminates the heat consumed in raising the temperature of the inert nitrogen and increases the heat available for melting the charge. The finely divided concentrate, projected as individual particles surrounded by the oxidizing gas, ignites at 400 to 500 °C and is fully molten by the time the particles are collected in a molten pool at the base of the furnace. The rate of reaction is very high. It is mass transport controlled, so the rate increases as the particle size is decreased. However, the dust losses in the exit gas also increase as the particle size is diminished. A balance must be struck between the maximum particle diameter that can be reacted completely before entrainment in the bath and the minimum diameter below which dust losses become unacceptable.

The heat released by oxidation of the sulfur is partially lost as sensible heat in the off-gas and only low-grade concentrates (i.e., high FeS content) can supply sufficient heat to melt the charge. Some hydrocarbon fuel is occasionally needed to supplement the heat supply in the Outokumpu smelter, but the INCO smelter with a pure oxygen blast is autogeneous. The SO_2 content of the gas ranges from about 20 vol% in the Outokumpu to as high as 75 vol% in the INCO process, thereby reducing the total volume of gas that requires cleaning and simplifying the gas treatment when compared to the traditional route. Both processes produce a matte containing up to 50% Cu and a slag containing about 0.6% Cu.

Insufficient heat is generated in flash smelting to permit complete conversion of the concentrate to metallic copper in one operation. The Kennecot modification of the Outokumpu process[73] achieves conversion in two flash smelting units. The concentrates are smelted to an intermediate grade in the first unit. The product is ground to a small particle size before flash smelting to metallic Cu in the second unit.

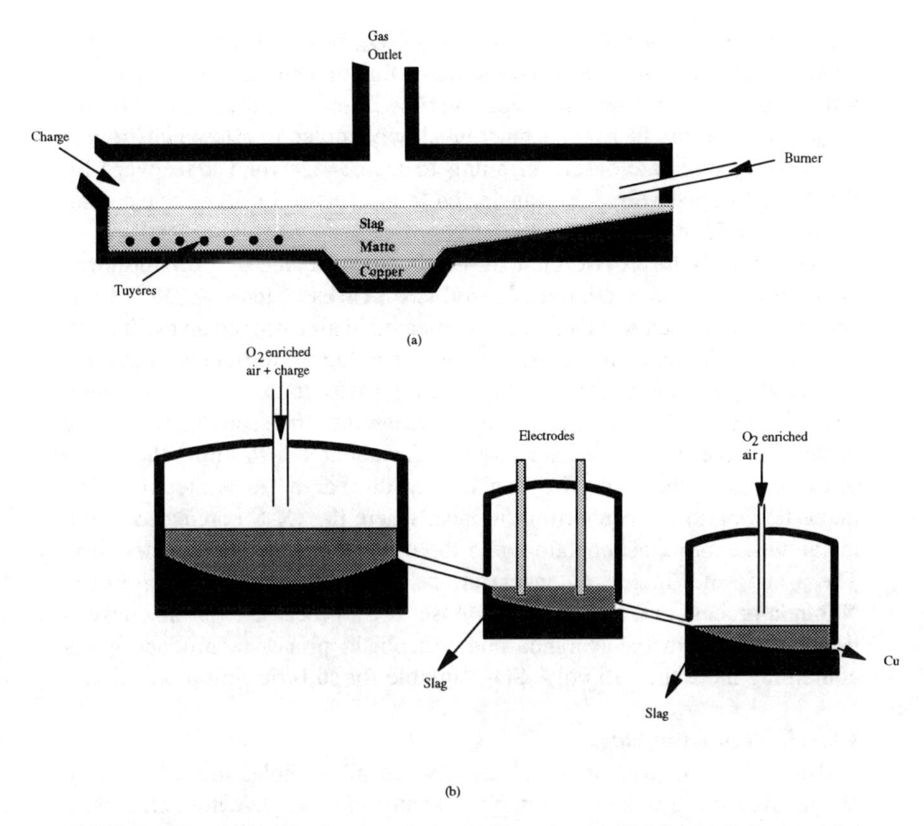

FIGURE 4.13. Schematic drawing of the (a) Noranda and (b) Mitsubishi furnaces for copper production.

The **Noranda process**[74] was designed as a nonflash smelting, continuous process to achieve all three stages in one operation. Concentrates, silica and flux are spread continuously from one end of the cylindrical furnace onto the slag that floats on a layer of almost pure Cu_2S, Figure 4.13a. A row of tuyeres submerged in the molten matte below the feed zone supplies the oxygen-enriched air required for the reactions, which raise the temperature in the furnace to 1200–1250 °C. Molten Cu collects below the matte in a well in the center of the furnace, from which it is tapped periodically. The furnace hearth slopes upward from the well to the opposite end of the furnace and rises to a height slightly above the top of the matte layer. This allows the slag to be discharged continuously without loss of the matte. The raised hearth also provides a quiescent zone in which Cu droplets and Cu_2S entrapped in the slag can settle out and drain back into the matte. However, the separation is not completed by the time the slag reaches the overflow and the discharged slag may contain more than 20% Cu. The Cu loss in the slag tends to increase with increasing grade of the ore. Consequently, the Noranda reactor is now used only to produce a rich Cu matte.

The slag may contain as much as 25% Fe_3O_4, however, and is not sufficiently fluid unless some lime is added as a flux or a burner is used to direct a flame onto the slag near the slag overflow. Since the slag cannot be recycled through the furnace, it must be slowly cooled to allow the Cu particles to agglomerate before grinding to a fine size for Cu recovery by flotation. Alternatively, it is transferred to an electric furnace for cleaning while it is still molten.

The inability to recycle the slag has been overcome by utilizing three furnaces in the semicontinuous **Mitsubishi process,** Figure 4.13b.[75] Concentrates, fluxes and recycled slag are injected in air enriched up to 50 vol% oxygen into the melt in the first furnace. The highly turbulent conditions produced by injection results in rapid heat transfer to the solids and a high rate of reaction. The matte and slag overflow into the second, electrically heated furnace where they are retained for about 1 h to allow the Cu to settle out from the slag. The slag is then discharged to waste while the matte flows into the converting furnace, where the Cu_2S is reduced to the metal with an air blast containing 25 to 30 vol% oxygen and injected flux. The S and other impurity concentrations in the Cu are lower than in the Noranda process and are similar to those in Cu processed via the conventional route. Both the Noranda and Mitsubishi processes produce gases containing more than 15 vol% SO_2, suitable for sulfuric acid production.

4.5.1.3. Foaming Slags

The volume of a melt is increased when gas bubbles are released by chemical reaction within a melt. The expansion is larger with increases in both the rate of formation of the gas bubbles and the average dwell time of the bubbles in the liquid before they escape to the atmosphere. The rate of escape is, in turn, decreased as the viscosity of the melt is raised (e.g., by lower temperature) and as the surface tension of the melt is lowered. A low surface tension results in a low rate of drainage of the liquid film surrounding the bubbles, retarding coalescence and, hence, reducing the rate of escape to the atmosphere. When the retention time is high, the melt is converted into a foam. In extreme cases, the volume of the foam exceeds the capacity of the containing vessel and the melt overflows.

Molten mattes generally have high surface tension and low viscosity, so they do not form a foam; but silica is surface-active and markedly lowers the surface tension of the ferrous silicate slags that are formed during Cu extraction. The complex, polymeric silica molecules are readily adsorbed at the bubble surfaces, resulting in a low rate of film drainage and a high foam stability.

The problem is exacerbated when a blast of air or oxygen is blown from above the melt with sufficient velocity to penetrate the slag and strike the liquid phase underneath. The matte is then emulsified at the point of impact and a slag-metal emulsion is formed in which the bubbles are trapped. The intimate contact between the slag and the metallic phase can result in an accelerated rate of chemical reaction (as in BOS steel refining), but can also cause violent gas release and the ejection of metal and foam from the reaction vessel.

Lime raises the surface tension of the slag and the foam stability is decreased as silica is substituted by lime. With a CaO/SiO_2 ratio of 1.25, the foam stability decreases rapidly as the FeO content of the slag is raised above 2%.[76] There is little or no risk of foam formation with CaO-FeO slags and silica is now being replaced by a lime flux for Cu smelting and converting.[77] These slags are capable of holding a higher concentration of the higher oxides of Fe in solution and are therefore also beneficial as a means of reducing the risk of build-up of solid iron accretions in the reaction vessel.

4.5.2. Treatment of Nickel Sulfides

A different procedure is required for the reduction of sulfide concentrates containing both Ni and Cu. It is apparent from Figure 4.11 that NiS can only be reduced directly to the metal at temperatures above 1500 °C, requiring a large input of thermal energy to supplement the heat released by oxidation of the sulfides, when all the components are present in their standard states. Also, Ni melts at 1453 °C, so a temperature of at least 1500 °C is required to produce a fluid melt when the product is pure Ni. Cu and Ni are completely miscible in metallic form, however, in both liquid and solid states. Thus the melting temperature of the alloy is reduced progressively with increasing Cu content when both elements are reduced simultaneously. An additional stage is then required to separate the Ni from the Cu.

The procedure actually adopted is similar to that used for Cu production up to the stage where the FeS has been oxidized and absorbed into the slag. The blow is continued until there is a sulfur deficiency with respect to the Cu and Ni sulfides and a small amount of a molten Ni-Cu alloy is formed, into which any precious metals in the ore are partitioned preferentially. Blowing is continued until the matte contains roughly 40% Ni, 30% Cu and 2 to 3% Fe. If equilibrium partitioning of the elements between the slag and the metal is attained, the Ni content of the slag increases rapidly as the Fe content of the matte falls below about 6%, Figure 4.14. Some matte may also be entrained as suspended particles in the slag. So, the slag is separated from the matte at this stage and is either recycled to the smelter or heated with sulfur and carbon in an electric arc furnace to recover the metal values.

Nickel is almost insoluble in solid Cu_2S. Nickel solidifies in the sulfur-deficient matte as the compound Ni_3S_2, in which the solubility of copper is about 6% at the melting point but falls to less than 0.5% as the temperature falls. Slow cooling of the matte allows time for the formation of relatively coarse crystals of the sulfide compounds and for the solubilities of Cu and Ni in the compounds to be adjusted by liquid and solid-state diffusion. The

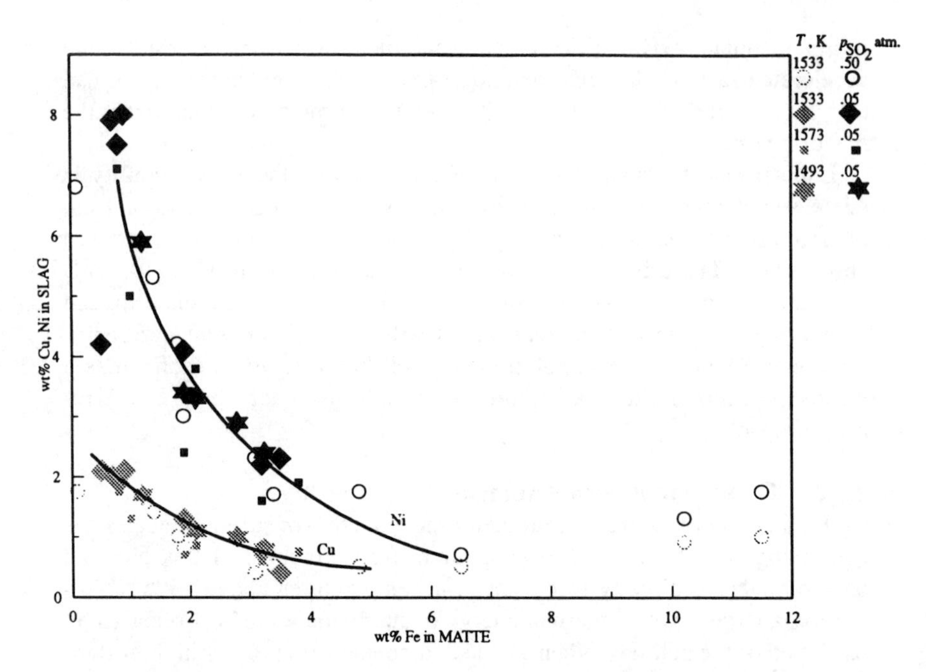

FIGURE 4.14. Equilibrium copper and nickel contents of smelting slags. (From Taylor, A. R. and Dinsdale, A. T., *User Aspects of Phase Diagrams,* The Institute of Materials, Petten, The Netherlands, 1991. With permission.)

solidified matte is crushed and ground to liberate the individual phases. The Ni-Cu alloy containing the precious metals is separated magnetically and the sulfides are parted by froth flotation. The Cu_2S recovered from the flotation circuit is fed into the reduction circuit for Ni-free Cu concentrates, while the Ni_3S_2 fraction is prepared for electrolytic extraction. Alternatively, Ni is extracted from the crushed matte with a chlorine leach, leaving a Cu_2S residue.

4.5.3. Treatment of Lead Sulfides

PbS can also be converted directly to the metal at temperatures above about 1200 °C, but Pb forms a series of basic sulfates with a wide range of stability at high partial pressures of O_2 and SO_2. If the gas atmosphere is in the range in which these compounds can form, the sulfates are absorbed into the slag that may contain up to 50% Pb. The amount lost in the slag is decreased if moderately reducing conditions are maintained during smelting.

In the **ISASMELT** process, the sulfide concentrate is injected with coke breeze in air enriched to 27 vol% oxygen through a lance and into a molten bath held at 1200 °C.[78] The lance orifice is submerged below the surface of the melt to reduce the amount of Pb escaping as fume from the reactor. A slag with a high Pb content is formed. This flows to a second reactor into which lump coal is fed in sufficient quantities to reduce the Pb to metallic form, while the Fe is retained as an oxide and fixed in the silicate slag. The lead content of the slag decreases with time in the second reactor, falling to 2% after a hold of 4 h.

4.6. REDUCTION OF THE MORE REACTIVE METALS

Some metals are not readily produced using the pyrometallurgical techniques considered thus far. There are various reasons for this limitation. For example, the oxides of aluminum, beryllium and uranium are very stable and extremely high temperatures are required for reduction by carbon. Alumina can be reduced by carbon above 2000 °C, but the metal carbide is formed in preference to the metal in the presence of excess carbon. Both Al and an aluminum suboxide, Al_2O, would be lost as volatile species at these temperatures. Some other metals, such as Nb, Ta and Ti, readily form carbides in the presence of carbon, and these metals are severely embrittled by small amounts of oxygen remaining in solution after reduction. Problems also arise when the ore contains minerals of two or more metals with similar stability, making separation by selective reduction more difficult. A pyrometallurgical route may still be feasible for these ores, using metallothermic or halide metallurgy techniques.

4.6.1. Metallothermic Reduction of Magnesium

The production of magnesium is an interesting example of a possible metallothermic route. This metal is primarily produced by electrolysis. In the past, attempts have been made to reduce MgO with carbon. This may seem surprising, since MgO is only a little less stable than Al_2O_3 and the Mg-MgO line intersects the C-CO line on the oxygen potential diagram (Figure 2.8) at about 1800 °C. But Mg boils at 1105 °C and, at higher temperatures, MgO is reduced by carbon to Mg vapor. Since the equilibrium constant for the formation of MgO at unit activity is given by:

$$\Delta G^{\ominus} = -RT \ln \left[\frac{1}{p_{Mg}^2 \cdot p_{O_2}} \right] \quad (4.59)$$

it is evident that the oxide will dissociate at a lower temperature if p_{Mg} is reduced. That is, a reduction in the total gas pressure moves the Mg-MgO line upward to less negative ΔG values on the oxygen potential diagram, Figure 4.15. Conversely, the equilibrium constant for the oxidation of carbon at unit activity to form CO:

$$\Delta G^{\ominus} = -RT \ln \left[\frac{p_{CO}^2}{p_{O_2}} \right] \quad (4.60)$$

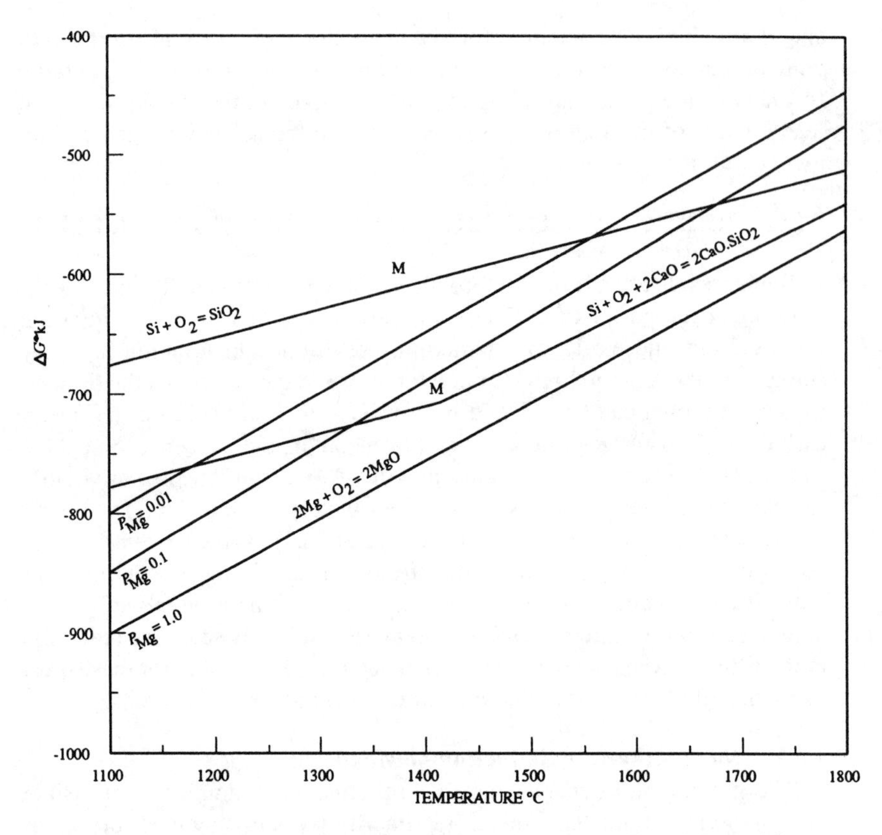

FIGURE 4.15. Free energy-temperature diagram illustrating the effect of the partial pressure of magnesium vapor and the formation of fayalite ($2CaO \cdot SiO_2$) on the minimum temperature for the reduction of MgO by silicon.

shows that the C-CO line moves downward on the diagram as the total pressure is reduced. The sum effect of a pressure reduction is a decrease in the temperature at which MgO can be reduced by carbon and the reaction can proceed if the pressure in the reactor does not exceed 10^{-2} atm at 1600 °C. But, as with the recovery of Zn from the blast furnace gas, the vapor must be shock-cooled to prevent reoxidation of the Mg as it is condensed. Since p_{CO} in equilibrium with the metal vapor decreases rapidly as the temperature is lowered, it is necessary to dilute the CO in the gas by quenching in a blast of hydrogen or inert gas. Mg is then recovered as a very finely divided powder which, in view of the stability of the oxide, is extremely pyrophoric. Very careful handling and storage in evacuated containers or under an inert gas is required to prevent oxidation prior to consolidation, and there is a serious explosion hazard.

This problem is avoided by eliminating carbon and reducing the oxide metallothermically, using Si or ferrosilicon as the reducing agent (**Pidgeon process**). The free energy of the reaction:

$$2MgO + Si = 2Mg + SiO_2 \tag{4.61}$$

is positive at all temperatures below 2000 °C, as is evident from Figure 4.15. However, the reaction can be made to proceed at much lower temperatures by operating at a reduced pressure in the temperature range where Mg vapor is formed and by adding CaO to combine with the silica and produce the compound called ''fayalite,'' $2CaO \cdot SiO_2$:

$$2MgO + 2CaO + Si = 2Mg + 2CaO \cdot SiO_2 \tag{4.62}$$

The equilibrium constant for this reaction is given by:

$$K = \frac{p_{Mg}^2 \cdot a_{2CaO \cdot SiO_2}}{a_{Si} \cdot a_{MgO}^2} \tag{4.63}$$

When pure Si is used as the reductant, $a_{Si} = 1.0$ and p_{Mg} is lowered by the reduced pressure. Alumina may be added to lower the melting temperature of the charge and form a molten mass, which is saturated with MgO to ensure that a_{MgO} is close to unity. Thus, reduction is possible even though, as with carbothermic Mg reduction, the standard free energy change for the reaction is positive. In fact, the reaction temperature required is only a little above 1200 °C if p_{Mg} is not greater than 0.01 atm, as is evident from Figure 4.15.

Reoxidation is avoided by removing the Mg vapor from contact with the melt and into an oxygen-free, evacuated chamber where the vapor is cooled slowly, allowing time for the nucleation and growth of a coarse, granular and less pyrophoric deposit to form before discharge to the atmosphere.

Since $\Delta G^{\ominus}$ for the reaction becomes less negative as the temperature is increased, it follows that the maximum pressure in the reactor at which reduction can occur is raised as the temperature is increased. The very high temperatures in the core of the ionized gas and the highly localized heating zone that can be achieved with plasma heating could be applied for this purpose.[79] However, the liquid phase field in the ternary system CaO-MgO-SiO_2 expands as the temperature is raised. An increase in temperature is counterproductive, therefore, if it results in a more rapid decrease in a_{MgO} than in a_{SiO_2} as the liquid solubility is increased.[80]

4.6.2 The Halide Extraction Route

The conversion of an oxide into a soluble or volatile halide compound is sometimes the starting point for the production of a metal. Bromide, fluoride and iodide compounds may be produced, but chlorination is usually the cheapest route and the one most widely adopted. In general, the melting and volatilization temperatures for the metal halides are lower than for the metal sulfides which, in turn, are lower than for the metal oxides. Hence, the metal can be separated as a volatile chloride at markedly lower temperatures than are normally required to achieve separation when it is necessary to melt the gangue materials.

4.6.2.1. Chlorination of Ores

The standard free energies of formation of the common metal chlorides from 1 mol Cl_2 gas are grouped closely in the range from -150 to -500 kJ mol^{-1} at 25 °C; only the alkalies and alkaline earths form more stable chlorides. The direct chlorination of an oxide:

$$2MO + 2Cl_2 = 2MCl_2 + O_2 \tag{4.64}$$

is only feasible if the free energy of the chloride is more negative than the free energy of formation of the oxide. $\Delta G^{\ominus}$ is a small negative quantity for most of the common metals, but the driving force is low and relatively high temperatures are required to obtain a reasonable reaction rate. A more negative free energy change, allowing chlorination to occur at lower temperatures, can be achieved if solid carbon or, at temperatures below about 800 °C, CO is present to combine with the oxygen in the ore:

$$MO + Cl_2 + CO = MCl_2 + CO_2 \tag{4.65}$$

Even lower temperatures can be used when the chlorine is supplied as CCl_4 or $COCl_2$ gas. The standard free energy change for Equation 4.64 is positive for some metals (including Al, Cr, Mn, Si and Ti) and the oxides of these elements can be chlorinated only in the presence of a reducing agent.

Soluble chlorides can be formed by adding NaCl to the mixture in a roasting furnace. Copper can be recovered in metallic form from, say, chalcopyrite concentrate by first roasting to remove most of the sulfur, then mixing the concentrate with finely divided coal or coke, silica and NaCl. When the mixture is heated in an inert atmosphere at 700 to 800 °C, the salt is hydrolyzed to form sodium silicate and HCl:

$$2NaCl + SiO_2 + H_2O = Na_2SiO_3 + 2HCl \tag{4.66}$$

and the Cu is volatilized by the HCl:

$$Cu_2O + 2HCl = 2CuCl + H_2O \tag{4.67}$$

The chloride dissociates on the coke, forming a surface deposit of Cu, and the HCl is regenerated to continue the extraction process:

$$2CuCl + C + H_2O = 2Cu + 2HCl + CO \tag{4.68}$$

Volatile chlorides of other elements, such as Be, Li, Nb, Ta, Ti, V and Zr, are usually formed by heating ore concentrates and coke fines in a Cl_2 or HCl gas in a fluidized bed, or by heating briquettes of ore and coke in Cl_2 in a shaft furnace. The volatile species are transferred to a condensing chamber. The chlorides of the alkalies and alkaline earths, which have a higher melting and volatilization temperature, remain in solid or liquid form. The liquids drain down to the bottom of the furnace from where they are periodically discharged. Excessive amounts of this liquid can cause particle agglomeration and interfere with gas access to the concentrate surfaces.[81] Dust carry-over in the effluent gas, which is prevalent with these treatments, is avoided when the ore is chlorinated in a bath of molten KCl-NaCl. $FeCl_3$ can be used as a catalyst. The chloride is reduced to $FeCl_2$ in contact with the concentrate and is reoxidized to $FeCl_3$ by the chlorine gas.[82] The viscosity of the melt increases with time, however, due to the accumulation of the chlorides of the gangue minerals and part must be drained off periodically and replaced with fresh salt.

4.6.2.2. Recovery of the Metal

The majority of the elements can form chloride compounds, and any species present in the ore that can form a volatile chloride under the imposed conditions may contaminate the vapor of the element being recovered. For example, the Cl_2-O_2 predominance area diagrams at 1200 K have been calculated for Al, Fe, Mg and Si[83] and these are superimposed in Figure 4.16. If all of these oxides are present at unit activity during chlorination, it is evident that iron oxides could be volatilized preferentially under mildly reducing conditions, leaving most of the other oxides in the residue. However, if the intention is to recover Mg as the chloride, the vapor is contaminated with the other three elements. The boundaries between the various phase fields move downward and to the left as the activities of the oxides decrease below unit. The extent to which any given species is volatilized is dependent, therefore, on the activity of the element in the ore, the gas composition, the pressure and the temperature.

Since several elements may be chlorinated simultaneously in the ore, extensive treatment is often required to separate only the chloride of the metal required. A variety of techniques is used for purification.[84] Differences in melting point can be used to separate some impurities, as in the rejection of the alkalies and alkaline earths during chlorination. The boiling points of the chlorides vary over a wide range, and fractional distillation is often used as a means of separation. Thus, $TiCl_4$ condenses at a lower

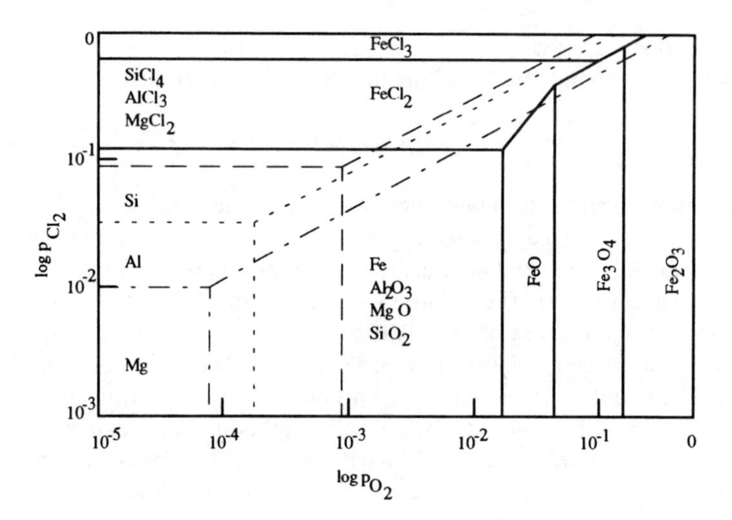

FIGURE 4.16. Chlorine-oxygen predominance area diagram for aluminum, iron, magnesium and silicon at 1200 °C.

temperature than the chlorides of Fe and Ta, so Ti can be retained as a vapor phase while the impurities are condensed by cooling the gas to below 200 °C. Repeated distillation is often required to obtain sufficient purity. Purification can also be achieved by selective reduction. For example, $ZrCl_4$ is often contaminated with hafnium chloride, $HfCl_4$. Hf contamination of metallic Zr is not acceptable when the Zr is used for containing the fuel for a nuclear reactor, since Hf has a very large neutron capture cross-section that would quench the reactor process. The Zr is separated from Hf by partial reduction of $ZrCl_4$ with $ZrCl_2$:

$$ZrCl_4 + ZrCl_2 = 2ZrCl_3 \tag{4.69}$$

At about 400 °C, $HfCl_4$ is not reduced, but remains as a gas and is separated from the liquid $ZrCl_3$; then, at temperatures above 450 °C, Equation 4.69 is reversed to release pure $ZrCl_4$ gas. Alternatively, Zr can be precipitated as an oxide while Hf is retained in the vapor phase by exposure of the gas to mildly oxidizing conditions at about 800 °C.

If the metal halide is readily soluble, it may be extracted into a solution and reduced to the metal by electrolysis, or it may be electrochemically reduced from a molten salt. A variety of pyrometallurgical techniques are also used for the reduction of the volatile salts. Carbon is not suitable as a reducing agent, since CCl_4 is one of the least stable chlorides; its free energy of formation is positive above 415 °C. However, carbon can be used in conjunction with some other element that forms a more stable chloride.

Thus, CuCl can be reduced to the metal in the presence of lime and carbon at about 1100 °C:

$$2CuCl + CaO + C = 2Cu + CaCl_2 + CO \tag{4.70}$$

and the chlorine is fixed as an insoluble compound. If the gas can be recirculated, however, it is preferable to reduce the Cu with hydrogen:

$$2CuCl + H_2 = 2Cu + 2HCl \tag{4.71}$$

The salt may be reduced metallothermically, as in the production of Ti. After cooling to condense out the impurities, the vapor is reheated and bubbled through a bath of molten Mg in the **Kroll process:**

$$TiCl_4 + 2Mg = Ti + 2MgCl_2 \tag{4.72}$$

(or molten Na in the Hunter process). The reaction is exothermic. When all the Mg has reacted, the $MgCl_2$ entrapped in the Ti metal sponge is extracted by leaching, any excess Mg remaining is removed by distillation and the Ti is consolidated by vacuum melting.

When a metal exhibits more than one valency, the effect of the entropy of formation on the temperature dependence of the stability of the compounds may be exploited for metal recovery. For example, Al forms two volatile chlorides: AlCl and $AlCl_3$. In general, the heat of formation of the compound formed between a metal and some other species increases with the valency exhibited by the metal in the compound. Thus, the heat of formation is much larger for $AlCl_3$ than for AlCl, and the former compound is produced by chlorination of alumina at low temperatures. However, in the reaction:

$$\tfrac{2}{3}Al + Cl_2 = \tfrac{2}{3}AlCl_3 \tag{4.73}$$

1 mol of gas is consumed in the formation of $\frac{2}{3}$ mol of gaseous compound. The overall entropy change is negative and the slope of the $\Delta G^{\ominus}$ vs T plot is positive (i.e., the stability of the compound decreases with increasing temperature). In contrast, 1 mol of Cl_2 is consumed in the production of 2 mol of the compound in the reaction:

$$2Al + Cl_2 = 2AlCl \tag{4.74}$$

The overall entropy change is positive and the slope of the $\Delta G^{\ominus}$ vs T plot is negative. At some temperature, the two lines intersect and $\Delta G^{\ominus}$ is then zero for the reaction:

$$2Al + AlCl_3 = 3AlCl \tag{4.75}$$

The intersection temperature of 1310 °C, which is found by equating the $\Delta G^{\ominus}$ relations for the two reactions, applies when all the species are in their standard states (i.e., all gases at 1 atm pressure), but moves to lower temperatures if the total pressure in the system is decreased.

This is the basis of the **Gross process** for the production of aluminum. The alumina is chlorinated at a temperature above the intersection of the two lines to form AlCl. The vapor is cooled to a temperature below the intersection in a separate chamber, causing Equation 4.75 to move to the left and deposit Al metal, and the $AlCl_3$ is recycled over the alumina. Unfortunately, alumina can only be chlorinated in the presence of a reducing agent. If carbon is used for this purpose, the Al is reoxidized by the carbon gases during cooling. An economically efficient method of exploiting this seemingly attractive route for extracting Al from the ore has not yet been devised. It has good potential, however, for removing Fe and Si from aluminum scrap to upgrade the metal, since these elements are not readily removed by other means.

4.6.2.3. Limitations of the Halide Route

If all the impurities are separated from the gas before the halide is reduced, a very pure metal can be produced by this route, with little or no need for further refining. The process, however, is expensive. Halide gases are expensive, toxic and aggressive. The reaction chamber must be completely gastight to prevent ingress of air and the escape of halide gases into the atmosphere. Few substances are completely immune to the halides at elevated temperature and the maintenance of the reactor lining is a major problem. The reaction temperature can be lowered by the presence of reducing agents, or halide gases containing a reducing agent can be used, but these are costly to produce. Sometimes, as with Zr, the temperature is lowered by first converting the mineral to a metal carbide, but this also raises the total cost of production. Hence, the adoption of this route is often a last resort, used only when other methods of production do not provide a sufficient yield or purity of product. It is also used when the very low concentration of the metal in the source material (ore, slag, flue dust, etc.) and/ or the small scale of production make other extraction methods uneconomic.

FURTHER READING

von Bogdandy, L. and Engell, H. J., *The Reduction of Iron Ores*, Springer Verlag, Berlin, 1971.

Boldt, J. R. and Queneau, P., *The Winning of Nickel*, Methuen, London, 1967.

Peacy, J. G., *The Iron Blast Furnace*, Pergamon, Oxford, 1979.

Walker, R. D., *Modern Ironmaking Methods*, Institute of Materials, London, 1986.

See also References 5 to 12.

Chapter 5

PYROREFINING

5.1. OBJECTIVES AND PRINCIPLES

5.1.1. Overview

The crude metal produced by pyroextraction usually contains several other elements as precipitates and/or in solution. The objective in refining is to remove the undesirable impurities to the extent that is necessary to produce those mechanical, physical and chemical properties in the metal required for the intended use (i.e., fitness for purpose), while retaining those elements beneficial to the properties.

A wide range of impurities may be present in the crude metal. As indicated in the previous chapter, the metal will be contaminated with any other elements present in the ore or in the concentrate that form compounds of a similar type (oxide, sulfide, etc.) with similar or lower stability to the compound of the metal that is reduced. It is difficult to control closely the oxygen potential in the reduction process, so a safety margin is allowed and the charge is overreduced to ensure that all the metal is recovered. Compounds that are a little more stable than the one of primary interest may then, simultaneously, be partially or even completely reduced and the elements released may also dissolve in the metal. Most of the sulfur and oxygen in the ore are removed during smelting, although sulfur may be picked up from the impurities in the coal or coke during carbothermic reduction. Small amounts of these elements often remain in the metal, however, in sufficient concentration to adversely affect the properties. Other impurities, such as carbon, hydrogen and nitrogen, may be absorbed from the reductant or from the atmosphere in contact with the melt.

Metal scrap is usually incorporated into the production cycle at the refining stage. This can introduce a further range of contaminating elements. The concentration of those impurities originating from the extraction stage should be low in the scrap metal but may vary over a wide range, depending on the composition of each of the pieces of scrap. The average concentration of these elements in a scrap metal charge may be higher or lower than the level required in the new material to be produced. Pure metals may have been melted together to form metallic alloys, which then introduce the alloy elements into the scrap melt. The bundles of scrap may also contain discrete components made from other metals (e.g., Cu wire looms and armature windings in steel scrap from automobiles that have not been separated by magnetic, levitation or other sorting methods, see Chapter 8). The elements thus introduced may not normally be present in the metal produced by pyroextraction (e.g., Cu is not normally present in Fe ores), and the refining processes must also remove or dilute the concentration of these elements if they are detrimental to the metal properties.

5.1.2. Standard Specifications

Metals are commonly purchased with reference to in-house, national or international standards of composition. Each standard specification defines a maximum, or a minimum and a maximum, content of each element that may affect the metal properties. Very low residual levels, often in parts per million (ppm), are quoted for some elements when high values of a specific property are required, such as electrical conductivity (solutes cause electron scattering), toughness, ductility or corrosion resistance.

The selling price of a metal increases as it is refined to meet progressively more stringent specifications of maximum impurity contents, and the value added by the refining operations increases accordingly. However, the production cost increases rapidly when very high levels of purification are required. It is evident from Equation 2.45 (i.e., $\overline{S}_A^M = -R \ln N_A$) that the entropy of the solution changes rapidly as the composition is changed at high dilution. Hence, it becomes increasingly difficult to remove more of a solute from a solution as its concentration approaches zero.

A wide range of refining techniques with markedly different operating costs are available. In general, a pyrometallurgical selective oxidation approach is the lowest-cost route because large quantities of metal can be treated in one operation and the total energy cost per ton is relatively low, since the molten metal product can be cast and hot-worked without the need for reheating. However, very high purity and the removal of some elements, such as Bi from Cu, is often difficult to achieve by this route and, as the purity required is increased, a stage is reached where it may become less expensive to use a more efficient but more energy-intensive route such as electrorefining to produce the metal.

The demand for metals of higher purity has increased over the last 2 or 3 decades. As the understanding has increased of the thermodynamic and kinetic factors controlling the refining processes, the metal refiner has often been able to modify the processes and reduce the residual levels of the impurities to meet this demand. Purity levels are being achieved today by pyrometallurgical methods that would have been regarded as impossible to attain only a few decades ago.

Steel is a good example of this trend. Until comparatively recent times, the majority of standard specifications for steels placed an upper limit of 0.05% on S and P contents. Charge materials containing only small amounts of these impurities were usually selected when a concentration of less than 0.025% was required. Today, significant quantities of steel are produced containing not more than 0.010% of each of these elements. Likewise, it was difficult to produce steel containing less than about 0.04% C, but the "interstitial free" deep-drawing grades of steel now marketed for the pressing of automobile body panels contain not more than 50 ppm carbon plus nitrogen. The metal may then contain more than 99.8% Fe. Since steel is defined as an alloy of iron and carbon, it is a misnomer when applied to metal of this purity. Commercial-purity iron would be a more appropriate name and would be consistent with classification of the corresponding grades of the nonferrous metals. "Microalloyed" steels, containing only very small amounts of alloying elements, coupled with new thermomechanical treatments and with very low concentrations of other residual elements, are now replacing steels which contain much higher concentrations of valuable alloy elements for many applications.

Product reliability legislation, which makes the manufacturer liable for any damages caused by the supply or use of defective goods, is also leading to the imposition of closer tolerance limits on the composition of the metal and a rising demand for "clean" metal as free as possible from harmful inclusions.

5.1.3. Partition Ratios

The aim in most pyrorefining operations is to obtain a high partition ratio of one or more impurities between a receptor phase (which may be a slag, another metallic mixture or a gas atmosphere) and the metal being refined. When a solute is transferred from one phase to another by a chemical reaction such as:

$$X_{(metal)} + O_{(metal)} = XO_{(slag)} \tag{5.1}$$

or

$$S_{(metal)} + 2O_{(metal)} = SO_{2(gas)} \tag{5.2}$$

the equilibrium partition ratio ($X_{(slag)}/X_{(metal)}$) for the solute between the two phases is determined by the equilibrium constant for the reaction. The numerical value of the constant is dependent on the temperature (*cf.* Equation 2.41). The ratio of the concentrations of the solute in the two phases can also be changed by altering the activities of the reactants or the products, by changing the partial pressure of a gaseous species (e.g., by applying a vacuum or by continuous purging of a gaseous product), or by changing the relative volumes of the two phases. These are all applications of the **Le Chatelier Principle** which, in effect, states that when one of the factors controlling a system in equilibrium is changed, the system will adjust itself and attempt to restore equilibrium under the new imposed conditions. A variety of methods can be used to manipulate these variables and change the partitioning ratio.

The receptor phase is often removed when the reaction has proceeded to the required extent and a new second phase is introduced to remove more of the first solute or to remove another impurity. This sequence may be repeated several times before the final end point is attained. There is a risk that some of the solute may revert to the metal when a reaction similar to Equation 5.1 occurs if the receptor phase is not removed completely before, say, the oxygen content in the metal is changed in a subsequent operation.

A high partition ratio is favored by a high activity of the solute in the metal being refined and a low activity of the solute in the second phase. The activity of the solutes in the refined metal cannot usually be raised without causing other deleterious changes in the composition of the metal. However, the solutes interact with each other, and it may be possible to remove one solute before a second solute is removed that raises the activity of the first solute. For example, the metal tapped from the iron blast furnace contains about 4% C. Carbon raises the activity of S in Fe, and h_S is 6.6 in carbon-saturated Fe at 1600 °C.[34] Although conditions are not ideal for refining in the ladle used to transfer the metal from the blast furnace to the steel refining unit, significant amounts of S can be removed at this stage while the C content is still high. The amount of sulfur which has to be removed in the refining furnace is thus markedly decreased, and a smaller volume of slag is required to complete the refining.

The common practice for achieving a high partition ratio is to select a composition for the phase into which the impurity is to be transferred which will lower markedly the activity of the solute. Ideally, this phase should not contain any other elements that can transfer into the metal and the metal should be insoluble in the second phase. It is usually difficult to avoid the partitioning of some of the metal into a slag and, when the metal has a high value, it is economically beneficial to recycle the slag through the earlier stages of the extraction process to recover the values.

Quite frequently, one or more stages during refining involve the partitioning of a solute into a liquid or a solid phase that is formed within the metal, as in the removal of dissolved oxygen by deoxidation or in the separation of a solute when the liquid solubility limit is exceeded during cooling (liquation). Action must then be taken to ensure that as much as possible of the second phase is removed before the metal solidifies and prevent it from forming inclusions with consequent detriment to the properties of the metal.

5.1.4. Refining Kinetics

Reaction kinetics are also important with respect to the cost of refining. Chemical reaction rates are rarely rate limiting at the elevated temperatures used for this purpose, and the overall rate is usually transport controlled. The solutes are often present in dilute solution at the start of refining, so the driving force for transportation is low in terms of solute concentrations and diminishes toward zero as the equilibrium partitioning is approached.

Many of the refining reactions involve transfer of a solute across an interface between the metal and a slag or a gas atmosphere. In the absence of induced turbulence, mass transport is dependent on diffusion aided by any convective circulation. A shallow bath of metal with a large interfacial area is conducive to increased reaction rates under these conditions. However, the area of contact with the refractory lining per unit volume of metal is raised as the depth of the metal bath is reduced, and the amount of heat lost through the lining is increased correspondingly. Transfer rates are raised dramatically when the bath is turbulent. In some cases, this occurs naturally when, for example, carbon is oxidized from molten Fe and the bubbles of CO agitate both the molten phases and the interface as they rise through the melt. In the absence of such natural phenomena, the reaction rates can be increased artificially by mechanically induced agitation or by bubbling an inert gas through the melt. Similarly, the rate of escape of dissolved gases (such as hydrogen, oxygen and nitrogen) across a clean, static metal surface into a high vacuum is very low, and much higher rates of transfer are obtained by flushing the melt with a stream of inert gas bubbles. Argon is usually used for this purpose. The bubbles create turbulence that shortens the diffusion distances within the melt and also act as a receiver phase in which the dissolved gas can build up a partial pressure and escape from the melt. The reaction rates are improved and, hence, the time required to reach equilibrium is lowered as the turbulence is increased, but, in practice, the agitation must be limited to prevent both ejection of the reactants from the refining vessel and excessive erosion of the furnace lining.

5.2. REFINING SLAGS

Slags can fulfill a variety of requirements. In smelting, the slag serves primarily as a means of removing gangue minerals and any other unreduced oxides from the ore and the ash left after the combustion of a solid fuel. Refining slags act as a reservoir to collect and immobilize the solutes removed from the metal. They may also be required to accommodate some gangue, as when direct reduced iron (DRI) is melted. The slag isolates the metal from contact with the gas atmosphere above the bath and thus prevents or retards the transfer of gaseous species (e.g., H_2, N_2, O_2) from the gas to the metal and the loss of volatile elements from the metal into the gas. Since an oxide melt usually has a much lower thermal conductivity than a molten metal, the slag acts as a thermal barrier that decreases the heat losses from the surface of the metal. Conversely, the slag retards heat transfer to the metal when heat is supplied by combustion above the bath, as in the reverberatory furnace.

Energy is required to raise the temperature and melt the slag-forming materials, which is only partially compensated by the exothermic heat of formation of the compounds that are formed in the slag. Thus, the thermal requirement rises with an increase in the amount of slag. If the average molecular weight of the slag is equal to the atomic weight of the metal, then equal weights of slag and metal are required to halve the concentration of a solute by partitioning between the slag and the metal (i.e., a partitioning ratio of 1) when the solute behaves ideally in both phases. The density of the slag is only about half the density of the metals, so the volume of the furnace required to accommodate this amount of slag is roughly twice the volume required to hold the metal. Both capital and maintenance costs, per ton of metal refined, increase as the furnace capacity is raised. In fact, the atomic weights of Cu and Fe are only a little greater, but the atomic weight of Pb is roughly three times greater than the molecular weight of the slag.

This simple approximation serves to emphasize the importance of a high partition ratio in minimizing the thermal requirements for refining. Since some of the metal being refined is invariably transferred into the slag, the metal loss also decreases as the partition ratios for the impurities are raised and the slag volume required is lowered. The cost incurred in recycling or disposal of the slag after it is removed from the furnace also rises with increasing slag volume. Thus, there are a number of incentives that encourage the formulation of slag compositions giving high partition ratios.

One other important factor to be considered is the effect of the time taken to form the molten slag on the total time required to achieve the desired change in composition. Maximum partitioning between two phases can only be realized when both phases are fully molten and homogeneous. Since the oxides that are the principal constituents of the slag have high melting temperatures, time is required for the solids to melt and form a homogeneous slag. Steps can be taken to accelerate the rate of slag formation. When lime is added to flux silica, for example, the time required for slag formation decreases with decreasing particle size and increasing ''reactivity'' of the lime. The lime is prepared by calcining calcium carbonate and is dead-burned (sintered) to reduce the rate of subsequent hydration. Stability against hydration is improved with increasing temperature and time of burning, but the reactivity, as measured by the time taken for the lime to dissolve in the slag, is also lowered. A variety of other techniques is used to accelerate dissolution, including careful control of the lime reactivity, the injection of very finely divided lime in a carrier gas into the slag and the addition of fluxes with a low melting point.

The slag remaining after all the economically recoverable metal values have been removed may still have a commercial value. It may be sold for use in civil engineering construction or as a component of fertilizers and high-temperature insulating materials (see Chapter 8).

5.2.1. Slag Structures

The rate of mass transport tends to decrease as the viscosity of the slag is raised. At constant composition, the viscosity tends to fall as the superheat above the liquidus temperature is increased. The temperature dependence of the viscosity is, in turn, dependent on the species present in the slag. Although in the calculation of equilibrium constants it is conventional to assume that a slag is composed of discrete molecules, a molten slag is actually a mixture of ions.

In oxide melts, elements such as Ca, Fe, K, Li, Mg, Mn, Na, Pb and Zn can ionize as electron donors (i.e., cations) and function as bases with the transfer of electrons to associated oxygen or other anions. The bond strength between these ions is relatively weak and the cations can readily interchange positions.

Silicon can shed four electrons, which are shared with four oxygen atoms in tetrahedral coordination. Since each mole of silica supplies one Si ion and only two oxygen ions, each oxygen ion is shared with an adjacent tetrahedron to form a three-dimensional network in pure silica. The covalent bonds between the tetrahedra are broken progressively as the concentration of the basic oxides is increased, and the three-dimensional network is changed into rings and linear or branching chains. The tetrahedra are separated completely when 2 moles of a divalent oxide or 4 moles of a monovalent oxide are present for each mole of the anion complex (i.e., at $2CaO \cdot SiO_2$ or $4Na_2O \cdot SiO_2$). This is classed as a *neutral composition.* Basic slags contain more and acid slags contain less oxygen ions than are required for a neutral composition. The presence of the polymeric anions in acid melts results in a marked increase in the viscosity. Thus, for a constant degree of superheat, the viscosity decreases rapidly as the composition is changed from silica saturation to the orthosilicate composition. It is little affected by structural considerations in more basic melts, where the oxygen ion requirements of the complex anions are satisfied and they exist as discrete entities with unattached oxygen ions. For cations with a common valency, the change in viscosity with composition in acid melts increases slightly with a decrease in the radius of the added cation. For example, additions of FeO, MgO and MnO produce a greater change in the viscosity than similar amounts of CaO which has a larger ionic radius.

Al and P can substitute for Si in the tetrahedra. Alumina is amphoteric. In acid melts, the Al atom sheds electrons and acts as a base, but it functions as an acidic oxide and an oxygen ion acceptor in basic melts. Ferric and chromic oxides behave similarly. The P ion can carry five electron charges, whereas Al can shed a maximum of three electrons. When these ions replace Si^{4+} in the tetrahedra, the cation-oxygen bond strength is increased accordingly by substitution of P and decreased by Al. These variations in bond strength have only a small effect on melt viscosity, but have a more marked effect on the thermodynamic properties of the melt. The activity of a basic oxide is lowered by the addition of anions and vice versa. The activity of the base is lowered to a greater extent when P is substituted for Si, and to a lesser extent when Al is substituted.

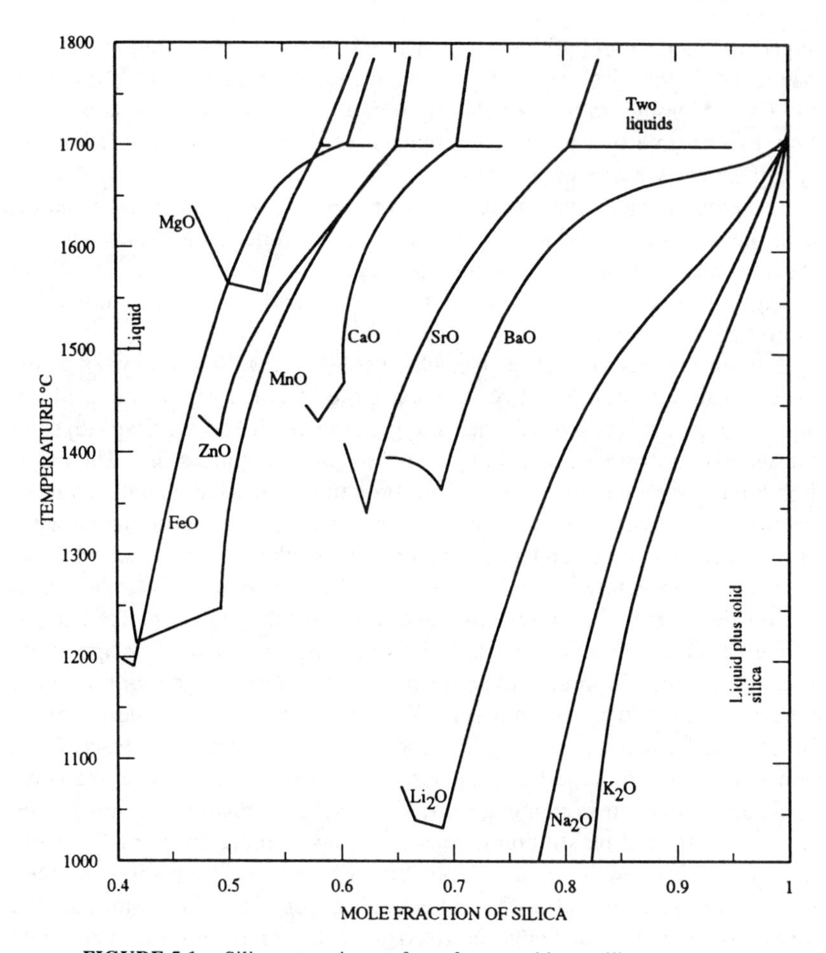

FIGURE 5.1. Silica saturation surfaces for some binary silicate systems.

Many basic oxides are not completely miscible with silica in the molten state, and a two-liquid region exists at high silica concentration. The maximum extent of this region is related to the cation-anion bond strength in the basic oxide. The bond strength is too weak to cause liquid immiscibility with the alkali oxides and is not quite sufficiently strong with BaO but, thereafter, the two-liquid region extends to increasing concentrations of the basic oxide as the bond strength is increased, Figure 5.1. Intermediate compounds, including metasilicates ($MO \cdot SiO_2$) and orthosilicates ($2MO \cdot SiO_2$), are often formed at still lower silica concentrations in the solid state.

A more detailed treatment of the structure and properties of fused oxides can be found in the first three texts listed as further reading at the end of this chapter.

5.2.2. Thermodynamic Properties

Separation of a melt into two phases is associated with a positive deviation from ideal behavior (i.e., $\gamma > 1$) by the solvent, whereas compound formation in the solid state is usually accompanied by a negative deviation from ideality in the melt. These trends are portrayed by binary mixtures of silica with a base that exhibit liquid immiscibility. The activity of silica remains close to unity across the immiscible region. It then falls rapidly as the concentration is lowered below the saturation limit and usually shows a marked negative deviation from ideal when the orthosilicate composition is reached, the negative deviation increasing as the bond strength of the basic oxide is raised. The activity of the basic oxide also shows marked negative deviation from ideality, falling with increasing silica content to near zero at the silica saturation composition. The melting temperature of some of the basic oxides is higher than the temperatures attained in the refining slag. The activity of the base then remains close to unity until sufficient anions are present to form a homogeneous melt and, thereafter, falls rapidly with increasing silica concentration. These trends are illustrated in Figure 5.2.

A two-liquid region, joining the miscibility limits in the binary silicate systems, is formed when silica is mixed with any two of the basic oxides that exhibit binary liquid immiscibility. This type of behavior is shown by the CaO-FeO-SiO_2 ternary system (Figure 5.3), the components of which are the major constituents of many smelting and refining slags. The iso-activity contours for CaO and SiO_2 in the ternary melts are roughly linear between points of equal activity of these species on the SiO_2-FeO and the SiO_2-CaO binary sides.[87]

Mixtures of CaO with FeO form a compound in the solid state, and a_{FeO} shows a marked negative deviation from ideal behavior in molten binary solutions of these oxides. However, a_{FeO} is raised when silica is added to the binary melts and the maximum values for γ_{FeO} coincide with the join from FeO to the composition, $2CaO \cdot SiO_2$. Iso-activity contours for FeO in the ternary melts are shown in Figure 5.4a at 1600 °C, corresponding to the temperatures encountered in steel refining, and in Figure 5.4b at the lower temperatures appropriate, for example, to the converting and refining of Cu. The free energy of formation of the CaO-SiO_2 compounds is markedly more negative than for the corresponding FeO-SiO_2 compounds. Consequently, the addition of CaO displaces FeO from association with the SiO_4^{4-} tetrahedra, raising the value of a_{FeO} and increasing the oxidizing power of the slag. This behavior can be also explained in terms of the strength of the cation-oxygen bond in CaO and FeO. When oxides of elements with a smaller cation radius (and hence weaker bond strength) but the same valency, such as MnO and MgO, are partially substituted for CaO in the refining slag, the positive deviation of a_{FeO} along the FeO-$2CaO \cdot SiO_2$ join is decreased. Deductions can be made on a similar basis about

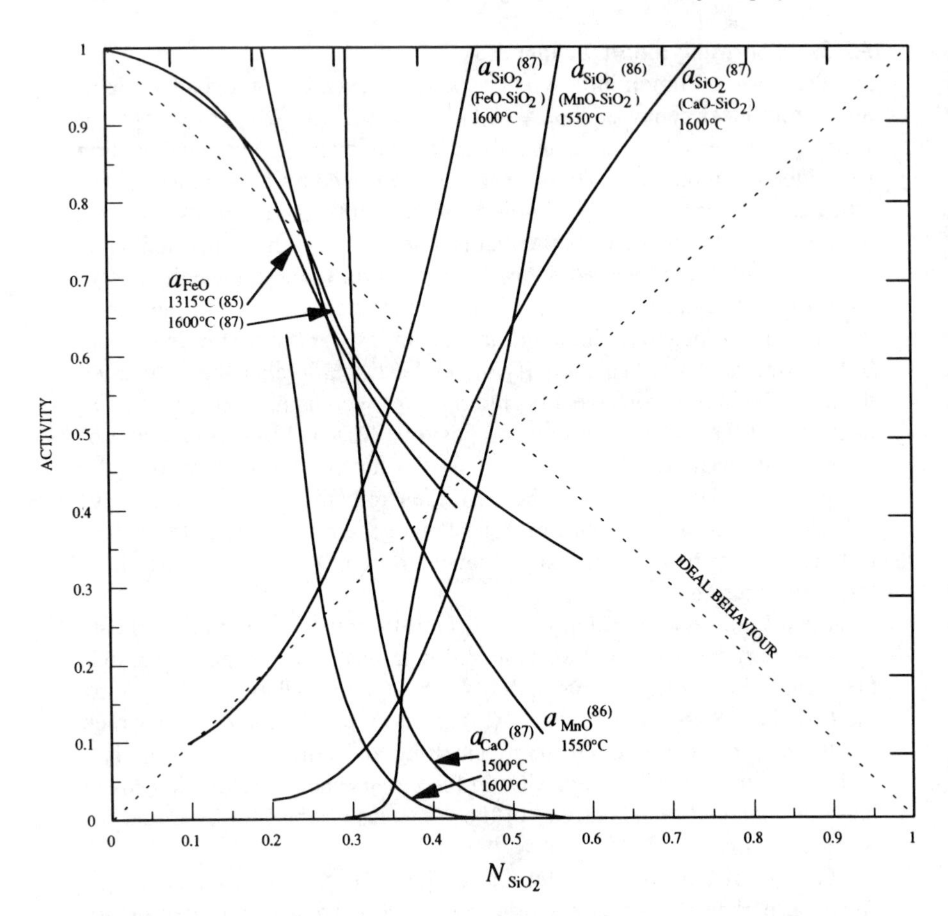

FIGURE 5.2. Activities in some binary silicate systems. Reference numbers for original data given in brackets.

the probability of deviations from ideal behavior of other constituents of oxide and sulfide melts.

Care is needed when assessing the activities of components in a slag for the evaluation of thermodynamic relationships such as equilibrium constants. The activities shown, for example, in Figure 5.4 apply only to homogeneous melts. Comparison of Figures 5.4a and 5.4b draws attention to the rapid increase in the composition range in which homogeneous melts can be obtained and the change in the activities as the temperature is raised. The temperature of the slag may change during refining and the composition of the slag is changing continuously with time as solutes are transferred from metal to slag, while other species (such as oxygen) may transfer in the opposite direction. When the molten slag is in contact with an oxide refractory lining (SiO_2, MgO, etc.), the slag attacks the lining and the composition changes progressively toward saturation with the oxides contained

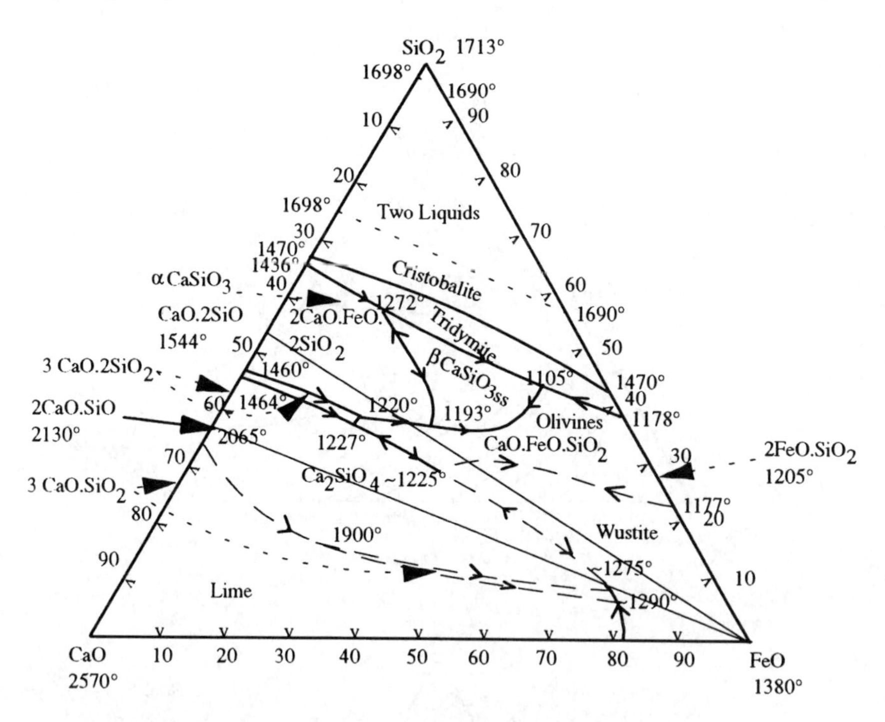

FIGURE 5.3. Phase diagram for the ternary system CaO-FeO-SiO_2. Tridimite (870 to 1470 °C) and cristobalite (>1470 °C) are high-temperature allotropic forms of silica. Olivines are an isomorphous family of minerals with the general formula M_2SiO_4. (From Bowen, Schairer, and Posnjak, *Am. J. Sci.*, 26, 204, 1933. With permission.)

in the refractory. Conversely, oxides in the slag may be reduced when in contact with a carbon lining. Chemical analysis of a slag sample gives the composition only at the time when the sample was taken, and the composition may have changed appreciably during active refining by the time the analysis is reported. The analysis may also be misleading if the slag is not homogeneous and contains droplets of metal or undissolved oxides.

As noted earlier, refining times are extended and costs are increased when the formation of a homogeneous slag is delayed by a slow rate of dissolution of high melting point constituents in the slag. Compounds with low melting points are often added as fluxes to increase the rate of slag formation. However, the flux may also change the activities of the slag components. For example, small amounts of fluorspar (CaF_2; m.p. 1385 °C) are used to accelerate lime solution. Each mole of CaF_2 added contributes two F^- ions to the melt. The fluoride ions break the covalent O^{2-} bonds between SiO_4^{4-} tetrahedra in acid melts, releasing O^{2-} ions and raising the oxidizing potential of the slag. In basic slags, where all the tetrahedra are separated, the fluoride ions cluster around the cations and lower the activity of the basic oxides.

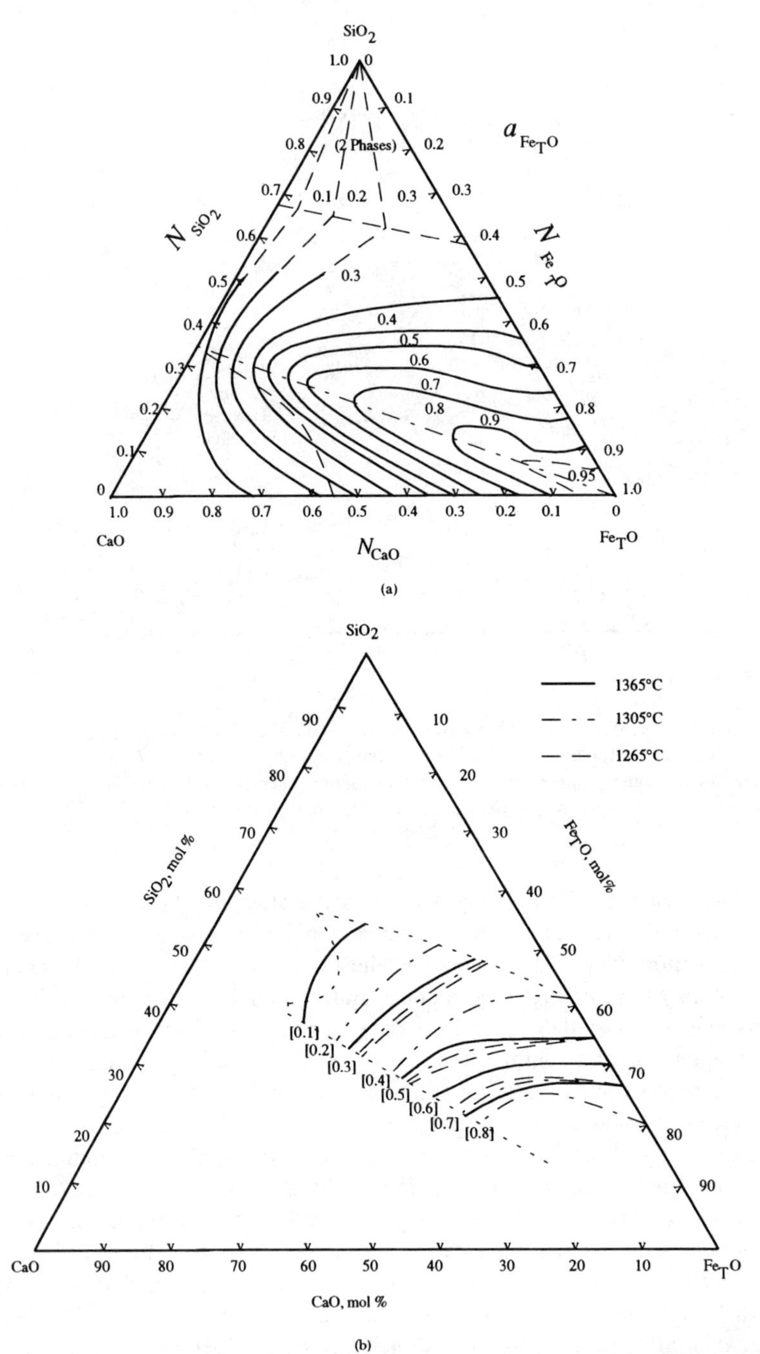

FIGURE 5.4. The activity of ferrous oxide in CaO-FeO-SiO_2 slags: (a) at 1600 °C; (b) at 1265 to 1365 °C. From Bodsworth, C., *J. Iron Steel Inst.*, 13, 193, 1959. With permission. (a) Reprinted with permission from *TRANSACTIONS OF THE METALLURGICAL SOCIETY, Vol. 203, (1955), pg. 485,* a publication of The Minerals, Metals & Materials Society, Warrendale, Pennsylvania 15086.

Low melting point halides, such as AlF_3, NaCl and KCl, are sometimes used as protective fluxes in nonferrous metal refining to prevent contamination of the metal by the gas atmosphere. Since halides tend to be hygroscopic, they must be dried before charging into the furnace to ensure that they do not introduce hydrogen into the melt. The halides help to flux any dross or other form of impurity that floats on the surface and aid the separation of this layer from the metal.

5.3. SLAG-METAL REACTIONS

When the metal is isolated from the atmosphere by a slag barrier, impurities that form more stable oxides than the metal being refined can be removed more or less completely from the metal by oxidation if they form constituents with low activity in the slag. Selective oxidation of impurities is the most common application of slag-metal refining. It can be used, for example, to reduce the concentration of elements such as Fe, Ni and Zn, which may be present in crude or scrap tin metal. However, the activities of the oxidation products in the slag cannot be lowered sufficiently to make this an effective method for the refining of metals that form very stable oxides, like Al. The application which has been studied most intensively is the production of steel.

5.3.1. Iron Refining

The iron produced in the blast furnace is saturated with C and contains variable amounts of Si and Mn, together with S and P as the most deleterious impurities. Since S and P form anions in the slag, they are removed most effectively when the composition of the slag is as close as possible to lime saturation, while still remaining homogeneous and fluid. Typically, this condition is achieved with a slag that is almost saturated with lime and has a basicity ratio $(CaO + MgO)/SiO_2$ of 2.8 to 3.5. Si is oxidized almost completely from the Fe during refining under a basic slag and, as the slag contains only about 20% SiO_2 at lime saturation, the slag bulk increases rapidly as the Si content of the blast furnace metal is raised.

The metallurgical load on the refining furnace can be eased considerably by pretreatment of the molten metal in the transfer ladle between the blast furnace and the refining unit. If Si is oxidized from the metal in the blast furnace runner by the addition of finely divided iron oxides (e.g., rolling mill scale) and the silica product is removed before the metal enters the ladle, then S and P can be removed extensively by forming a $CaO\text{-}CaF_2$ slag in the transfer ladle. Or, the concentrations of Si, P and S can be lowered simultaneously by addition of Na_2CO_3, for example, to form silicate, phosphate and sulfide complexes with the sodium oxide.[88] With appropriate control of the blast furnace slag composition and temperature in the reduction of a high-grade ore, followed by treatment in the transfer ladle, it is not difficult to limit the concentration of these elements in the metal charged to the refining furnace to less than 0.1% Si, 0.05% P and 0.05% S.

Steel is recycled extensively as scrap metal, which costs less to procure and refine than to produce metal from the ore via the blast furnace and the converter. The scrap metal contains much lower concentrations of carbon and usually less S and P, but it may contain a wide variety of alloy elements. Some of these (such as Cr, V and Ti) are readily removed but, if they are present in significant amounts, their oxidation products may adversely affect the basicity, viscosity, etc. of the refining slag. Other elements like Cu, Ni and Sn cannot be removed by selective oxidation from Fe, and the refining processes must be modified to cope with these impurities. Nonferrous metals are similarly recycled as scrap material and the procedures may have to be modified to deal with alloy elements introduced into the melt from the scrap. On the other hand, many of these alloy elements are expensive to produce, and it is often financially more rewarding to segregate the scrap containing significant amounts of the alloy elements and process it separately to reproduce the alloy compositions. The refining process must then be modified to prevent the oxidation of the alloy elements and retain them in the melt.

5.3.1.1. Equilibrium Considerations

If oxygen gas is blown onto molten iron, the order in which the impurity elements start to oxidize is determined initially by the free energies of formation of the oxides of those elements which, in turn, determine the values of the equilibrium constants for the oxidation reactions. It is useful to express the latter in terms of the dilute, wt% standard state for the solutes dissolved in Fe. For example, the equilibrium constant for the oxidation of Mn in terms of the alternative standard state is obtained by summation of three free energy relations:[89]

$$Mn_{(l)} + \tfrac{1}{2}O_{2(g)} = MnO_{(l)}$$

$$\Delta G^{\circ} = -343{,}400 + 56.1T \text{ J mol}^{-1} \tag{5.3}$$

$$Mn_{(l)} = Mn_{(\%\text{ in Fe})}$$

$$\Delta G^{\circ} = -28.6T \text{ J mol}^{-1} \tag{5.4}$$

$$\tfrac{1}{2}O_{2(g)} = O_{(\%\text{ in Fe})}$$

$$\Delta G^{\circ} = -118{,}000 + 2.4\,T \text{ J mol}^{-1} \tag{5.5}$$

Hence,

$$Mn_{(\%\text{ in Fe})} + O_{(\%\text{ in Fe})} = MnO_{(l)} \tag{5.6}$$

$$\Delta G^{\circ} = -225{,}400 + 97.1T \text{ J mol}^{-1}$$

and

$$\log K_{Mn} = \log \frac{(a_{MnO})}{[a_{Mn}][a_O]} = -\frac{\Delta G^{\circ}{}_{Mn}}{19.15T} = \frac{11,770}{T} - 5.07 \qquad (5.7)$$

where square and round brackets signify activities in the metal and in the slag, respectively.

The equilibrium constants for the oxidation of other solutes can be derived in a similar manner to yield the following relations:

$$Si_{(\% \text{ in Fe})} + 2\, O_{(\% \text{ in Fe})} = SiO_{2(l)}$$

$$\log K_{Si} = \log \frac{(a_{SiO_2})}{[a_{Si}][a_O]^2} = \frac{29,660}{T} - 11.23 \qquad (5.8)$$

$$C_{(\% \text{ in Fe})} + O_{(\% \text{ in Fe})} = CO_{(g)}$$

$$\log K_C = \log \frac{p_{CO}}{[a_C][a_O]} = \frac{1168}{T} + 2.07 \qquad (5.9)$$

$$2P_{(\% \text{ in Fe})} + 5\, O_{(\% \text{ in Fe})} = P_2O_{5(l)}$$

$$\log K_p = \log \frac{(a_{P_2O_5})}{[a_p]^2[a_O]^5} = \frac{35,700}{T} - 30.3 \qquad (5.10)$$

$$S_{(\% \text{ in Fe})} + 2\, O_{(\% \text{ in Fe})} = SO_{2(g)}$$

$$\log K_S = \log \frac{p_{SO_2}}{[a_S][a_O]^2} = \frac{300}{T} - 3.2 \qquad (5.11)$$

and, for the oxidation of Fe:

$$Fe + O_{(\% \text{ in Fe})} = FeO_{(l)}$$

$$\log K_{Fe} = \log \frac{(a_{FeO})}{[a_{Fe}][a_O]} = \frac{7154}{T} - 2.92 \qquad (5.12)$$

The variation with temperature of these equilibrium constants is illustrated in Figure 5.5. This would appear to indicate that C, Mn and Si can be oxidized preferentially in the working temperature range, but P and S cannot be removed before Fe is oxidized.

Relative to the dilute wt% standard state, a_O would equal unity at 1 wt% O in solution in Fe if oxygen conformed to Henry's Law up to that concentration. However, oxygen is only slightly soluble in molten Fe. The solubility in pure Fe is 0.16% at the melting point (1535 °C) and increases slowly with rising temperature according to the relation:[90]

$$\log [\%O] = -\frac{6320}{T} + 2.70 \qquad (5.13)$$

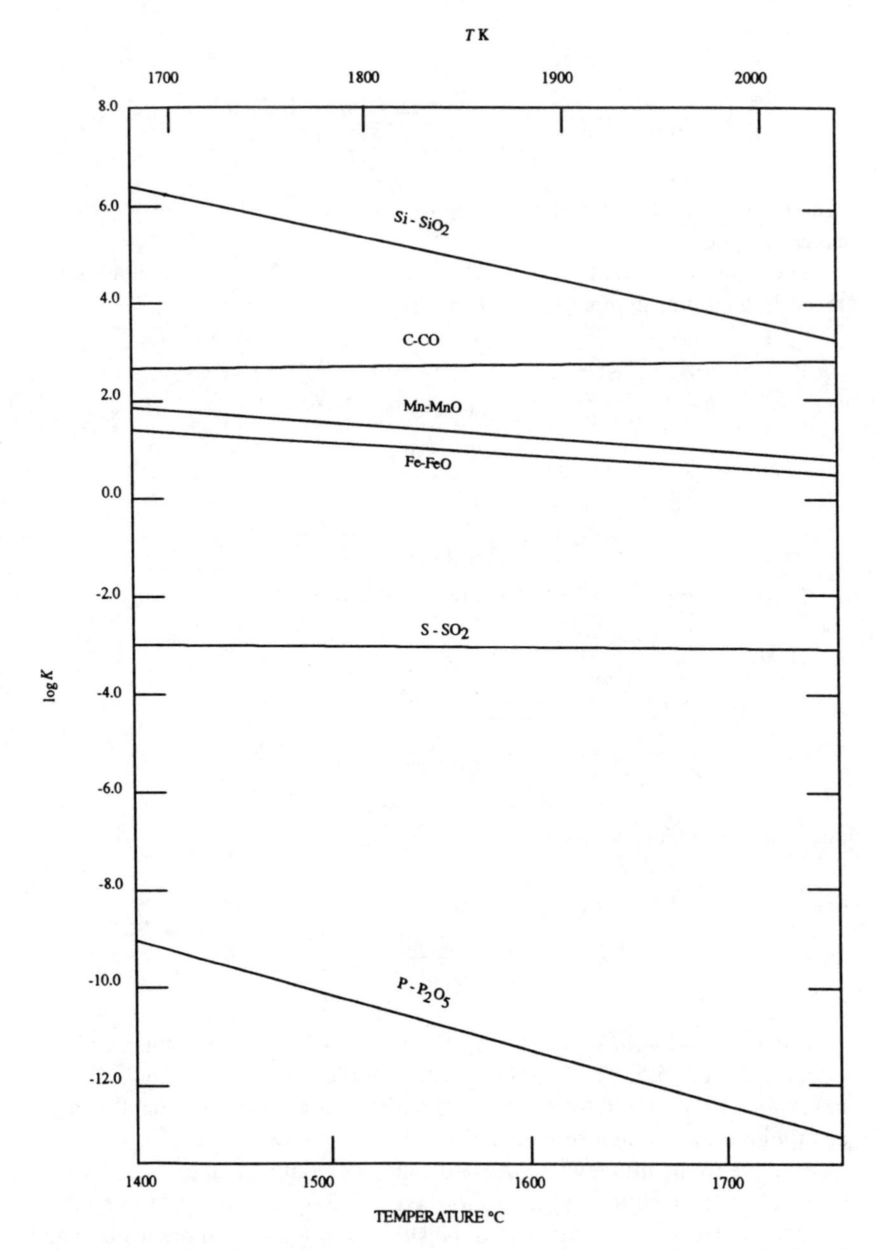

FIGURE 5.5. Temperature dependence of the equilibrium constants for the oxidation of solutes from molten iron.

Within the solubility range and in the absence of other solutes, a_O shows a negative deviation from Henry's Law (i.e., $e_0^0 < 1$). The interaction coefficients (e_O^X) for C, Mn, Si and S are all negative, indicating that the negative departure from Henry's Law behavior is increased (f_O is decreased) by these solutes. P has a negligible effect on a_O. Hence, the actual values of a_O inserted when solving the above equations are markedly lower than unity. The effect of this low value of a_O on the equilibrium ratio of the activities of the other two species in a reaction becomes progressively greater as a_O is raised to powers greater than unity in the equilibrium constant. Furthermore, if equilibrium is maintained between the various reactions during refining, then, at any instant, the value of a_O must be the same in all the above equations.

The activity coefficient for FeO in refining slags is greater than unity (*cf.* Figure 5.4a) and is typically in the range 2.0 to 2.5. γ_{MnO} values are similar, but γ_{SiO_2} is very low and is usually in the range 0.1 to 0.01. Inserting these values into the appropriate equations (remembering that $a_M = \gamma_M \times N_M$), it is readily apparent that Si is oxidized almost completely into the slag before Mn and Fe start to oxidize. Most of the Mn can also be oxidized before the Fe when the initial Mn content in the metal is less than about 2% and N_{MnO} in the slag is, correspondingly, only about 0.03 to 0.04.

It would seem from the data in Figure 5.5 that carbon would oxidize before Mn. However, if the carbon is oxidized within molten metal that is isolated from the atmosphere by the slag barrier, the effective partial pressure of CO is greater than unity due to the difficulty of nucleation of gas bubbles within a molten melt (see Section 3.4.2 and Worked Example 12). This shifts the equilibrium position for Equation 5.9, relative to that for Equation 5.6, and makes carbon removal more difficult. Depending on the actual operating conditions, the start of Mn oxidation may precede or coincide with the start of C removal.

Each element can continue to oxidize until the ratio of its activity in the slag to its activity in the metal is in equilibrium with the activity of the oxygen dissolved in the metal. The reaction would then cease if a_O and the temperature remained constant. However, a_O rises as the concentrations of C, Mn, Si, etc. decrease, so the reactions can continue in order to maintain equilibrium with a_O in the Fe. Eventually, a stage is reached, however, where the Fe will begin to oxidize rapidly if oxidation is continued. Refining is halted, therefore, before this stage is reached, leaving a small but finite concentration of each of the impurities in the metal. Only those elements (such as Al) forming very stable oxides relative to FeO can be virtually eliminated by oxidation. Similar restrictions apply to the limiting partition ratios for all metals.

5.3.1.2. *Dephosphorization*

Phosphorus is conventionally assumed to exist in the slag as the acid radical, P_2O_5. The activity of this species is high, therefore, in acid slags, but is low in basic melts. Few direct measurements have been made of the activity of P_2O_5, but insertion into Equation 5.10 of experimental data for P partitioning between slag and metal yields values for $\gamma_{P_2O_5}$ in the range 10^{-14} to 10^{-18} in basic steelmaking slags. Since a_O is raised to the fifth power in Equation 5.10, it is evident that dephosphorization is also facilitated by a high oxygen content in the metal and, because C, Mn and Si lower h_O, P partitioning becomes greater as the other solutes are removed. Figure 5.5. shows that the equilibrium constant for P oxidation is markedly dependent on the temperature when compared with the other refining reactions; and, in practice, under otherwise constant conditions, the residual phosphorus content remaining in the metal is increased by at least 0.01% if the metal temperature is raised by 50 °C. So, P removal is favored by the rapid formation of a highly basic and oxidizing slag at a relatively low temperature, and P partition ratios of at least 150 can be attained under these conditions. Phosphorus can then be removed simultaneously with carbon.

The lack of direct measurements for $\gamma_{P_2O_5}$ hampers the use of the equilibrium relations and one must sometimes resort to the use of empirical equations to express P partitioning. Phosphorus is actually present in the slag as the anion, (PO_4^{3-}):

$$P_{(\%\ \mathrm{in\ Fe})} + \tfrac{5}{2}O_{(\%\ \mathrm{in\ Fe})} + \tfrac{3}{2}O_{(\mathrm{slag})} = PO^{3-}_{4(\mathrm{slag})} \tag{5.14}$$

This reaction can be fitted empirically to dephosphorization by the relationship:[91]

$$\begin{aligned}\log k_P &= \log \frac{(\%P)}{[\%P][\%O]^{5/2}} \\ &= 19{,}872/T(\mathrm{K}) - 8.566 + 0.0667(\%CaO + \%CaF_2)\end{aligned} \tag{5.15}$$

The fit of this equation to experimental and plant data at various temperatures is illustrated in Figure 5.6.

Ironically, while the demand for steels containing less than 0.010% P for use at low temperatures is growing, a demand has now arisen for P contents up to 0.07% to improve the strength and deep drawability of sheet steel for forming into shaped components. This emphasizes the comment made at the beginning of this chapter that impurities need to be removed during refining only to the extent required for the intended use.

5.3.1.3. *Desulfurization*

The equilibrium constant for the oxidation of sulfur to SO_2 has a very low numerical value at 1600 °C and, if typical values for a_O and a_S are

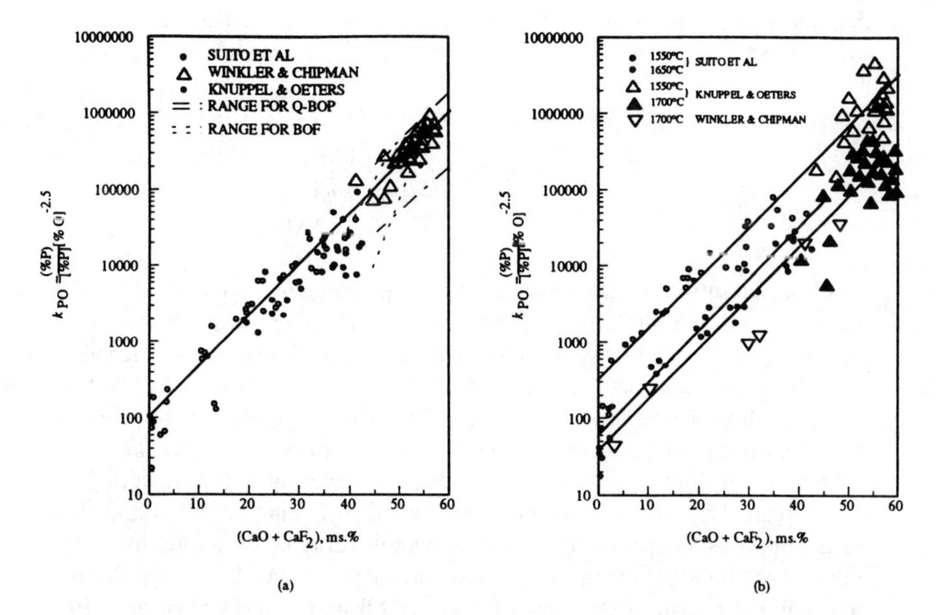

FIGURE 5.6. Variation of the phosphate capacity, k_{PO}, with slag composition (a) at various temperatures and (b) at 1600 °C. (From Turkdogan E. T., *Perspectives in Metallurgical Development,* The Metals Society, London, 1984, 49. With permission.)

inserted in Equation 5.11, it is found that the maximum value of p_{SO_2} is very low. Since SO_2 is removed as a gas, the pressure generated by the reaction is insufficient to nucleate bubbles of SO_2 within the metal. A small amount of SO_2 could dissolve in the bubbles of CO as they rise through the bath, but the amount removed in this way is not significant and S cannot be removed effectively from steel by oxidation.

Desulfurization is achieved by the formation of anions in the slag:

$$S_{(\text{in Fe})} + 2e^- = S^{2-}_{(\text{slag})} \tag{5.16}$$

Electroneutrality is preserved by simultaneous release of electrons, either by transfer of oxygen to the metal:

$$O^{2-}_{(\text{slag})} = O_{(\text{in Fe})} + 2e^- \tag{5.17}$$

or by the oxidation of other elements (such as Fe, Mn and Si) that form cations in the slag:

$$Mn_{(\text{in Fe})} = Mn^{2+} + 2e^- \tag{5.18}$$

By summing Equations 5.16 and 5.17, the desulfurization reaction is written as:[92]

$$
\begin{aligned}
S_{(\text{in Fe})} + O^{2-}_{(\text{slag})} &= S^{2-}_{(\text{slag})} + O_{(\text{in Fe})} \\
\log K_S &= \log \frac{(a_{S^{2-}})[a_O]}{(a_{O^{2-}})[a_S]} \\
&= -3750/T + 1.996 \qquad (5.19)
\end{aligned}
$$

Desulfurization is thus promoted by a high a_S value and a low a_O value in the Fe and by a high oxygen ion activity in the slag. Sulfur forms compounds with Fe in the solid state and, correspondingly, shows negative deviation from ideal behavior in the melt, but the activity is raised by carbon and a_S is high in carbon-saturated Fe. The activity of the oxygen ions is raised by increasing concentration of the basic oxides in the slag. FeO is a base and contributes O^{2-} ions to the slag. However, an increase in a_{FeO} is accompanied by an increase in a_O in the metal (Equation 5.12) and this effect opposes desulfurization. Hence, S removal is favored by a low FeO content in the slag and a high C content in the metal. These conditions apply in the hearth of the blast furnace, and S partition ratios of over 100 can readily be achieved during extraction of the iron from the ore. But oxidizing conditions prevail when impurities are being oxidized during refining and partition ratios of 1 to 5 are then more common.

All basic oxides do not have an equivalent effect on desulfurization. The differing behavior can be assessed qualitatively from the difference in the free energies of formation of the oxide and sulfide compounds of the elements. In the relevant temperature range, the free energy of formation is more negative for sodium sulfide, Na_2S, than for the oxide, Na_2O. This higher affinity of sodium for S explains why Na_2O is so effective for desulfurization in the blast furnace transfer ladle. The affinity of both Ca and Mg is greater for O than for S, but the difference in the stability of the compounds is very much greater for Mg than for Ca. Correspondingly, the ratio $(a_{S^{2-}})/(a_{O^{2}})$ decreases as Ca is replaced by Mg in a slag of constant basicity ratio. High partition ratios can be obtained with $CaO\text{-}CaF_2$ slags containing little or no FeO.

5.3.1.4. Removal of Other Elements

As noted earlier, elements such as Cu, Ni and Sn that form less stable oxides than FeO cannot be removed by selective oxidation. These impurities have adverse effects on the workability of the metal and on properties such as weldability. They originate primarily from the scrap metal and their concentrations increase slightly when the scrap is remelted because some

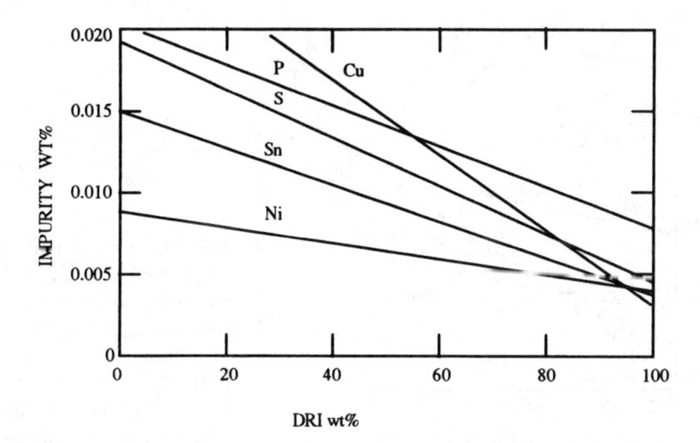

FIGURE 5.7. Dilution of the concentration of tramp elements in iron by the addition of direct reduced iron, DRI.

of the Fe is lost into the slag. Cr, V, Ti and other elements that form somewhat more stable oxides than FeO are almost entirely transferred into the slag, but even these elements cannot be oxidized preferentially when their concentrations (and hence their activities) fall to very low levels in the Fe. The residual amount remaining in the metal is sometimes sufficient to affect adversely the properties of the metal for the intended use.

When not required as alloy elements in the steel, these species are usually referred to as "tramp elements." Procedures other than those conventionally adopted in slag-metal refining must be adopted when they must be removed to very low levels;[93] but more commonly, the requirement is merely to reduce the concentrations to below some acceptable level. Direct reduced iron (DRI) usually contains only small amounts of impurities and a cost-effective method is to incorporate a sufficient quantity of DRI in the charge to ensure that the content of tramp elements falls below the critical level, as shown in Figure 5.7.

Metallic chromium is more expensive than Fe and, when scrap metal with a high Cr content (e.g., stainless steel containing 18% Cr) is remelted, it is often desirable to prevent the loss of Cr by oxidation. However, oxidizing conditions are required if C is to be removed. The numerical value of the equilibrium constant for Cr oxidation decreases with increasing temperature, whereas K_C is almost independent of temperature; thus, carbon can be oxidized in preference to Cr if the temperature is raised sufficiently. The high slag and metal temperature may then result in severe attack of the refractory lining of the furnace.

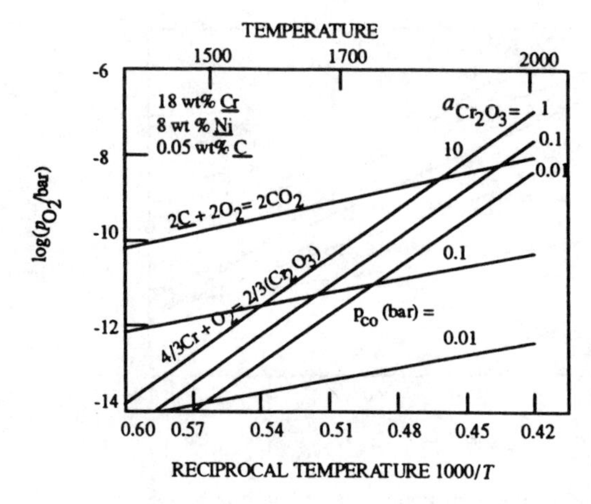

FIGURE 5.8. Equilibrium relations for C-CO and Cr-Cr_2O_3 in stainless steelmaking. (From Neuschutz, D., in *User Aspects of Phase Diagrams,* Hayes, F. H., Ed., Institute of Metals, Petten, The Netherlands, 1991, 199. With permission.)

An alternative approach is adopted that allows the preferential removal of C at lower temperatures. The oxidation of Cr can be described by the reaction:

$$\tfrac{2}{3}\mathrm{Cr}_{(\text{in Fe})} + \mathrm{O}_{(\text{in Fe})} = \tfrac{1}{3}\mathrm{Cr_2O}_{3(\text{slag})}$$

$$K_{Cr} = \frac{(a_{Cr_2O_3})^{1/3}}{[a_{Cr}]^{2/3}[a_O]} \tag{5.20}$$

Combining this with Equation 5.9 for C oxidation at constant a_O:

$$[a_O] = \frac{p_{CO}}{K_C[a_C]} = \frac{(a_{Cr_2O_3})^{1/3}}{K_{Cr}[a_{Cr}]^{2/3}} \tag{5.21}$$

Therefore,

$$\frac{[a_C]}{[a_{Cr}]^{2/3}} = \frac{[\%C \cdot h_C]}{[\%\mathrm{Cr}]^{2/3}[h_{Cr}]^{2/3}} = \frac{K_{Cr}p_{CO}}{K_C(a_{Cr_2O_3})^{1/3}}$$
$$= (K_{Cr}/K_C) \cdot p_{CO} \tag{5.22}$$

if $a_{Cr_2O_3}$ is held constant and only p_{CO} is changed. The partial pressure of CO can be lowered by bubbling an inert gas through the bath into which the CO oxidation product can diffuse. This avoids the supersaturation required to nucleate bubbles within the melt and lowers p_{CO} by dilution in the inert gas, thus shifting the balance of the reaction in favor of C removal.

The partial pressure of oxygen in equilibrium with both C and Cr in a steel containing 18% Cr, 8% Ni and 0.05% C is shown in Figure 5.8. If the slag is saturated with Cr, so that $a_{Cr_2O_3}$ is unity, the two reactions are in equilibrium at about 1850 °C when p_{CO} is equal to 1 atm and C can only be removed to lower concentrations without simultaneous oxidation of Cr if this temperature is exceeded. The equilibrium temperature is raised further if $a_{Cr_2O_3}$ is lowered by dilution in the slag while p_{CO} remains at 1 atm. However, if p_{CO} is lowered to 0.01 atm, the equilibrium temperature is decreased to about 1520 °C even when $a_{Cr_2O_3}$ is as low as 0.01. In the **AOD process,** a mixture of oxygen and argon is bubbled through the bath and the O/Ar ratio is decreased progressively as the C is removed. It is possible to lower the C concentration to about 0.02% in this way, while retaining most of the Cr in the metal, at temperatures in the normal refining range (e.g., 1600 to 1650 °C). Any Cr oxidized into the slag can be recovered by adding ferrosilicon to the bath when the oxidizing reactions have been completed.

5.3.2. Steelmaking Practice

About three quarters of the steel produced today is refined in the basic oxygen (L.D., BOS, QBOP, etc.) converter which is a squat, cylindrical vessel, closed at the bottom, with a restricted opening at the top (Figure 1.7c) that can hold up to 400 t metal. The vessel is vertical during refining, but can be rotated around a central trunnion to decant the refined metal and the slag. Some metal is still refined in the open-hearth furnace, which is similar to a reverberatory furnace but with heat exchangers at each end that are used to raise the temperature of the combustion gases and enable the required temperature to be attained in the bath. Most of the remainder is refined, primarily from scrap metal, in an electric arc furnace.

The oxidation of P, Si and (to a lesser extent) Mn from solution in Fe is strongly exothermic. In the original Bessemer converter process, the heat liberated when air was blown through molten metal from the blast furnace containing relatively high concentrations of these elements was sufficient to raise the temperature of the metal from about 1350 °C at the start of the blow to over 1600 °C at the end of refining, without any other extraneous heat supply. The high-grade iron ores smelted today have only a low P content and the blast furnace is operated to produce a metal with low Si content. Consequently, the heat liberated by the oxidation of these impurities during refining is not sufficient to achieve the required temperature rise. The only metalloid present in appreciable quantity is C, but the heat released by the oxidation of C to CO made only a small contribution to the energy balance in the Bessemer process. Sufficient heat is supplied by this reaction alone, however, if the air blast is replaced by commercially pure oxygen gas, thus saving the energy consumed in raising the temperature of the nitrogen in the air up to the reaction temperature.

Oxygen gas is jetted into the metal in the modern converter process. The heat released by oxidation of the C plus small amounts of the other impurities is sufficient not only to raise the temperature of the molten metal by the required amount, but also to melt cold metal scrap equal to about 20%

of the total charge weight. The raw material cost, per tonne of refined metal, is lowered by the scrap addition, while the tramp elements in the scrap are diluted by the hot metal from the blast furnace. The amount of solids that can be melted can be increased by partial combustion of the CO to CO_2 within the vessel and/or by injecting coal fines in the oxygen blast. The only energy requirement is for the production and compression of the oxygen gas and, as with the Bessemer process, no extraneous energy is needed for heating or for the refining reactions.

5.3.2.1. ***The Top Blown Converter Process***

Oxygen gas is admitted through a lance at supersonic velocity to impact on the surface of the melt in the top blown BOS process. When the jet first strikes the metal surface, oxygen dissolves in the metal, and Si and some Fe are rapidly oxidized. These oxides rise to the surface of the melt and flux the solid lime in the charge to form a fluid, basic slag. Very high temperatures, in excess of 2000 °C, can be generated by the oxidizing reactions in the jet impact zone. Impurities such as Pb and Zn have a high vapor pressure at these temperatures and are more or less completely volatilized. They condense as oxides in the gas off-take, from where they can be recovered and sold to a nonferrous metal extractor.

The turbulence in the impact zone ejects droplets of metal into the slag layer. When carbon starts to oxidize, the bubbles of CO leaving the metal may be temporarily trapped in the slag to produce a foaming condition. The tendency of the slag to form a foam rises with increasing viscosity of the slag and with increasing rate of formation of the gas bubbles. It also rises as both the surface tension and the temperature of the slag are lowered. SiO_2 and P_2O_5 are both surface active and lower the surface tension, P_2O_5 having a markedly greater effect than silica. Consequently, the presence of the latter species in the slag during Fe refining increases the foam stability relative to that encountered during Cu converting.

As the bubbles of CO leave the metal and enter the slag, they carry with them a thin film of metal. The adsorbed slag layer stabilizes this film and retards drainage back into the metal bath, meanwhile, the metal spray, ejected by the blast turbulence, further increases the amount of metal held in the slag. The suspended metal droplets react with the FeO in the slag to produce small CO bubbles that remain attached to their surfaces, increasing the buoyancy of the droplets. A foaming slag can thus be regarded as a slag-metal emulsion in which a very high surface area per unit volume of the metal is in intimate contact with the slag. This condition is ideal for the oxidation reactions. If lime with a high reactivity is dissolved rapidly to form a basic slag, P can transfer from metal to slag while the droplets are in suspension, and the P content of the metal bath can be reduced to a low level while the temperature is relatively low and the C content is still high. The formation rate of CO bubbles decreases as the metal C content falls and the foam eventually collapses as the end point of refining is approached.

5.3.2.2. Bottom Blowing Practice

When oxygen is blown onto the top of the metal, some turbulence is created by the energy of the blast, but circulation within the bath is caused primarily by the bubbles of CO rising through the melt. Circulation decreases, therefore, as the rate of bubble formation falls at low carbon contents. As a result, it is difficult to lower the C content below about 0.05% without a rapid increase in the amount of Fe oxidized into the slag. Air was blown through tuyeres in the bottom of the vessel in the original Bessemer process and this caused vigorous circulation within the metal. Bottom blowing with oxygen instead of air is now practiced in processes such as the OBM (oxygen, bottom, Maxhutte) and QBOP (quick basic oxygen process). However, intense heat is generated by the oxidation reactions in the vicinity of the tuyeres when oxygen gas is injected. It is necessary to protect the refractories in this zone by injecting a shroud of hydrocarbon gas through annular tuyeres surrounding the oxygen stream. The temperature at the tuyere mouth is thus reduced by the endothermic dissociation of the hydrocarbon. Since Fe atoms are more numerous than other species in the melt, the oxygen reacts initially with the iron to form FeO and the FeO is then reduced as it comes into contact with the more readily oxidizable elements.

Injection of the gas through the metal allows continued refining until carbon has been almost completely eliminated from the Fe. Very finely divided (pulverized) lime can be injected into the metal in the oxygen blast without risk of dust loss into the atmosphere. The lime is then rapidly fluxed by the FeO and reacts with the phosphorus as it rises through the bath. Thus, P is again removed at the same time as C is oxidized, but at a lower rate than with top blowing since slag foaming is suppressed and most of the P is transferred to the slag toward the end of the C oxidation period. Since the oxygen is consumed by reaction as it rises through the metal, the amount of Fe lost in the slag is less than with top blowing, particularly at very low C contents. Thus, the minimum P content obtainable with a constant volume ratio of slag/metal is correspondingly decreased. Less heat is available for melting scrap metal in the charge, and hydrogen is absorbed by the metal from the hydrocarbon shroud gas, requiring additional treatment for its removal. It is also more difficult to halt the refining at the appropriate stage if a higher C content of, say, 0.2 to 0.4% is required in the steel.

5.3.2.3. Combined Blowing

Today, the separate advantages of top and bottom blowing are frequently combined in one vessel. Oxygen is jetted onto the surface of the melt to obtain rapid lime solution and the formation of a basic, foaming slag to aid P removal; but up to 10% of the oxygen is blown through basal tuyeres. CO_2 is sometimes substituted for the bottom blow to reduce the amount of heat generated at the tuyeres and avoid the need for a hydrocarbon shroud.

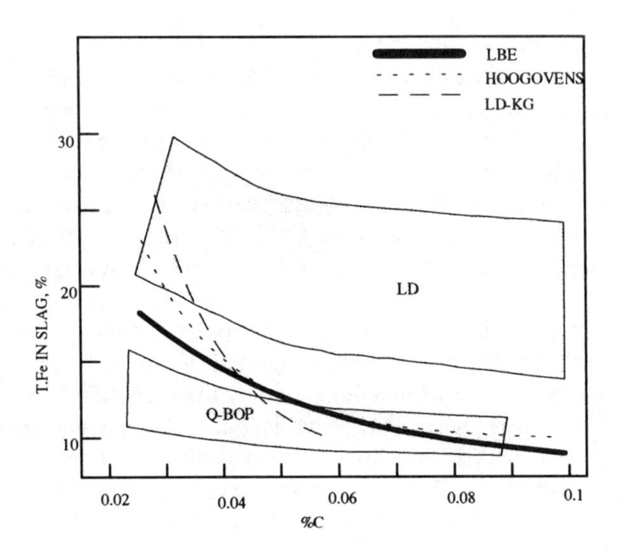

FIGURE 5.9. Effect of blowing method on the relation between %C in iron and the total iron oxides in the converter slag. LBE, LG-KG and Hoogovens are combined blowing processes. (From Singh, R. P. and Ghosh, D. N., *Ironmaking and Steelmaking,* 17, 333, 1990. With permission.)

Since CO_2 is not stable in the presence of C at elevated temperatures, it acts as an endothermic source of oxygen in the metal. Alternatively, an inert gas (argon) is injected through the tuyeres to increase bath agitation, the amount supplied increasing as the C content falls to low levels. However, argon is expensive and the refining costs rise in line with the increase in gas consumption, so nitrogen may be substituted in the early stages of the blow and argon used only in the later stages when the nitrogen content of the steel is not too critical. At constant carbon concentration, the amount of iron lost in the slag with combined blowing is greater than when all the oxygen is admitted from the bottom of the vessel, but is less than with only top blowing, Figure 5.9. Since, from Equation 5.12, the equilibrium oxygen activity in the metal is determined by a_{FeO} in the slag, it follows that the metal contains less oxygen at the end of refining with bottom blowing or combined blowing practice, and a smaller quantity of deoxidants are required to finish the refining than are needed when all the oxygen is top blown (see later).

A modern converter can refine 400 t blast furnace metal in about 35 min, giving a production rate of about 700 t h^{-1}. No other refining process approaches this rate of production. The final metal temperature can be held to within ±10 °C of the target and the final C content can be controlled within very narrow limits at any required concentration.

5.3.2.4. The Arc Furnace Process

The electric arc furnace is primarily used for refining a solid charge of recycled scrap metal, and the time taken to melt the charge results in a much lower rate of production (typically 30 to 40 t h^{-1}) than is obtained with the converter. The oxygen required for refining is supplied through a lance after the solid charge is submerged below a layer of molten metal, so melting and refining proceed simultaneously.

Very high temperatures are generated in the arc plasma and volatile species are very effectively removed from the metal. However, a scrap steel charge has a much lower carbon content than blast furnace metal, so a smaller volume of CO is released. The CO bubbles scavenge nitrogen dissolved in the Fe as they rise through the bath. The smaller volume of gas released from the lower carbon content in the charge results in a higher residual nitrogen content in steel refined in the arc furnace (about 0.007%) than in steel produced from blast furnace metal in the converter (0.003 to 0.005%). The smaller gas volume also decreases the tendency of the slag to form a foam in the arc furnace, and dephosphorization proceeds primarily by transfer across the slag-metal interface, so the reaction rate is lower. The lower volume of CO bubbles also results in less intensive stirring of the molten bath and extends the refining time. This can be countered by admitting a stream of small gas bubbles of argon (or nitrogen when the nitrogen content of the steel is not critical) through a submerged diffusion bubbler. The slag composition is usually adjusted to encourage some foam formation to cause the slag to rise above the electric arcs and act as an insulating blanket that prevents overheating of the furnace roof.

The rate of oxidation is more readily controlled in the arc furnace and this is an advantage when the loss of expensive alloy elements must be restricted. It is also possible to produce the reducing conditions that are more conducive to sulfur removal by removing the slag when the oxidizing reactions have been completed and forming a new basic slag containing a reducing agent, such as ferrosilicon or calcium carbide, to remove any remaining FeO. The reaction rate is very slow, however, in the absence of CO gas agitation, and it is necessary to induce turbulence by bubbling an inert gas or by electromagnetic induction to accelerate S transfer from metal to slag.

Whereas the BOS converter derives all the energy required from the sensible heat in the molten metal charge and from the exothermic oxidation reactions, the electric arc furnace melting cold scrap metal requires a large electrical energy input. Since the efficiency of energy conversion to produce electricity is less than 40%, the total energy consumption of the EAF process is very large. The electricity demand can be reduced by using high-intensity oxy-fuel burners, inserted through the furnace roof, to distribute the heat more uniformly over the entire surface area of the bath and accelerate the melting of the charge. Preheating the cold scrap charge by transferring sensible heat from the hot gas leaving the furnace also reduces the energy requirement in the furnace. The modern, three-phase AC electric arc furnace with burner-assisted melting consumes 350 to 400 kWh electrical energy per ton of refined metal, together with the equivalent of 40 to 50 kWh t^{-1} energy supplied by the burners. The oxidation reactions supply only about 25% of the net energy requirements. The energy demand can be reduced by using some of the sensible heat in the exhaust gas from the furnace to preheat the scrap metal charge. A recent development is the direct current (DC) arc furnace, which has one electrode above the bath and one incorporated in the hearth below the bath. The strong magnetic field that is then developed through the molten metal results in strong agitation and rapid mass transport in the bath. The electrical power requirement is reduced by up to 10% by DC operation, but problems arise in maintaining the furnace lining around the bottom electrode.

5.3.2.5. Ladle Refining

The electric arc furnace is energetically most efficient when full power is applied to raise the temperature and melt the charge. The electrical efficiency of the transformers, etc. falls when the power input is reduced to maintain the temperature constant during refining. Hence, there is a growing trend to use the furnace primarily for melting and preliminary refining. When melting is complete, the slag is separated and the metal is transferred to a ladle in which the refining reactions are completed and any alloy additions are made. The ladle may be heated by a separate arc furnace hood or by induction, and the metal may be circulated by induction. The metal can be tapped from the arc furnace at a lower temperature when the ladle is heated, thus allowing more P to be held in the preliminary refining slag and lowering the amount that remains to be removed in the ladle.

The sulfur content of the metal can be reduced to below 0.005% by forming a suitable slag in the ladle. If the desulfurization reaction is written as:

$$S_{(metal)} + O^{2-}_{(slag)} = O_{(metal)} + S^{2-}_{(slag)} \tag{5.23}$$

then the equilibrium constant for the reaction can be expressed as:

$$K \cdot \frac{(a_{O^{2-}})[h_S]}{(\gamma_S)} = [a_O] \cdot \frac{\%S_{(slag)}}{\%S_{(metal)}} \tag{5.24}$$

where h_S and γ_S represent the activity coefficients of the sulfur in the metal and in the slag, respectively. The left-hand side of this equation is termed the *sulfide capacity*, C_S, of the slag.

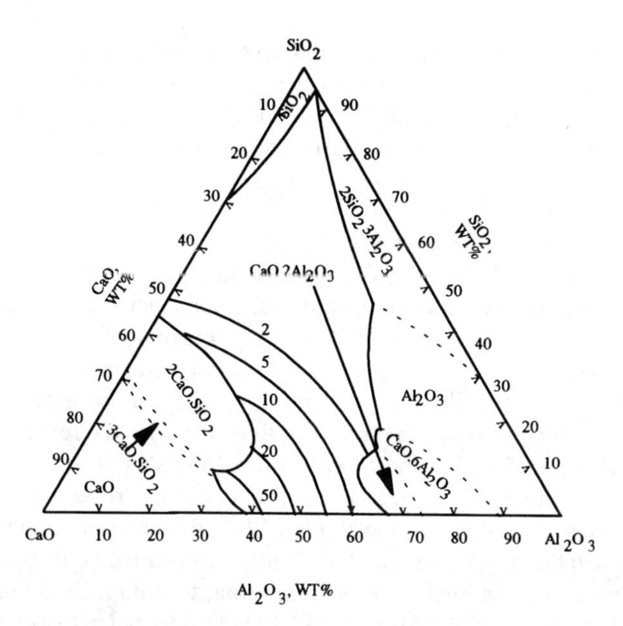

FIGURE 5.10. The sulfide capacity of lime-alumina-silica slags at 1600 °C. (From Ohnishi, T., *Secondary Steelmaking for Product Development,* The Institute of Metals, Petten, The Netherlands, 1985, 139. With permission.)

The sulfide capacities of slags in the $CaO-Al_2O_3-SiO_2$ ternary system are shown in Figure 5.10. At constant lime content, C_S is increased as Al_2O_3 is substituted for SiO_2 in these melts and the highest values of C_S are found along the $CaO-Al_2O_3$ binary edge. However, the melting temperatures of the binary melts are high and the molten slags are too viscous at typical ladle refining temperatures of 1600 °C, so calcium fluoride, CaF_2, is added as a flux. If all the oxidizing slag is removed during the transfer of the metal from the furnace to the refining ladle, a calcium aluminate slag can be formed in the ladle that is virtually free from FeO such that a_O is very low and the conditions are ideal for sulfur removal.

A typical slag might contain 50% CaO, 15% CaF_2 and 35% Al_2O_3. Lime-based powders are also injected into the metal to aid the reaction. Residual S levels of less than 5 ppm can be realized when CaSi or CaC_2 is injected.

The phosphorus content can also be lowered to 0.005% under a slag containing, say, 65% CaO, 20% CaF_2 and 15% FeO. Both slags have a near-maximum oxygen ion activity, and the essential difference between them is the oxidizing potential. The concentration of both impurities in the melt can be lowered by first forming a slag suitable for dephosphorization and then, when the concentration of P has been lowered sufficiently, decanting the slag completely before forming a new slag suitable for the removal of S. Argon is usually admitted for 2 to 3 min through porous plugs in the base of the vessel to improve temperature and composition control when the refining reactions have been completed. Care must be exercised to avoid the introduction of S and/or P into the metal from the additions when alloy elements are added to the steel after the ladle treatment.

5.3.2.6. Reversion From Slag to Metal

When refining is finished, a slag layer is left on top of the metal to reduce the rate of contamination from the atmosphere and act as a thermal blanket while the steel is poured from the ladle. It can take up to 30 min to empty the ladle when the metal is poured into a continuous casting machine. During that time, the slag may absorb oxygen from the atmosphere and oxygen may transfer from slag to metal. Reaction with the refractory lining of the ladle may also alter the composition of the slag. These changes can lower the stability of the solutes in the slag and cause reversion of S and P from the slag to the metal as the slag/metal ratio increases during the emptying of the ladle. Care is required, therefore, to ensure that the composition of the slag cover is adjusted, prior to pouring, to minimize the possibility of reversion occurring. This is particularly important when the specification calls for a maximum concentration of only a few parts per million of these elements.

5.4. METAL-METAL EQUILIBRIA

The refining reactions considered thus far exploit the differences in the attraction between oxygen and the various elements present in the melt, relative to the attraction between the solvent and the solutes. The slag acts primarily as a means of lowering the activities of the oxidation products and thereby increases the differences in the tendencies to associate with oxygen. In some cases, a solute can be separated from the solvent without the necessity of forming a slag to change the activities of the products.

5.4.1. Liquation and Drossing

5.4.1.1. Restricted Solubility

When the solvent metal melts at a very much lower temperature than a solute, or when there is extensive liquid immiscibility between the two elements, it is sometimes found that the solubility of the solute in the melt decreases to very low values as the temperature is lowered toward the melting point of the solvent. Some examples of this behavior in binary systems are shown in Figure 5.11. These variations with temperature in liquid solubility can be exploited to achieve separation because the activity of the solute in the melt is raised as the solubility decreases.

In binary mixtures of Pb and Zn, the solubility of Zn in molten Pb is 2% at 417 °C. So, a_{Zn} is close to unity at the saturation surface and falls rapidly

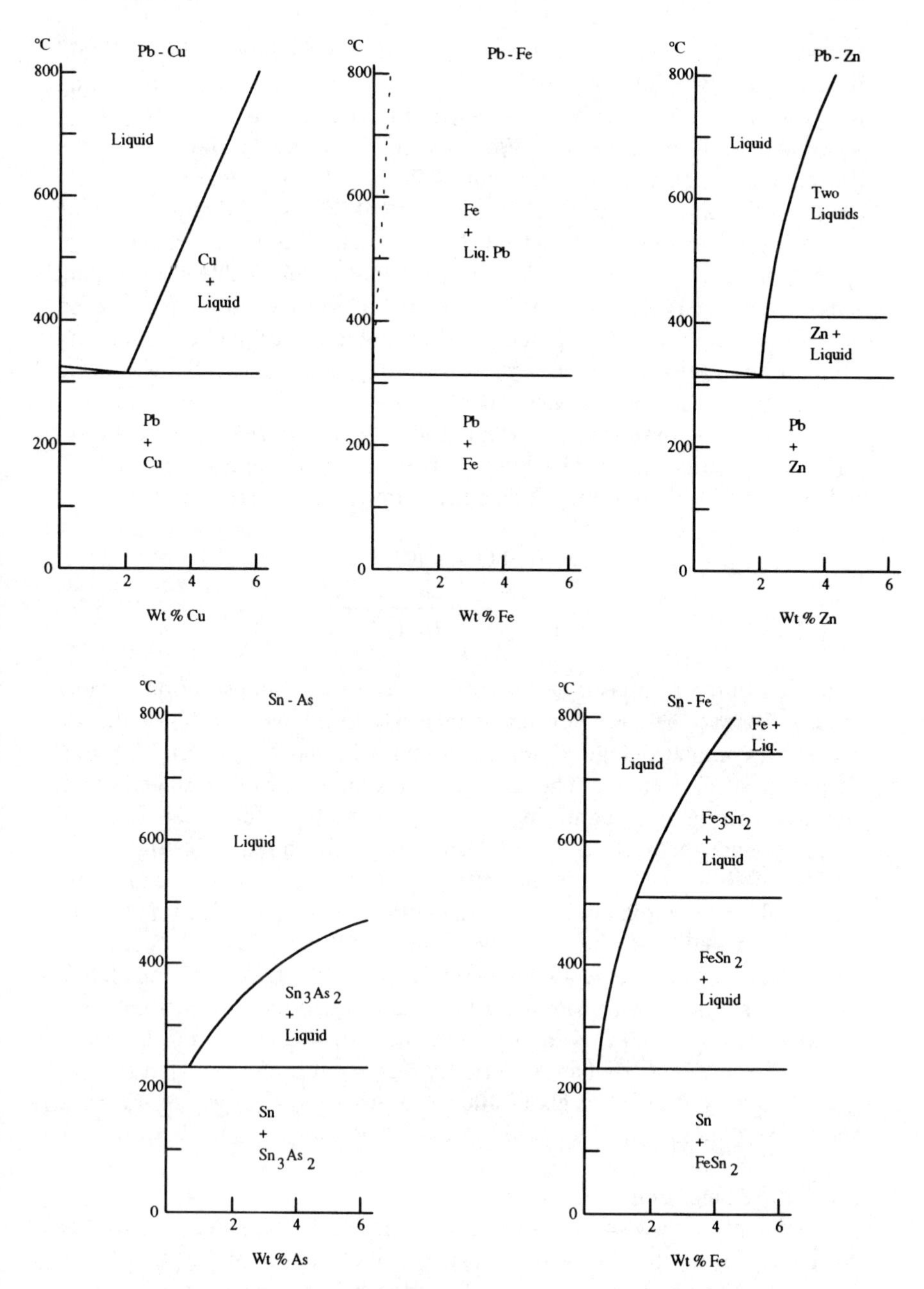

FIGURE 5.11. The solubility of solutes in some binary metallic systems.

to zero as the composition changes across the narrow homogeneous liquid field to pure Pb. The solubility of Zn continues to decrease with falling temperature to about 1.5% at the eutectic temperature of 318.2 °C. If the solubility limit for Zn in liquid Pb is exceeded during cooling, the excess Zn, containing only a small amount of Pb in solution, separates out as a phase having a lower density than Pb and floats on top of the melt. This is called **liquation** and is the basis of the process used to recover Zn when it is condensed from the vapor phase in the Zn blast furnace process. In that application, the Pb is not cooled below 417 °C in order to recover the Zn as a liquid phase. In refining, the metal is cooled to just above the melting point of Pb to achieve further separation as the temperature falls, with solid Zn separating between 417 and 318 °C.

The large increase in a_{Zn} in the liquid Pb, as a consequence of the restricted solubility, raises the attraction between Zn and oxygen atoms, relative to the Pb-O attraction. The equilibrium position in the reaction:

$$Zn + PbO = ZnO + Pb \tag{5.25}$$

$$K = \frac{(a_{ZnO})[a_{Pb}]}{(a_{PbO})[a_{Zn}]} \tag{5.26}$$

is moved to the right as a_{Zn} is increased, so Zn in dilute solution is more readily removed as an oxide phase than would be the case if Pb and Zn formed a complete range of liquid solutions in which the activities conformed to ideal behavior. The removal of a solute showing high activity at low concentrations by admitting small amounts of oxygen as air, steam or as compounds that dissociate to release oxygen, is called **drossing.**

The solubility of Fe in molten Pb is very restricted and Fe can be removed almost completely by liquation without need for oxidation. Cu can also be removed from Pb by liquation, the solubility of Cu in Pb decreasing to 0.2% at the binary eutectic temperature of 350 °C. Liquation can also be used to separate a metal with a very low melting point from one that has a much higher fusion temperature when the two metals are mechanically mixed. For example, tin can be partially recovered from tin-plated steel by heating the solid metal to about 400 °C, a temperature at which the Sn is molten and drains out of the solid mass.

5.4.1.2. Compound Formation

Frequently, when the liquid phase saturation is exceeded, the molten melt is not in equilibrium with a liquid or solid solution of the pure solute, but with an intermetallic compound of the two elements as, for example, in binary mixtures of Sn with Fe or As. When the solubility limit is exceeded, the excess solute then separates out as the compound, often containing more solvent than solute atoms. The activity of the solute is less than unity in the compounds, but its activity in the melt at compositions close to the saturation limit is higher than for an ideal solution, and separation by precipitation is still possible.

Sometimes, however, a solute can be removed more effectively by adding another metal with which the solute forms a more stable compound. The metal added for this purpose must show either negligible solubility or must be easily removed from solution in the host metal. For example, arsenic and antimony form simple eutectic systems with Pb, so they would be expected to show positive deviations from ideality in the liquid. However, they form compounds with Fe, and a negative deviation occurs when these elements are mixed with Fe. Correspondingly, when Fe is present in the dross, As and Sb partition selectively into the Fe to form a separate phase of arsenate and antimonate compounds called a **speiss.** These impurities can be removed more effectively from the Pb by adding NaOH and $NaNO_3$ at about 450 °C to form sodium arsenate and antimonate compounds (**Harris process**). Sn is removed simultaneously as a stannate compound:

$$5Sn + 6NaOH + 4NaNO_3 = 5Na_2SnO_3 + 3H_2O + 2N_2 \quad (5.27)$$

Both As and Sb form compounds with Sn in the solid state and hence are expected to show negative deviation from ideality, but Al forms even more stable compounds with these impurities. If finely divided Al is stirred into the molten tin, the Al compounds are formed and float to the surface, from where they can be skimmed off. The excess Al is then removed from the Sn by preferential oxidation.

Silver and gold in Pb and Pb-Zn ores are transferred almost completely into the Pb during smelting and they can be recovered from the Pb in a similar manner. The Pb-Ag binary system is a simple eutectic, but Ag forms a series of intermetallic compounds with Zn. When excess finely divided Zn is stirred into the melt at about 480 °C, the Ag partitions preferentially into the Zn to form Ag_2Zn_3 (m.p. 664 °C) as the Pb is cooled slowly to just above the melting point (**Parkes process**). After separation of the dross, the excess Zn is separated by liquation, oxidation or by the formation of $ZnCl_2$ (see Worked Examples 8 and 9). The dross removed from the melt contains only Ag and Zn in a large amount of Pb if the other impurities are removed before the Zn is added. The Ag is recovered from the dross by liquation under a chloride flux (to minimize oxidation) to remove most of the Pb before the Zn is removed by distillation. The remaining Pb is then rejected by oxidation (**cupellation**) to leave a residue of Ag with a purity of over 99.9%.

The attraction between oxygen and Cu atoms is considerably weaker (i.e., ΔG° for oxide formation is less negative) than the attraction between oxygen and Pb or Sn. Consequently, these elements and others showing even stronger metal-oxygen bonding can be refined almost completely from Cu by oxidation to form an oxide dross. However, the removal of Cu from

Pb cannot be achieved by oxidation. Furthermore, Pb and Sn are completely miscible in the liquid state and show similar attraction for oxygen; thus Pb cannot be removed from Sn or Sn from Pb without severe loss of the host metal. So, recourse must be made again to association with other species with which the required partitioning can be achieved.

The bonding of S atoms with Cu is stronger than with Pb, and the residual Cu left in the melt after liquation can be removed by adding sulfur as PbS or by bubbling a sulfur gas through the melt to form a dross of Cu_2S (see Worked Example 10). Similarly, chlorine is more strongly bonded to Pb than to Sn, and Pb can be removed from Sn by injecting a Cl_2 gas into the melt or by adding $SnCl_2$:

$$SnCl_2 + Pb = Sn + PbCl_2 \tag{5.28}$$

This is similar to the technique that can be used for removing Zn as a chloride from Pb, referred to above. In that case, the chlorine is more strongly bonded to Zn than to Pb and this demonstrates again the importance of the relative stability of compounds in refining.

Bismuth is another impurity often present in Pb and Sn ores and which cannot be removed by selective oxidation. The only effective way for complete removal of this impurity is electrorefining, but the concentration can be lowered sufficiently for some applications by stirring Ca and Mg metal into the melt (**Kroll-Betterton process**). These metals are not attracted by Pb or Sn, but form insoluble compounds with Bi that separate in a dross. The Ca or Mg are added when the metal is about 150 °C above the melting point and the compound separates as the melt is slowly cooled down toward the melting point. The excess Mg and Ca can be removed from the melt by chlorination and oxidation, respectively.

These examples illustrate how knowledge of solubility-temperature relations, compound stabilities, etc. can be exploited to aid refining. Only a small amount of scum is formed on the melt during drossing operations and it occupies a much smaller volume than a refining slag. However, the scum often contains large amounts of the host metal. In some cases, it consists primarily of the oxide of that metal in which the impurities are held in fairly dilute solution. It is often economically profitable either to recycle the scum through the earlier stages of the extraction process to recover the host metal or to use other specialized treatments to recover both the host metal and the other metallic elements.

5.4.2. Deoxidation

5.4.2.1. Oxygen Solubility in Metals

Oxygen is slightly soluble in most molten metals and, in general, the solubility is raised as the concentration of other impurities is lowered during refining. Only very small amounts dissolve in molten Pb, Sn and Zn (e.g., about 0.02% just above the melting point of Pb) and this does not seriously affect the properties of the metal. Larger amounts can dissolve in some other metals. The solubility is 0.16% in pure molten Fe at the melting point, but it decreases rapidly when small amounts of carbon remain in the metal, falling to about 0.05% at 0.05% C and continuing to fall at a decreasing rate as the C content is increased further. The actual amount in solution may be much higher if equilibrium is not attained during refining due to the difficulty of nucleating CO bubbles in the melt. Molten Cu can absorb up to 4% O when the metal is oxidized to remove the last traces of sulfur, but the solubility again falls rapidly with increasing residual S content.

The solubility of oxygen in a metal decreases sharply on solidification and continues to drop as the metal is cooled to values of only a few parts per million at room temperature. In pure metals, the oxygen rejected from solution on cooling forms an oxide of the metal. When the surface tension is high between the metal and the oxide, the oxide tends to adopt a globular form. This is less detrimental than when, as with Fe, the surface tension is low and the oxide readily forms grain boundary films that cause severe loss in toughness and ductility.

The globular oxides formed in pure Cu are less detrimental to the mechanical properties, but hydrogen introduced into the solid lattice during bright annealing of Cu can readily diffuse to the oxide particles, where it can react to produce steam if the temperature is high enough. The H_2O molecule is too large to escape by diffusion through the solid lattice. Thus, a pressure builds up at the reaction site that may be sufficient to cause blister formation on the surface from the reaction of hydrogen in solid solution with subcutaneous oxide particles. For these reasons, one of the final stages in metal refining is deoxidation to lower the residual oxygen content sufficiently to avoid problems in subsequent fabrication and use.

5.4.2.2. Selection of Deoxidizers

Any element can be used as a deoxidizer that shows a higher affinity for oxygen at the reaction temperature than does the metal being deoxidized, the efficiency of the deoxidizer increasing as the difference in the affinities increases. In this context, efficiency can be regarded as the equilibrium atom percent of the deoxidizer remaining in the metal when the chosen residual concentration of oxygen in the metal has been attained.

Hydrogen can function as a deoxidizer for some metals by combining with oxygen to form steam. In the traditional process, Cu was deoxidized by "poling" the metal with freshly cut, green timber that released hydrogen when immersed in the hot metal. In modern practice, the hydrogen is provided by bubbling ammonia or a hydrocarbon gas through the metal. These treatments lower the oxygen content to about 0.05%, which is low enough for many applications. In this condition, the metal is marketed as tough pitch Cu, since the metal is very ductile and tough in comparison with oxygen-saturated Cu.

Hydrogen could be used to remove more oxygen from the metal, but large volumes of gas would be required and the solubility of H_2 in Cu increases rapidly at very low oxygen contents. Thus, an alternative reagent is used to remove the remaining oxygen when lower residual levels are required. P_2O_5 is markedly more stable than Cu_2O, and Cu can be deoxidized very effectively by the addition of a relatively cheap Cu-P alloy containing about 15% P. The residual level of P in the metal is around 0.02%. This has little effect on the mechanical properties, but does adversely affect the electrical conductivity of the metal, so lithium is used as the deoxidizer when the metal is destined for high conductivity applications.

Hydrogen is not used to deoxidize iron because the residual hydrogen that would be left in solution would cause severe embrittlement, but a variety of other elements can be used for this purpose. Al, Ca, Mn or Si are most commonly selected, the residual equilibrium concentration of the added element decreasing in the order Mn > Si > Al > Ca. Thus Ca is the most powerful deoxidizer of this group and, if a sufficient amount is added under a suitable slag cover, both S and O contents of the metal can be lowered simultaneously to less than 10 ppm. But it is also the most expensive element in this group and the treatment cost is lowered by preliminary deoxidation, usually with Al, to remove most of the oxygen before the Ca is introduced. It is important that the vessel is lined with stable refractories such as MgO or dolomite when the oxygen content of the metal is lowered to only a few parts per million, since less stable refractories such as fireclay may be reduced by the metal at very low oxygen potentials and feed oxygen back into the metal (see Worked Example 11).

The deoxidizer is added to the bath either in granular form or as a continuous wire feed. It dissolves rapidly in the metal, producing numerous small volumes wherein the transient enrichment of the concentration of the added element is sufficient to overcome the nucleation barrier (Equation 3.39) and a shower is formed of fine precipitates of the oxide of the added element.

If a pure solid oxide is precipitated, its activity is unity. When iron is deoxidized, the surface energy term (γ) in the nucleation equation is the energy of the interface between molten Fe and the solid oxide. This is higher when Al_2O_3 or CaO are formed than for MnO and SiO_2 reaction products,[95] but the difference is swamped by the much larger value of ΔG° for the formation of CaO and Al_2O_3, and a finer dispersion of precipitates is formed when the metal is deoxidized with Al or Ca.

The relationship between the residual amounts of oxygen and the deoxidizer when equilibrium is established can be calculated from the appropriate form (i.e., solid or liquid reaction product) of the equilibrium

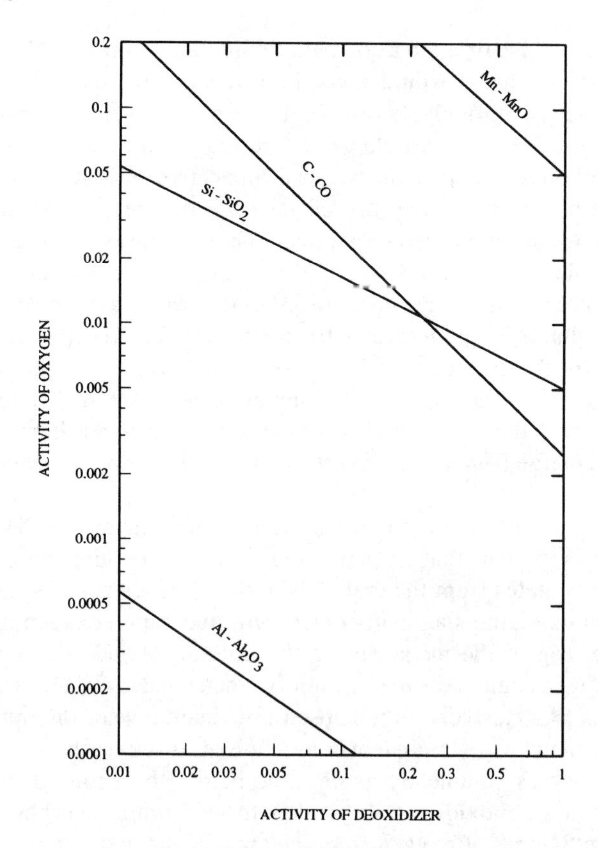

FIGURE 5.12. Equilibrium relations for the deoxidation of molten iron at 1600 °C.

constants given by Equations 5.7 and 5.8 for Mn and Si oxidation and similar equations for the other deoxidizers. The results obtained are shown in Figure 5.12. From inspection of the temperature dependence of the equilibrium constants, it is evident that the lines move downward and to the left as the metal temperature decreases, indicating that deoxidation continues as the metal is cooled toward the solidification temperature. This shift in the equilibrium position with falling temperature may be accommodated by continued growth of the precipitates formed at higher temperature or by nucleation of additional precipitates.

5.4.2.3. Removal of Deoxidation Products

Deoxidation converts the dissolved oxygen into a less innocuous form, but the mechanical properties and the formability of the metal are still adversely affected if the deoxidation products remain suspended in the metal. The oxides have a lower density, however, than the metal and tend to rise to the melt surface. The rate of escape to the surface can be approximated from Stokes' Law (Equation 1.3) if it is assumed that the oxides are spherical and rise through a stagnant liquid. Substitution of the appropriate data indicates that it would take about 6 min for a particle 0.02 mm in diameter to rise through 300 mm in the furnace bath, for example, or about 1 h for the same size particle to rise through 3 m in a ladle. But, Stokes' Law is not directly applicable because convective currents cause circulation in the bath even in the absence of gas evolution or induced agitation and the particles are not always spherical. Thus, alumina forms as needles or plates, which rise less rapidly than predicted by Stokes' Law, but the individual precipitates may cluster together to form an aggregate with a larger effective diameter and increase the rate of rise. The escape velocity is also affected by the surface tension between the deoxidation product and the molten metal.[96] On arrival at the metal surface, the inclusion should be absorbed into the slag, but if the surface tension is not high, the particle will not emerge from the melt surface and may continue to circulate within the metal.

Although Stokes' Law gives only a rough estimate of the escape velocity, it is evident that long holding times are required to remove very small precipitates from the melt. The cost increases with the holding time and, if an oxidizing slag is in contact with the metal, oxygen may transfer from the slag to the metal during the holding period. Consequently, the metal oxygen content decreases rapidly when the deoxidizers are added and then rises progressively with elapsed time back toward the initial level. If very low metal oxygen contents are required, it is necessary to remove the refining slag as completely as possible before the strong deoxidizers are added. A new nonoxidizing slag is then formed, which acts both as a barrier to prevent ingress of oxygen from the atmosphere and as a receiver for the deoxidation products.

When the deoxidizing elements are added individually, the product is usually the oxide of that element at unit activity. When two or more deoxidizing elements are added together, however, the reaction product is a mixture of oxides, which has two important effects. When Mn and Si are added together, the product is a mixture of FeO, MnO and SiO_2 in which the activities of each of the oxides is very much lower than unity. The equilibrium position in the deoxidation reactions, shown in Figure 5.12, is then shifted downward and to the left, lowering the amount of oxygen remaining in solution for any chosen residual concentrations of the deoxidizers. Deoxidation is thus more efficient than when either element is added alone. Similarly, the effectiveness of Al as a deoxidizer is enhanced by first adding Si and/or Mn to form a complex deoxidation product.

The reaction product may be molten at the refining temperature when two or more deoxidants are added together. Surface tension can cause the coalescence of the molten products if they come into contact with each other, increasing the diameter and hence increasing also the rate of rise in the metal bath and escape into the slag cover. Less inclusions then remain to be entrapped as the metal solidifies. When Ca and Al are added together, the calcium aluminate product is less abrasive than alumina inclusions on the cutting tools during the machining of the solid metal. If the steel contains much sulfur, the calcium aluminate is surrounded by a softer layer of Ca-Mn-S.

5.5 METAL-GAS REACTIONS

5.5.1. Volatilization

Condensed substances exert a gaseous or vapor pressure in the surrounding atmosphere that increases with rising temperature until the substance becomes completely gasified by boiling or sublimation. Equilibrium at constant temperature below the boiling point is established when the rate at which atoms or molecules transferring from the condensed species into the gas phase is equal to the rate of transfer in the opposite direction. The rate of approach to equilibrium is affected by mass transport in the gas and in the condensed substance, by adsorption, desorption, etc. (see Equations 3.26 to 3.31).

5.5.1.1. Vapor Pressures

The vapor pressure exerted by most metals is very low in the solid state, but it increases rapidly between the melting and boiling points. The temperature interval from the melting point to the boiling point varies markedly from one metal to another. This interval is only about 400 to 500 °C for some metals with low melting temperatures, such as K, Na, Mg, and Zn. However, the temperature range is not a consequence of a low melting point, for it is 1413 °C for Pb, 1740 °C for Al and 2500 °C for Sn. The reason for this behavior can be explained by considering the volatilization of some species, M:

$$\begin{aligned} M_{(liquid)} &= M_{(gas)} \\ \Delta G^{\ominus} &= -RT \ln K \\ &= -RT \ln p_M \end{aligned} \tag{5.29}$$

if $a_{M(liquid)} = 1$. Therefore,

$$\ln p_M = -\frac{\Delta G^{\ominus}}{RT} = -\frac{\Delta H^{\ominus}}{RT} + \frac{\Delta S^{\ominus}}{R} \tag{5.30}$$

However, the entropy of vaporization is approximately 92 J $K^{-1}mol^{-1}$ for all substances (**Trouton's rule**) and can be regarded as a constant. Hence,

$$\ln p_M = -\frac{L_e}{RT} + C \tag{5.31}$$

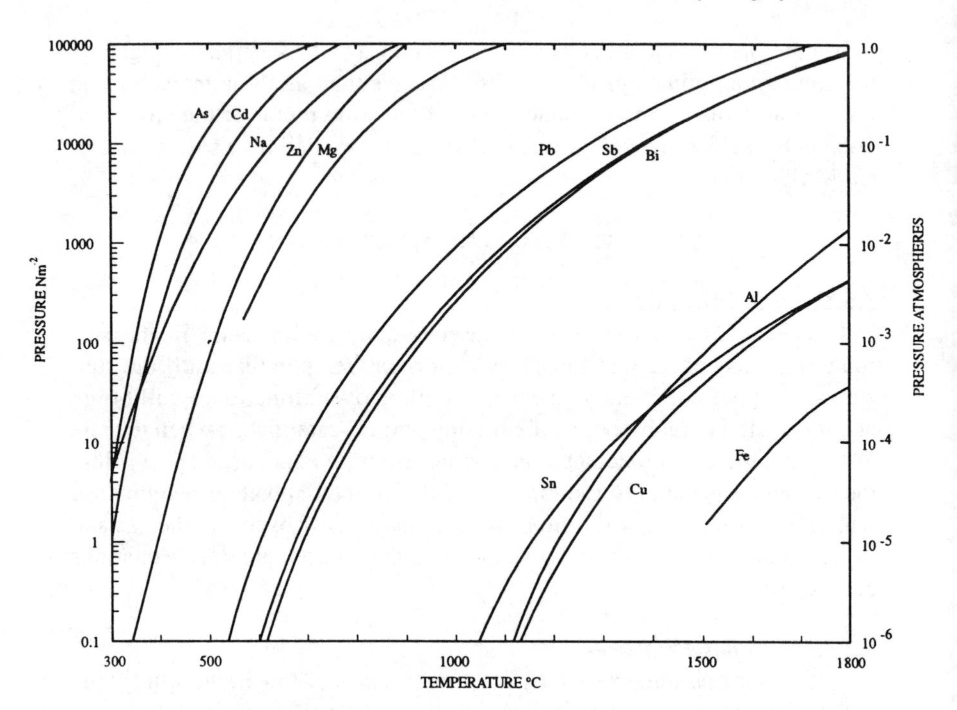

FIGURE 5.13. The variation with temperature of the vapor pressure of some pure elements.

where L_e is the latent heat of evaporation and C is a constant. The differences in volatilization temperatures can thus be attributed to L_e, which tends to increase with increasing Group Number of the elements in the Periodic Table (i.e., L_e is much larger for transition series elements than for alkalies and alkaline earths). Differences in the melting points of metals are similarly dependent on the latent heats of fusion, but the latter vary less systematically with position in the Periodic Table.

The differences in the temperature interval between the melting and the boiling points result in marked differences in the vapor pressures exerted by the metals at, say, 100 °C above their melting points and hence in the rates of evaporation from the melt. The temperature dependence of the vapor pressure of some pure metals is shown in Figure 5.13. From inspection of this diagram, it might be assumed that all of the elements that lie on the left-hand side could be removed completely by volatilization from the metals that lie further to the right and exhibit relatively low vapor pressures in the relevant temperature range. This assumption would be erroneous. Assuming ideal behavior of the gas, the activity of an element in a liquid solution, relative to the pure substance standard state, is defined by the relation:

$$a_M = \frac{p_M}{p_M^0} \tag{5.32}$$

where p_M^0 the vapor pressure of the pure element at the temperature considered. If the solute exhibits ideal behavior in the melt, then, at an atom fraction concentration of 0.001 (corresponding to 0.1 atom%), the vapor pressure of the solute above the melt is 10^3 smaller than p_M^0. The maximum concentration of many impurities tolerated in a metal is often much less than 0.1 atom percent, and the vapor pressure is correspondingly lowered even further at the acceptable residual level. Deviations from ideal behavior in the melt also modify the vapor pressure of the solute. Thus, As and Sb form solid compounds with Fe, their activities show negative deviation from ideality in the melt and it is difficult to remove these elements completely from Fe by volatilization. Conversely, Bi, Mg, Pb and Zn show positive deviations from ideal behavior and are more readily removed from Fe.

5.5.1.2. Distillation from Melts

From the kinetic theory of gases, Langmuir derived an expression for the mass of vapor molecules, ω_i, of a species, i, striking unit area of surface in unit time:

$$\omega_i = p_i \left(\frac{M_i}{2\pi RT} \right)^{1/2} \tag{5.33}$$

where M_i is the atomic or molecular weight of the species in the gas phase. Substituting Equation 5.32 for p_i:

$$\omega_i = a_i p_i^0 \left(\frac{M_i}{2\pi RT} \right)^{1/2} \tag{5.34}$$

At equilibrium, the rate of condensation is equal to the rate of evaporation, so that this equation also gives the rate at which atoms or molecules are transferred from the melt to the gas. Hence, if two species, A and B, can volatilize at a given temperature, the relative rates of volatilization are given by:

$$\frac{\omega_A}{\omega_B} = \frac{a_A p_A^0 (M_A)^{1/2}}{a_B p_B^0 (M_B)^{1/2}} \tag{5.35}$$

If A and B represent the solute and the solvent, respectively, the extent to which the solute can be removed without excessive loss of the solvent increases as the ratio ω_A/ω_B is increased. This is the basis of refining by distillation. By careful selection of the distillation temperature, most of the more volatile species can be removed while retaining most of the solvent in the condensed form (fractional distillation). Or, the melt may be completely volatilized and the elements separated by fractional condensation of the gas at suitable temperatures. This technique is used, for example, in the separation of impurities from volatile halides. Repeated distillation may be necessary to obtain sufficient purity. Another example of this technique is found in the refining of Zn. Lead and cadmium are the major impurities present in Zn produced in the blast furnace. p^0_{pb} is much lower and p^0_{Cd} is much larger than p^0_{Zn} at any given temperature. Thus, when the impure Zn is completely distilled in a refluxing unit, Cd is also distilled, but, by controlling the maximum temperature, Pb can be retained as a liquid that can be drained off. Most of the Cd remains in the vapor phase if the vapor is then cooled just sufficiently to condense the Zn. Complete removal of the impurities by this method is difficult, and higher purity Zn is produced more effectively by electrolytic refining.

The rate of removal of a solute by volatilization can be increased by lowering the pressure of the gas phase above the melt and, in vacuum melting, the pressure above the melt is readily lowered to about 10^{-3} atm. The rate can also be increased by providing a cooler surface onto which the evaporated species can condense, thus lowering the partial pressure of that species in the gas and decreasing the rate of return of gas molecules to the metal surface.

Several factors can lower the rate of removal of the impurities. Surface-active solutes absorbed onto the surface of the melt reduce the number of sites available for occupancy by the volatile species, while complete coverage of the surface by an oxide film or by a slag layer may reduce the rate of escape almost to zero. In the absence of forced circulation of the melt, the surface layer may be depleted of solute by evaporation more rapidly than it is replenished by mass transport from the bulk melt, and the activity term in Equations 5.32 and 5.34 then relates to the activity of the solute at the concentration existing in the surface layer. A large surface area per unit volume and induced circulation of the melt are essential to obtain reasonable rates of volatilization.

Vacuum melting is expensive. A less costly technique, which is sometimes equally effective, is to bubble an inert gas through the melt. If the bubbles are very small and retained in the melt for sufficient time, the partial pressures of the impurities in each bubble closely approach the equilibrium values at the melt temperature. The bubbles also create turbulence in the bath, reducing or eliminating concentration gradients and accelerating the rate of solute transfer into the gas phase, relative to the rate obtained when the melt is stirred only by convective currents.

Commercially pure inert gases contain traces of oxygen and other gases, however, and these may be transferred from gas to metal during passage through the melt. For example, p_{O_2} in equilibrium with Fe and FeO is about 10^{-9} at 1600 °C and decreases as the oxygen content of the iron is decreased below the saturation limit. A ''pure'' inert gas containing oxygen at 10^{-6} atm pressure is thus oxidizing to molten Fe at this temperature. Since the partial pressure of oxygen in equilibrium with a metal and its oxide decreases with decreasing temperature, the inert gas becomes more oxidizing as the temperature is lowered. Very high purity is required, therefore, in injected gases. Injection of a cold gas also cools the bath and compensating heat must be supplied when large quantities of gas are used. For these reasons, in addition to the cost of the gas, the quantity of gas injected should be just sufficient to effect the required impurity removal, and the bubble size should be as small as practicable to increase the probability that the equilibrium partial pressures are attained in the bubbles.

5.5.1.3. Volatilization from Solids

Since diffusion is very much slower in the solid than in the liquid state, volatilization is not an efficient method for the removal of impurities from a solid metal, although significant amounts of some species (such as As and Sb) may be removed during the sintering of finely divided ores. However, solid-state diffusion is not a problem when the impure metal is completely volatilized. This technique is used, for example, in the carbonyl process for the refining of nickel.

The impure Ni produced in the extraction stage may contain up to 20% Fe, together with variable amounts of Co and Cu, depending on the composition of the ore. Nickel carbonyl is formed when CO gas is passed over the impure metal at a low temperature:

$$Ni + 4CO = Ni(CO)_4 \tag{5.36}$$

and the boiling point of the carbonyl is only 43 °C at a pressure of 1 atm. Fe also forms a carbonyl, $Fe(CO)_5$, but with a higher boiling point temperature of 105 °C. Cobalt carbonyl, $CO_2(CO)_8$, has very low volatility at low temperatures, while Cu and any precious metals present in the impure Ni do not form volatile species under these conditions. Hence, the Ni, together with a little of the Fe, can be volatilized by passing CO over the impure metal at, say, 80 °C, leaving the other impurities as a solid residue.

The formation of nickel carbonyl is only slightly exothermic, but 4 mol gas and 1 mol solid are consumed in the formation of 1 mol of the carbonyl, so the entropy contribution to the free energy of the reaction becomes dominant as the temperature is increased. Hence, if the vapor is transported to a second chamber in which the temperature is raised to about 160 °C, the direction of the reaction is reversed and the Ni carbonyl decomposes to deposit pure Ni. A higher temperature is required to cause reversal of the Fe carbonyl reaction, so the iron remains in the gas phase. Finely divided Ni powder is placed in the decomposition chamber to act as a nucleant. The gas decomposes on the powder surfaces and the particle size increases with time to produce pellets with a purity of 99.99%. The CO gas released from the chamber is cooled to condense the iron carbonyl before recirculation over the impure Ni.

The standard free energy change for Equation 5.36 is low and the value of K is close to unity at the reaction temperature, so the yield (e.g., % CO reacted) is low. The rate of reaction between a gas and a solid is also low at temperatures near to ambient, but the rate increases with rising temperature. More Fe will be volatilized, with increased risk of contamination of the product, if the temperature is raised in the gasification stage while maintaining the total pressure at 1 atm. If the pressure is increased, however, then according to Le Chatelier's principle, the equilibrium in Equation 5.36 is moved to the right and the dissociation temperature is raised. Since 5 mol CO are consumed in the formation of the Fe carbonyl and only 4 mol CO are consumed for the Ni compound, an increase in operating pressure has a greater effect on the dissociation temperature for $Fe(CO)_5$. In practice, the reactor is pressurized to about 20 atm and the carbonyl is dissociated at about 250 °C, leaving almost all of the Fe carbonyl in gaseous form.

Environmental regulations are placing increasingly tighter limits on the amount and composition of pollutants that can be discharged into the atmosphere. When metals are refined by volatilization and the impurities are removed in the exit gas, the volatile species condense as the gas cools. Low temperatures, however, may be required to ensure complete condensation. When carbonyls are present, for example, it may be necessary to cool the gas to ambient temperature before discharge to the atmosphere. Some of the condensate accumulates on cold surfaces in the exit gas ducts. The remainder of the suspended solids are removed in bag filters, electrostatic precipitators, etc. and the residue is often sufficiently rich in metal values to justify processing for their recovery.

5.5.2. Degassing

Most molten metals can dissolve hydrogen, oxygen and nitrogen when in contact with the atmosphere. Oxygen is transferred to the metal from oxidizing slags and hydrogen is introduced when any moist or hydrated compound (e.g., partially hydrated lime) is fed into the bath. The solubility of these species is often only a few parts per million in the solid but is higher in the liquid state. The decrease in solubility on solidification may result in the formation of gas bubbles that can become trapped in the solidifying mass (i.e., blowholes), but most of the excess gas is precipitated as oxide and nitride compounds. Some metals (e.g., Nb, Ta, Ti, V, and Zr) also form hydrides. These precipitates, together with the small amounts of the gases remaining in solid solution, can have a very deleterious effect on the ductility and toughness of the metal. Thus, the final stage of refining is often designed to lower the concentration of the dissolved gases to an acceptable level. Metals refined by electrolysis may also be contaminated by hydrogen deposition at the cathode and require this treatment.

The solubility of a diatomic gas, X, in a metal is given by Sievert's Law, Equation 2.61, which can be expressed as the wt% of the element in solution by using the wt% standard state scale. Assuming conformity to Henry's Law in solution and ideal behavior of the gas:

$$\%X = K(p_{X_2})^{1/2} \tag{5.37}$$

It follows from this that a gas is evolved from the molten metal if the partial pressure of that species above the melt is lowered by exposure to a vacuum. In practice, this is usually an effective way for the removal of hydrogen, since the small hydrogen atom has a very high diffusion rate through the melt, but it is less effective for the removal of the somewhat larger nitrogen and oxygen atoms, which diffuse at a lower rate. In some cases, however (e.g., with Al), the very low levels of hydrogen that can be tolerated necessitate the bubbling of chlorine through the melt, or the addition of a compound that can dissociate to release Cl_2, to remove hydrogen as HCl gas.

A reduction in the pressure above the melt is very useful when dissolved oxygen can react with some other species to escape as a volatile species. For example, the equilibrium position in the reaction between C and O dissolved in molten Fe to form CO is moved to the right when the pressure is lowered, as illustrated in Figure 5.14. If equilibrium is established for this reaction during the oxidizing stage of refining and under a pressure of 1 atm, the metal contains oxygen at an activity of 0.04 when a_C has decreased to 0.05. (At these low activities and in the absence of other solutes, the activities on the wt% standard state scale are approximately equal to the wt% concentrations.) If the partial pressure of the gas above the melt is lowered to, say, 0.1 atm, the reaction is restarted. Since 16 g oxygen combine with every 12 g C removed from the metal to form 1 mol CO, a decrease of 0.01 wt% in the carbon concentration is accompanied by the removal of 0.0133 wt% oxygen. Thus, the activities (and concentrations) of both species are lowered until equilibrium is established at the new value of p_{CO}. The quantity of deoxidizing elements required to combine with the residual oxygen is reduced and the volume of inclusions formed by deoxidation is decreased accordingly. This method can also be used to produce a metal containing only a few parts per million carbon. Very small amounts of iron oxide may then be injected into the melt if the metal oxygen content is not sufficient to combine with all of the carbon.

The true equilibrium in the decarburizing reaction at reduced pressure is not achieved in a static melt because of the difficulty of nucleating bubbles in the melt and, in practice, argon is usually bubbled through the metal in the low pressure chamber to assist the release of the CO. In a similar way, most of the S and O remaining in refined Cu can be removed as SO_2

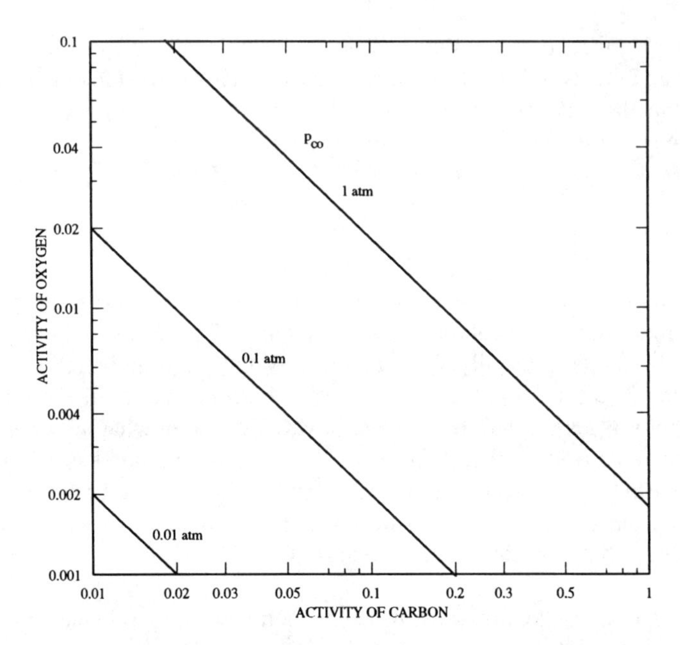

FIGURE 5.14. The effect of the partial pressure of carbon monoxide on the equilibrium between carbon and oxygen in molten iron at 1600 °C.

by melting the metal under vacuum or argon at low pressure,[97] prior to final deoxidation with phosphorus.

A variety of techniques is used in practice for degassing. The melt may be held in an evacuated container, with induced circulation of the melt to transport the gas atoms to the surface. Or, the melt may be poured through an enclosed orifice into an evacuated chamber to facilitate gas escape from the stream of metal as it flows into the chamber (vacuum stream degassing). A large evacuated chamber equipped with expensive vacuum pumps is required for these types of treatment. A less expensive technique is used with the RH type treatment, where the ladle containing the melt is open to the atmosphere. Two hollow legs extend from a small evacuated chamber to below the surface of the melt in an RH unit. Injection of a stream of inert (argon) gas bubbles at the bottom of one leg causes the metal to rise, flow through the evacuated chamber and return down the other leg. The argon flow also stimulates the release of CO or SO_2 gas as described above. The residual oxygen content in molten steel can be lowered from, say, 50 ppm to about 10 ppm by circulation for 10 min through the degasser. The metal is losing heat throughout the treatment, however, and it must be superheated, prior to degassing, to ensure that it is still sufficiently fluid for casting when the treatment is completed.

5.5.3. Melt-Refractory Reactions

The extent to which the oxygen content of the melt can be lowered is limited by the stability of the refractories used to contain the melt. A suitable refractory for lining the furnace or the ladle should have adequate heat and thermal shock resistance to withstand the temperature and the temperature changes encountered in the process. It should have good resistance to attack by the slag and the metal and a low thermal conductivity. The cheapest materials that come nearest to satisfying these requirements for containing the less reactive metals are the common oxides having high melting points. Refining furnaces using acidic slags are normally lined with SiO_2 or mixtures of SiO_2 and Al_2O_3. Lime hydrates too readily for general use, but MgO, dolomite ($CaO \cdot MgO$) and chrome-magnetite ($MgO \cdot Cr_2O_3$) are suitable for use with basic slags.

These oxides can dissociate, however, and supply oxygen to the melt if the partial pressure of oxygen in equilibrium with the melt falls below the level at which the oxides can be reduced, and the elements then dissolve in the metal or escape as volatile species such as SiO or Mg (see Worked Example 11). The probability of a metal-refractory reaction decreases with increasing stability of the refractory oxide. Pure alumina is the most stable of this group, but it is much more expensive than the common refractories and is still susceptible to attack under very high vacuum. Carbon has good heat and thermal shock resistance, but it has a high thermal conductivity. If this can be tolerated, a chilled layer of slag is formed on the surface of the refractory that retards reaction between the carbon and the oxides in the slag. It cannot be used, however, when carbon is soluble in the metal and a low C content is required in the product. Nitride and boride refractories are very expensive in comparison with the common oxides and their use is restricted to containing the very reactive metals.

5.5.3.1. Production of Clean Metals

The presence of some oxide inclusions is unavoidable in the solidified metal produced by normal pyrorefining methods. More expensive techniques must be adopted when a low inclusion content and/or high purity is essential. Thus, reactive and refractory metals can be melted in a water-cooled Cu container (cold hearth melting) using an electron beam or a plasma torch as the heat source.[98,99] Very high temperatures can be attained in the melting zone, increasing the rate of removal of volatile impurities, but a chilled layer of metal prevents any reaction with the Cu container. Inclusions with a density higher than the density of the metal become entrapped in the chilled layer, while low-density inclusions coagulate and float to the upper surface.

In electroslag refining, one end of a cylinder of solid metal is immersed in a molten slag.[100] Electrical resistance heating causes the metal to melt and droplets of molten metal fall through the slag to resolidify below the slag. A low melting temperature and adequate electrical conductivity can be obtained with a slag containing 60 to 80% CaF_2 together with additions of CaO, MgO and Al_2O_3. Inclusions emerging on the surface of the cylinder during melting, or on the droplets of metal, are transferred into the slag. Some inclusions may dissolve in the metal and are then removed by a slag-metal reaction as the droplets pass through the slag. The slag volume is small, however, in comparison with the volume of metal passed through it and the extent to which the metal can be purified, other than by inclusion removal, is very limited.

FURTHER READING

Paul, A., *Chemistry of Glasses,* 2nd ed., Chapman and Hall, London, 1990.

Rawson, H., *Properties and Applications of Glass,* Vol. 3, Glass Science and Technology, Elsevier, Amsterdam, 1981.

Strnad, Z., *Glass-Ceramic Materials,* Vol. 8, Glass Science and Technology, Elsevier, Amsterdam, 1986.

Alcock, C. B., *Principles of Pyrometallurgy,* Academic Press, New York, 1976.

Coudourier, L., Hopkins, D. W., and Wilkomirski, I., *Fundamentals of Metallurgical Processes,* 2nd ed., Pergamon, Oxford, 1985.

Deo, B. and Boom, R., *Fundamentals of Steelmaking,* Prentice-Hall, Englewood Cliffs, NJ, 1992.

Moore, C. and Marshall, R. I., *Steelmaking,* Institute of Metals, 1990.

Moore, J. J., *Chemical Metallurgy,* 2nd ed., Butterworths, London, 1990.

See also References 6–12 and 53.

Chapter 6

HYDROMETALLURY

6.1. OVERVIEW

Many metals will dissolve in acidic and alkaline solutions and the corrosion of metallic artifacts is an unwanted demonstration of this phenomenon. The rate of dissolution is usually very low at ambient temperatures, but rises with increasing temperature and with increasing acidity or alkalinity of the solution in contact with the metal. Wet assay methods for the chemical analysis of metals start by dissolving the solid in an acid and a boiling or near-boiling solution is used to accelerate the reaction. The concentration of each of the elements in the metal is then determined by converting them into discrete chemical compounds that can be titrated or precipitated.

Similar principles are applied in hydrometallurgical extraction of minerals from an ore. Many metallic minerals can be leached (dissolved) in aqueous solutions, and examples were given in Section 1.2.3 of the formation of ore bodies by solution and reprecipitation of minerals as a result of acidified water percolating through the strata. The rate of solution of the mineral again increases with rising temperature and with the acid or alkaline strength of the liquid phase. With suitable selection of the conditions, the gangue and possibly some of the other metallic minerals can be retained as insoluble residue, while those elements that do go into solution can be separated from each other by exploiting the differences in the stabilities of their chemical compounds.

In comparison with the pyrometallurgical route, the capital costs for hydrometallurgical extraction are very low, although the costs escalate when pressurized containers are used to allow operation at temperatures above the boiling point of the aqueous solution at atmospheric pressure. Costs also rise when expensive materials such as titanium or stainless steel are required to line the container to resist attack by very aggressive solvents. The chemicals or **lixiviants** used for the dissolution of the minerals are mostly inexpensive, but a higher cost is incurred for the chemicals that have to be used when the metal is recovered from the leach solution by techniques such as ion exchange or solvent extraction. After the metal has been recovered from the solution, the exhausted liquor (raffinate) is usually recirculated to the leaching stage to conserve the chemical reagents. When the raffinate is finally rejected, it must be neutralized and any noxious species must be converted into insoluble compounds before they can be discharged safely into the environment. Since only low temperatures are involved, there is no problem of atmospheric contamination by sulfur-containing gases, CO_2, volatiles or other noxious species.

The major disadvantage of the hydrometallurgical route is the very low production rate when compared with pyrometallurgy. When both of these routes can be used for the extraction of a particular metal from an ore body, there is a financial break-even point where similar costs arise from production by either route. This point is often determined by the concentration of the metal in the ore. The cost of beneficiation treatments to produce a feed sufficiently rich for the smelting processes increases rapidly as the mineral content of the ore falls to a low level. However, an ore can be leached after only limited preparation and metals can even be recovered directly by leaching from an ore body *in situ* in the earth. For example, it is generally less costly today to extract Cu from an ore containing not more than than 0.5% Cu by leaching than by other techniques, particularly when the metal is present as an oxide and not as a sulfide, which is more amenable to smelting treatments. About 10% of the world production of this metal is now produced by leaching of low-grade ores. There is, of course, another break-even point, when the mineral content of the ore is so low that extraction by leaching is only just capable of producing metal at a profit. Both break-even points are affected by the market price of the metal, which fluctuates with supply and demand, and by progressive changes in the efficiencies of the various extraction processes. They are also dependent on the amount of other valuable elements and deleterious impurities in the ore that may be recovered or rejected more effectively by one method than by another.

The ore composition is not the only critical factor. Zinc can be extracted in the blast furnace, and this is an important route for processing Zn ores containing significant amounts of Pb, since both metals can be recovered simultaneously. The energy costs per ton of pure metal produced are high because:

1. Zn vapor must be maintained at a high temperature up to the stage where it is converted to a liquid in the Pb condenser
2. Multistage refining is required to purify the condensed Zn
3. Some metal is lost as ZnO dust since condensation is not complete

Production costs for ores containing only small amounts of Pb via the leaching-electrowinning route are low in comparison, and about 80% of the total annual Zn production is extracted by hydrometallurgy.

Many of the less common metals are found only in very low-grade deposits requiring extensive beneficiation to produce a suitable pyrometallurgical feed. Most of them cannot be reduced economically to metallic form with carbon or carbonaceous gases. The slag, dross, flue dust and other residues from smelting and refining processes are sometimes valuable sources of these metals. Leaching of these ores and residues is often a more cost-effective means of recovery of the elements than alternatives such as halide metallurgy.

Leaching is also used to remove impurities from an ore prior to extraction of the metal, and Al production is an important example of this application. Most of the impurities have already been dissolved by percolating rainwater from the naturally enriched bauxite deposits. However, significant amounts of iron oxides, silica and sometimes titania remain *in situ* and these oxides are reduced preferentially before alumina. The remaining impurities can be leached almost completely from the ore to leave a pure alumina feedstock suitable for production of the metal.

6.2. LEACHING

The lixiviant used for dissolving the metallic mineral should be cheap and readily available. Abundant supplies of sulfuric acid are obtained from the treatment of the exhaust gas produced during roasting and smelting of sulfide ores. This acid satisfies the requirements for the leaching of oxidic ores of elements such as Co, Cu, Fe, Ni and Zn at ambient or slightly elevated temperatures, i.e.,

$$CuO + H_2SO_4 = CuSO_4 + H_2O \tag{6.1}$$

and more or less complete separation can be achieved from other elements such as Ag, Au, Pb, Pt and Sb, which are only sparingly soluble or completely insoluble under these conditions.

Pourbaix diagrams (see Section 2.3.4) can be used to identify the conditions required to dissolve a metallic compound in an aqueous solution at a specific temperature and pressure. Water is the predominant species in the solution, so the potential-pH diagrams for the relevant metal-water systems can be applied to examine the equilibrium conditions. The phase fields shown on the diagram for any chosen metal are dependent on the solid and ionic phases considered as possible participants when constructing the plot, and care must be taken to select a diagram that indicates the stable domains of all the forms in which the metal can exist under the conditions to be examined.

Figures 6.1 and 6.2 show the potential-pH diagrams at 25° C for the Cu-H_2O and the Fe-H_2O systems, respectively, when the solid phases indicated in the legends are taken into account in evaluating the boundaries. The stable ions Cu^{2+}, Fe^{2+} and Fe^{3+}, which can coexist with these solid phases, appear in the upper left-hand domains. A family of lines, designated 0, −2, −4 and −6, separate the ionic species from the adjoining solid phases. These lines indicate the phase boundary for solutions in which the

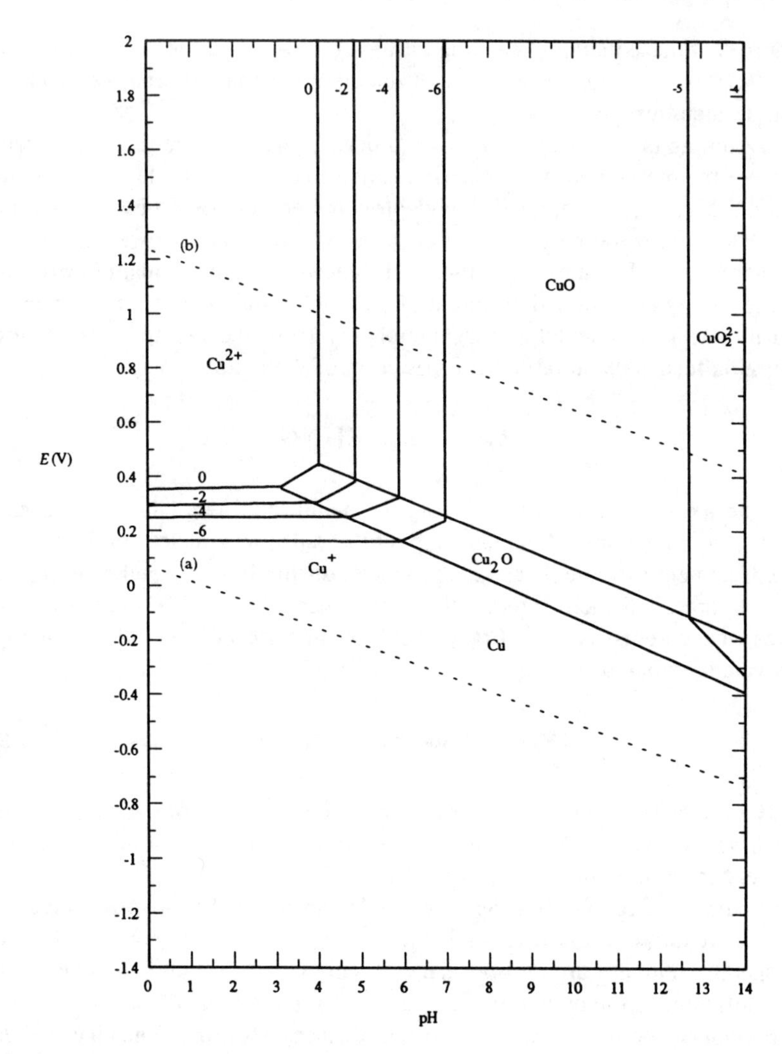

FIGURE 6.1. Pourbaix diagram for the copper-water system at 25° C, considering only solid Cu, Cu_2O and CuO.

activity of the ion in solution is 1 (i.e., 10^0), 10^{-2}, 10^{-4} and 10^{-6}, respectively, in equilibrium with the appropriate solid phase at unit activity. They show that an ion can exist over a wider range of *E* and pH as the activity of the ion in solution decreases. The horizontal lines separating the Cu-Cu^{2+} and the Fe-Fe^{2+} phases at unit ion activity are located at the standard electrode potentials for the reduction of these ions as listed in Table 2.1. A horizontal line at +0.77V similarly separates the Fe^{2+} and Fe^{3+} domains when both ions are present at the same activity. There is no corresponding

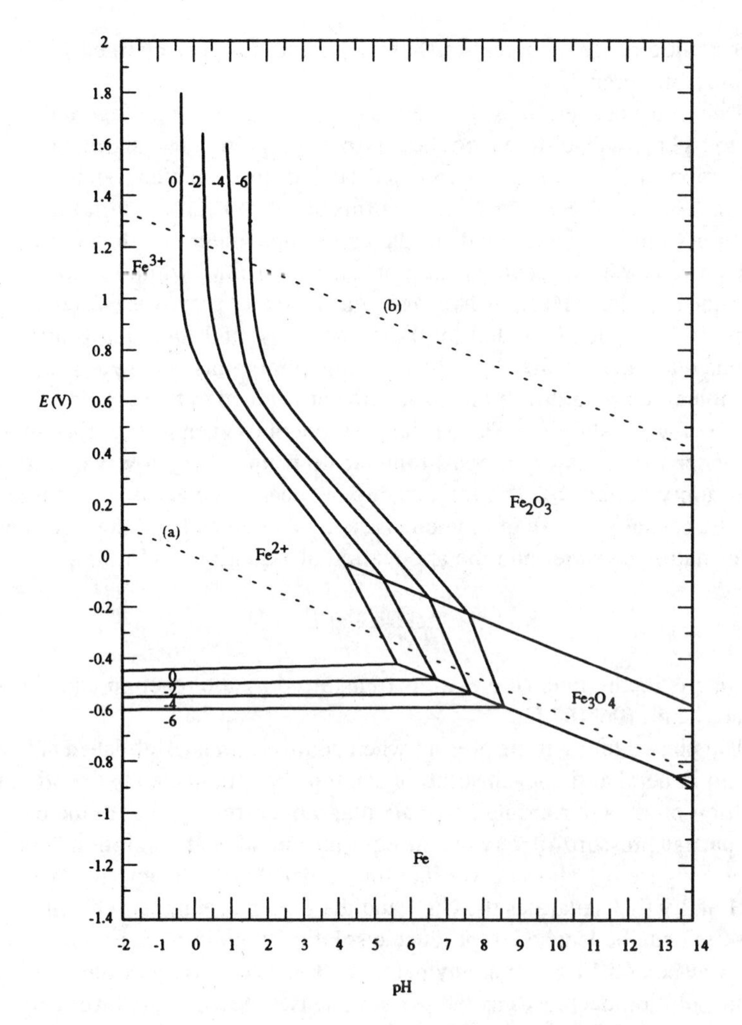

FIGURE 6.2. Pourbaix diagram for the iron-water system at 25° C considering only solid Fe, Fe_3O_4 and Fe_2O_3.

domain for Cu^+. This may seem surprising, since Cu_2O is thermodynamically more stable than CuO. The apparent anomaly arises because the ionization potential for the formation of Cu^+ ions (0.52 V) is not very different from the value for forming Cu^{2+} (0.34 V). The two ions are of similar size, so the charge density (i.e., charge/volume) is greater on the Cu^{2+} ion, and it is more capable of breaking (solvating) the polymeric bonds holding the water molecules together. In marked contrast, two electrons are readily shed by Fe to form Fe^{2+} ions, but a much greater energy must be applied to remove three electrons to form Fe^{3+} ions. As a result, ferrous ions can exist over a wide range of potentials before the more highly charged ferric ion becomes dominant.

Two other dashed lines, labeled (a) and (b) and falling diagonally from left to right, are also shown on the Pourbaix diagrams. The significance and derivation of these lines is explained later in this chapter (Section 6.3.2) and in Chapter 7 (Section 7.2.1). At this stage, it is sufficient to note that the upper line (b) corresponds to the conditions under which water can be ionized to release oxygen gas at a pressure of 1 atm, while the lower line (a) relates to the release of hydrogen gas at 1 atm pressure. Water is only stable in the space bounded by these two lines, and the equilibrium conditions that can be utilized for the leaching of minerals in aqueous solutions at ambient temperature and pressure are constrained to those defined by the region of water stability. The partial pressure of oxygen in equilibrium with the water increases on a logarithmic scale from a very low value at line (a) to unity at line (b). The relationship between the electrode potential, E, and the partial pressure of oxygen is readily evaluated from the free energy of formation of water and the application of Equation 2.115, i.e.,

$$\Delta G = -zFE$$

where z is the number of electrons transferred by the reaction and F is the Faraday unit (96,487 C).

The phases that will be present when equilibrium is established between a solid mineral and a leach solution can thus be established by locating the position on the appropriate Pourbaix diagram corresponding to the pH and the partial pressure of oxygen in equilibrium with the solution. For example, Figure 6.1 shows a vertical line, labeled with an activity of 10^0, at a pH of 3.9 that separates the Cu^{2+} and the CuO phase fields. This indicates that CuO can be leached to produce a solution containing Cu^{2+} ions at unit activity (i.e., 63.57 g l^{-1}) at any pH up to 3.9. The activity of the Cu^{2+} ions in the solution decreases as the pH is increased beyond this level and falls to 10^{-6} at a pH of 6.9. A diagonal line joins the 10^0 line separating the Cu^{2+}-CuO phases to the horizontal 10^0 line separating Cu metal from Cu^{2+}. This diagonal line is the boundary between Cu^{2+} and Cu_2O. It indicates that a lower pH is necessary to produce a 1 m Cu^{2+} solution from Cu_2O than from CuO, the pH decreasing as the partial pressure of oxygen in equilibrium with the solution is decreased.

Figure 6.2 indicates that a 1 m solution of Fe^{2+} ions could be produced at a maximum pH of about 5.5, where line (a) is coincident with the Fe_3O_4 boundary, but this also corresponds to equilibrium with hydrogen at 1 atm pressure. The pH for a 1 molal solution of this ion decreases rapidly with increasing partial pressure of oxygen in the solution. At a potential, E, equal to or greater than 0.77 V, corresponding to equal stabilities of Fe^{2+} and Fe^{3+} ions, the equilibrium activity of the metal ions is only about 10^{-3} at

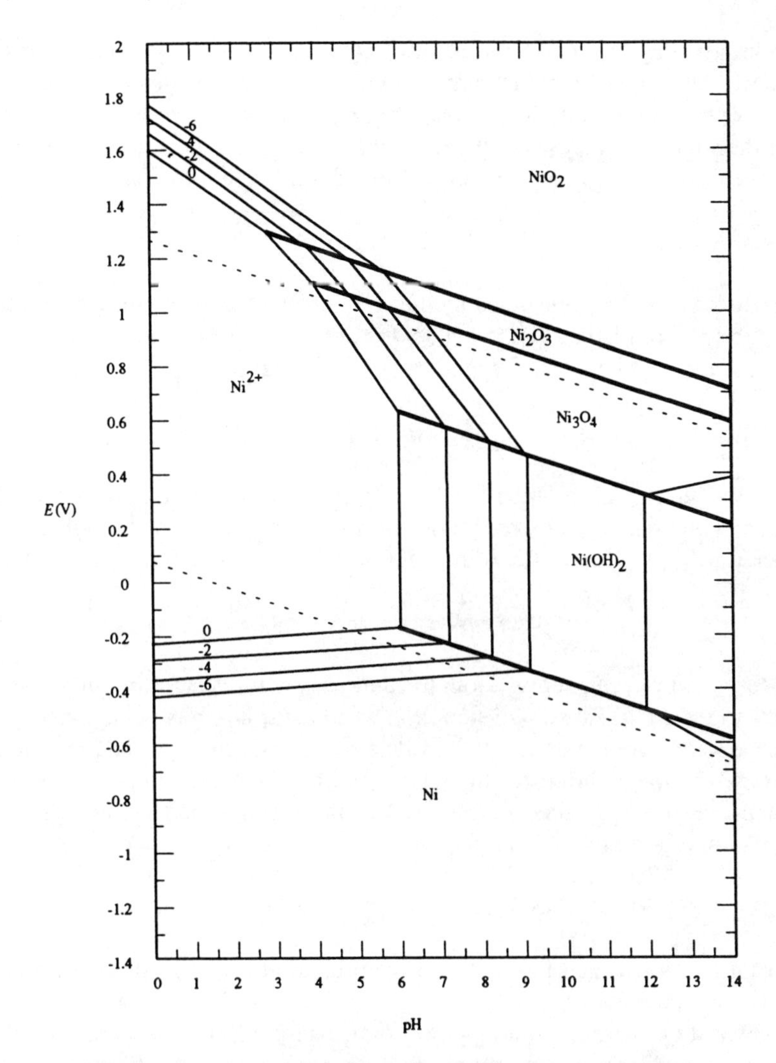

FIGURE 6.3. Pourbaix diagram for the nickel-water system at 25° C.

pH 1.0 and decreases to even lower levels as the pH is raised above unity. (The Fe^{2+}-Fe^{3+} boundary is not usually included on Pourbaix diagrams.) Thus, by controlling the partial pressure of oxygen in equilibrium with the solution, it is possible to leach Cu from an oxide ore while most of the Fe remains in the insoluble residue and only a small amount of Fe is dissolved in the solution. Conversely, Cu cannot be leached from an oxide ore in acidic solutions at ambient temperature and pressure without simultaneous leaching of nickel oxides, since the Ni^{2+} phase field extends to higher pH values than the Cu^{2+} boundary at similar activities of the ions, Figure 6.3.

Pourbaix diagrams can be used similarly to examine the possibility of selective leaching of other elements in the presence of impurities in the ores.

Copper is normally leached at a pH in the range 1.2 to 2.2, but the pH of the solution increases as the Cu is dissolved. The reason for this is more readily apparent when Equation 6.1 is written in the ionic form:

$$CuO + 2H^+ = Cu^{2+} + H_2O \tag{6.2}$$

Hydrogen ions are consumed by the reaction and the pH increases until the solubility limit is reached if the acid is not continuously replenished.

Sulfide ores consume oxygen during leaching:

$$CuS + \tfrac{1}{2}O_2 + 2H^+ = Cu^{2+} + H_2O + S \tag{6.3}$$

but the sulfur can be recovered in elemental form from cold or slightly warm solutions, thereby reducing environmental problems. At higher temperatures (100° C), all the sulfur may appear in the liquor as sulfate ions:

$$CuS + 2O_2 = Cu^{2+} + SO_4{}^{2-} \tag{6.4}$$

The rate of oxygen supply from the atmosphere to the solution may be too low to match the rate of consumption when the solution is not agitated, and the amount supplied may not provide a sufficiently high partial pressure of oxygen in the solution for the oxidation of the more stable sulfides. Oxidizing agents may then be required in the solution and ferric sulfate is commonly used for this purpose:

$$CuS + 2Fe^{3+} = Cu^{2+} + 2Fe^{2+} + S \tag{6.5}$$

or the ore is leached under a positive pressure of oxygen in an autoclave (see Section 6.2.1.2).

Variations in the solvating power of acids for different minerals can also be exploited to achieve separation of elements from a feedstock partially processed along a pyrometallurgical route. Thus, Cu can be leached in an oxidizing sulfate solution from a Cu-Pb matte, leaving Ag, Pb and S in a residue that can be processed further by pyrometallurgy.

A more expensive reagent is sometimes required to extract the metallic mineral. For example, PbS is practically insoluble in H_2SO_4 but is readily dissolved in HCl. The higher reagent cost raises the break-even point, however, and it is generally less expensive to extract Pb by smelting. Many other sulfides can be leached in chloride solutions. Ferric chloride is frequently used as an oxidizing agent, e.g.,

$$CuFeS_2 + 4FeCl_3 = CuCl_2 + 5FeCl_2 + 2S \tag{6.6}$$

and the cation concentration in the liquor is often much higher than with a sulfate leach. However, the solution may contain other elements less readily dissolved as sulfates (e.g., Ag, As, Bi, Cd and Pb), and the difficulty of solution purification from a chloride solution is increased. Chloride leaching under oxidizing conditions is useful, however, for the treatment of ores and flue dusts contaminated with arsenic, for the arsenic can then be retained in the residue as an insoluble ferric arsenate. A few minerals like the Sn oxide, cassiterite, are insoluble in both H_2SO_4 and HCl and these acids can be used to remove other metallic minerals to leave a purer ore for processing by other routes.

The maximum acidity of the leach liquor is limited not only by the increasing reagent cost and the risk of attack on the container lining as the pH decreases, but also by the risk of dissolution of some of the gangue that would raise the consumption of acid and further complicate the purification of the pregnant liquor. The minimum acidity is often limited by the risk of the precipitation of oxides or hydroxides that could adsorb a coating of the mineral salt being extracted and thus decrease the yield.

Carbonate ores release CO_2 gas in acid solutions, which consumes more acid and can cause extensive frothing. This is one reason why some ores are treated to yield anions in alkaline solutions. However, high pH conditions may also be chosen for other reasons. ZnS can be leached, for example, either in a strong oxidizing acid solution to yield Zn^{2+} cations or in a strong alkali:

$$ZnS + 2NaOH = Na_2ZnO_2 + H_2 + S \tag{6.7}$$

CuS and NiS are also dissolved in the acid solution but remain in the residue with the alkaline leach, and the Zn liquor then requires less work for purification. Ammoniacal ammonium carbonate is sometimes used to produce soluble ammines from ores, drosses, mattes, slags and from mixtures in which the element to be extracted has first been reduced to the metallic form.

Gold is not soluble in mineral acids but can be leached (together with impurities such as Ag, Cu, Fe, etc.) in an alkaline cyanide solution with $pH > 10$ to prevent formation of HCN:

$$4Au + 8NaCN + O_2 + 2H_2O = 4NaAu(CN)_2 + 4NaOH \tag{6.8}$$

(or it can be leached to form a bromide, $AuBr_4$). Uranium can be leached in sodium carbonate:

$$UO_2 + 3Na_2CO_3 + 2H_2O + O_2 = Na_2UO_2(CO_3)_3 + 4NaOH \tag{6.9}$$

Oxidizing conditions are required to drive both of these reactions. Bauxite is leached similarly in caustic soda:

$$Al_2O_3 + 2NaOH = 2NaAl_2O_3 + H_2O \qquad (6.10)$$

leaving the oxides of Fe, Si and Ti in the residue. The reaction rate is negligible at atmospheric temperature, so the ore is leached under pressure at an elevated temperature.

6.2.1. Leaching Practice

As in all extraction operations, the aim in leaching is to recover the element from the ore at the lowest cost and in a form requiring minimum cost for further processing to produce a pure metal. These two requirements are often in conflict. If only capital costs are considered, there is an inverse relationship between cost and the rate of extraction.

6.2.1.1. Mine, Dump and Heap Leaching

Some low-grade underground ore bodies cannot be processed profitably by mining, but the mineral can be recovered economically by leaching the ore *in situ* in the ground. Ores containing less than 0.5% Cu as CuS can be leached simply by diverting surface water down boreholes to the top of the deposit and pumping the extract up to the surface from the bottom of the bed. Uranium has been leached similarly, but an aqueous solution of ammonium carbonate, together with an oxidant, is normally injected as the lixiviant for the recovery of this element. The ore body must be permeable, or shattered by explosives, and resting on an impermeable rock for efficient extraction *in situ.* The only capital costs involved are for the drills to form the boreholes and the pumps for recovery of the liquor, but the concentration of the metal in the extracted solution is very low and it may take up to 20 years to recover most of the metal values. The pillars and walls left as roof supports in old mine workings can also be leached by this method. There is a risk, however, that some of the leach liquor may escape through faults in the surrounding rock strata and contaminate other underground waters, causing stream and river pollution which may only be detected at some considerable distance from the ore body.

Waste dumps of rock left from mining operations in earlier times, when beneficiation treatments were less efficient than present practice, often contain sufficient metal values to justify leaching the dump. A more expensive but still low-cost route for low-grade deposits is to mine the ore and, following primary crushing, pile it into a large heap that is exposed to percolation by the lixiviant. This is particularly useful when a low-grade ore contains large amounts of pyrites, which is difficult to remove during beneficiation and would increase the thermal load if the ore was smelted. In both dump and heap leaching, the ore must again rest on an impermeable bed that allows the liquor to be drained to suitable collection points without risk of contamination of other surface waters.

All of these leaching methods are very slow, but they are accelerated by the presence of a particular bacterium, ***Thiobacillus ferrooxidans,*** which occurs naturally in some mine waters. These microorganisms use oxygen and CO_2 for growth and derive their energy from the oxidation of both sulfides to sulfates and ferrous sulfate to ferric sulfate. The oxidation reactions provide both the sulfate ions required to hold the metal in solution and the oxidizing conditions necessary for the dissolution of the sulfide minerals, according to Equation 6.5, so acid additions are not essential. However, it is evident from Figure 6.2 that the pH of the solution must be held at or below 2.0 to hold sufficient Fe^{3+} ions in solution. The microorganisms require a pH in the range 1.5 to 4.0 at 20 to 40° C to sustain their growth, and forced air cooling may be needed to prevent the temperature rising too high from the heat released by the exothermic oxidation of the sulfate minerals. The rate of dissolution by microorganisms can be enhanced by the application of a positive potential of 400 to 600 mV to the solution.[101] Bioleaching was originally applied primarily for the recovery of Cu from low-grade deposits, but it is now used also for other purposes such as the acid leaching of uranium, the recovery of Ni and Cu from laterite minerals, and the destruction of sulfide, antimonide and arsenide compounds in refractory Au ores before they are leached in cyanide solution.[102]

6.2.1.2. Vat and Pressure Leaching

The pregnant liquor obtained from *in situ,* dump and heap leaching rarely contains more than about 5 g l^{-1} metal and the average concentration is frequently not more than 1 g l^{-1}. When the metal is isolated from the leach solution in the next treatment stage, the difficulty of metal recovery becomes greater as the concentration of metal ions in the solution is diminished. There is an obvious advantage, therefore, if more concentrated solutions are derived from the ore. This can be achieved by either vat or pressure leaching, and the rate of extraction is also markedly enhanced. However, both the capital and operating costs are increased and the minimum grade of ore that can be exploited economically is raised accordingly.

For vat leaching, the ore is first crushed and ground to produce a feed usually in the size range 5 to 20 mm to increase the mineral exposure to the lixiviant and it is then loaded into large open containers. Several vats are connected in series and are emptied and filled in rotation. Fresh leach solution is introduced into the vat containing the ore nearest to exhaustion. It is allowed to percolate through the ore and then pumped to the next container. The process is repeated until the liquor is finally recovered from the vat that was the last to be loaded with fresh ore. Each tank may contain up to 10,000 t ore and the metal is completely extracted in a few days, producing a liquor containing as much as 40 g l^{-1} of the metal ions. Alternatively, the ore may be ground to about 100-μm particle size and fed into the vat as a pulp containing about 40% solids in suspension. The pulp is agitated continuously by impellers immersed in the solution.

It may be economically advantageous to use a beneficiation treatment, prior to leaching, when the ore has to be ground to a fine size to expose the mineral and increase the rate of solution. Sulfide concentrates can also be roasted to produce a water-soluble sulfate compound (e.g., $CuSO_4$), or a more acid-soluble oxide (e.g., ZnO) and to convert Fe compounds into less-soluble Fe_2O_3 as with the treatment of Au ores. However, roasting introduces a further stage in the process, with increased costs for capital equipment, thermal energy, etc. and for gas cleaning when toxic volatiles and SO_2 are evolved.

Pressure leaching in an autoclave, first applied to sulfide ores about 40 years ago, avoids the need for a roasting treatment and gives even more rapid dissolution. The leaching rate of a metallic mineral is accelerated as the temperature is raised, but the solution boils if the temperature is raised to about 100° C under atmospheric pressure. Much higher temperatures, up to about 350° C, can be used if the pressure above the solution is increased. Furthermore, sulfide minerals can be oxidized directly to sulfates at elevated temperature if the solution is pressurized with oxygen gas. The exothermic oxidation reaction can provide all the thermal energy required and the process is autogenous when more than about 5% S is oxidized.

The most common form of autoclave is a horizontal, cylindrical pressure vessel subdivided internally into three or more compartments with an agitator in each compartment. Fresh concentrate is pumped in at one end of the vessel and overflows from one compartment to the next, while the lixiviant is fed from the opposite end and passes through in countercurrent flow to the ore. It is still a relatively novel technique, but the number of applications is increasing.[103] One of the first applications was for ammonia leaching of Ni concentrates. Ni-Cu mattes are now treated by a two-stage sulfate leach. Most of the Ni is extracted at about 130° C in the first autoclave and the remainder, together with the Cu, is recovered from the second vessel operating at about 160° C. ZnS concentrates can also be leached directly in this way, without prior roasting, and base metals can be removed from Au ores prior to cyanide treatment. With suitable control of the conditions, Fe minerals are precipitated in the autoclave as Fe_2O_3 or as compounds such as the mineral jarosite, $NH_4Fe_3(SO_4)_2(OH)_6$, which also contains any Pb present in the ore. If arsenic is present in the concentrate, it can also be precipitated as ferric arsenate.

The earliest and most extensive application of pressure leaching was for the purification of bauxite by the Bayer process for Al production. The whole of the aluminum extraction industry is still dependent on this process. The bauxite ore is digested in a strong solution of NaOH (150 to 350 g l^{-1}) at about 160° C under a pressure of 0.5 to 1.0 MN m^{-2} to form soluble sodium aluminate according to Equation 6.10. Bayerite, $Al(OH)_3$, is precipitated when the solution is cooled and diluted in a separate container. The insoluble residue remaining in the autoclave contains Si and Ti, but is primarily hydrated iron oxides that impart a reddish-brown color and give rise to the name of ''red mud'' for the residue. Pressure leaching with NaOH is also more attractive than acid leaching for low-grade ores of tungsten and some of the other less common metals.

6.2.1.3. Liquid-Solid Separation

The pregnant liquor must be separated from the insoluble residue and any precipitated phases before the metals are recovered from the solution. Traditionally, this is accomplished in settling tanks, which are shallow bowls with a large diameter and a conical bottom draining to a solids outlet at the center of the base. Slowly rotating rakes move the solids toward the discharge point while the liquor and the wash water, used to recover the remaining liquor adsorbed on the solids, overflows at the periphery. Flocculents are often added to coagulate the finer particles. More rapid separation is now obtained with mechanical centrifuges. The solution is then filtered to lower the solids content to about 10 to 20 mg l^{-1}.

6.2.2. Leaching Kinetics

The liquid-solid reactions that occur during leaching can be considered from a kinetic viewpoint as a series of consecutive steps analagous to those described for gas-solid reactions in Section 3.3.2. The interface between the leached and unreacted portions of a particle of ore or concentrate advances toward the center of the particle by a topochemical type of progression. So, four stages could be rate controlling if the liquid is stirred to eliminate concentration gradients in the bulk solution. These are:

1. Diffusion of the lixiviant through the liquid phase boundary film to the surface of the particle
2. Diffusion of the reactants and the products within the solid
3. Chemical reaction at the unreacted solid interface
4. Diffusion of the reaction products through the liquid phase boundary film into the bulk solution

The rate of leaching is accelerated by vigorous stirring when either steps 1 or 4 are rate controlling, but agitation has no effect if the rate is limited by steps 2 or 3. Insoluble reaction products, such as elemental sulfur and ferric hydroxides, may be produced during dissolution. These products may be adsorbed onto the surface of the mineral and retard steps 2 and 3 by impeding, or even preventing completely, access of the lixiviant to the reaction surface.[104] Care is required, therefore, to ensure that conditions are maintained that do not cause suppression of the chemical reactions.

When the reaction interface within the solid advances by a topochemical mechanism the rate can be fitted by equations of the type:[105]

$$1 - (1 - \alpha)^{1/3} = k\frac{C}{d}\exp\left(-\frac{E_A}{RT}\right)^t \qquad (6.11)$$

where α is the fractional amount of the metal that has been leached and k is the linear rate constant. This shows that leaching is accelerated as the molal concentration of the lixiviant, C, and the temperature, T, are increased and as the average particle diameter, d, is decreased. The equation must be modified to include the positive effect of oxygen pressure when it is applied to pressure leaching.

6.3. RECOVERY OF THE METAL FROM SOLUTION

By suitable selection of the lixiviant, it is sometimes possible to leach a metal and leave as insolubles other elements in the ore that would have been extracted simultaneously if a pyrometallurgical route had been followed. Iron oxides are soluble in all the common acidic solvents (although not in neutral or alkaline ones), and the pregnant liquor obtained by acid leaching is inevitably contaminated by Fe. Other elements may also dissolve and contaminate the extract, depending on the conditions imposed. Some of these impurities can be precipitated (hydrolyzed) from solution simply by increasing the pH, but a nucleation barrier may intervene and kinetically prevent precipitation occurring if the pH is raised only just sufficiently to shift the equilibrium position to where precipitation could occur, as indicated by the relevant Pourbaix diagram. A higher pH is usually required in practice to provide the driving force for the reaction.

It is standard practice to recirculate the raffinate to the leaching circuit after the metal has been recovered in order to conserve reagent costs. Greater quantities of lixiviant are required to restore the leaching potential if the pH is altered to precipitate the impurities. Additionally, the greater quantity of water required to recover ions adsorbed on the surface of the precipitate dilutes the average concentration of the ions in the solution. Consequently, most present-day treatments are designed to separate and recover the required metal or metals from solution, leaving the other elements in the liquor from which they may be subsequently removed by other means before recirculation to the leach circuit. A variety of techniques is used for this purpose.

6.3.1. Cementation

The precipitation of metallic Cu from solution by contact with solid Fe has been practiced for about 500 years and is still in use. The overall chemical reaction can be written as:

$$Cu^{2+} + Fe = Cu + Fe^{2+} \qquad (6.12)$$

This is the reaction that causes a coating of Cu to form over the surface of an iron nail when the nail is immersed in a copper sulfate solution. The mechanism of the displacement reaction is more readily apparent from electrochemical considerations. As explained in Section 2.3.4, the half-cell reaction for the reduction of Cu^{2+} to the metal can be written as:

$$Cu^{2+} + 2e^- = Cu \tag{6.13}$$

and the reversible electrode potential for the reaction is given by the Nernst Equation 2.122:

$$E = E^{\ominus} + \frac{RT}{2F} \ln a_{Cu^{2+}} \tag{6.14}$$

if pure metallic Cu ($a_{Cu} = 1$) is precipitated. The value of $E^{\ominus}$ for this reaction is +0.34V for a 1 *M* solution (i.e., $a_{Cu^{2+}} = 1$) at 25° C as listed in Table 2.1. Similarly, the Nernst Equation for the Fe half-cell reaction:

$$Fe^{2+} + 2e^- = Fe \tag{6.15}$$

takes the form:

$$E = E^{\ominus} + \frac{RT}{2F} \ln a_{Fe^{2+}} \tag{6.16}$$

and $E^{\ominus}$ is −0.44 V. Subtraction of Equation 6.15 from 6.13 gives the overall cell reaction, Equation 6.12, for which the reversible electrode potential at 25° C is:

$$E = E^{\ominus}_{Cu} - E^{\ominus}_{Fe} - \frac{RT}{2F} \ln \frac{a_{Fe^{2+}}}{a_{Cu^{2+}}} \tag{6.17}$$

$$= 0.34 + 0.44 - 0.0296 \log \frac{a_{Fe^{2+}}}{a_{Cu^{2+}}} \tag{6.18}$$

The reversible potential is positive and hence, from Equation 2.115 (i.e., $\Delta G = -zFE$), the free energy change for the overall reaction is negative. So, even in the absence of an applied potential, electrons are removed from the solid Fe to discharge the Cu ions, and the Fe simultaneously goes into solution to preserve electroneutrality. The reaction continues until either equilibrium is attained, all of the Fe is dissolved or all of the Cu ions are discharged.

If the liquor initially contains only Cu^{2+} cations, then the activity of the Cu ions decreases and the activity of the Fe ions rises from zero as the displacement reaction proceeds. Correspondingly, E decreases for Equation 6.13 and increases (becomes less negative) for Equation 6.15. Eventually, in the presence of an excess of solid Fe, E reaches the same value for both half-cells and equilibrium is attained. When the two half-cell electrode potentials are equal, then (from Equation 6.18) the activity quotient is equal to $[-\text{antilog}\ (0.78/0.0296)] = 4 \times 10^{-27}$, indicating that virtually all the Cu^{2+} ions should be precipitated. The raffinate is usually recirculated to the leaching circuit before this stage is reached. The Fe^{2+} ions in the solution can then be oxidized to Fe^{3+} by the microorganisms and promote the dissolution reaction, but some iron must be removed periodically to prevent excessive build-up of the concentration of this element in the liquor.

The copper obtained by cementation is not a high-purity metal. Any other element present in the solution for which the standard electrode potential is more positive than for Fe (e.g., Co^{2+}, Ni^{2+}) can also be precipitated by cementation. However, the residual amount of an element remaining in the solution when equilibrium with the iron is established is raised as the difference decreases between the standard half-cell potentials for that element and for Fe. When the potential difference is small, the equilibrium concentration may be greater than the initial amount of the impurity in the solution and all of the impurity is then retained in the solution. The Cu precipitate is also contaminated with small particles of occluded Fe, which have been protected from dissolution by a coating of Cu, and by Fe ions adsorbed onto the surface. Thus, the purity is usually lower than that obtained by pyrorefining, and further treatment is necessary to produce commercially pure grades of Cu.

Iron is used for cementation of Cu because relatively cheap scrap steel can be used. However, any metal that is markedly electronegative, relative to the metal to be recovered, can also be used for cementation. Thus, Zn can be used to precipitate Ag or Au from cyanide or bromide solutions:

$$Zn + 2Au(CN)_2 = Zn(CN)_2 + 2Au \tag{6.19}$$

and the excess Zn is removed by oxidation when the precious metals are melted for casting into ingots. The reaction is not selective and almost all the other metallic elements present in the solution are also precipitated by the Zn. This behavior can be useful for the purification of solutions of strongly electronegative elements. For example, addition of finely divided Zn to a $ZnSO_4$ solution causes precipitation by cementation of impurities such as Co, Cd, Cu, Ni, Sb and Th to leave a pure solution from which the Zn can be recovered by other means. The solid Zn particles are replaced periodically and the cementation coating can provide a valuable source of the less common metals.

6.3.2. Hydrogen Reduction

The standard electrode potentials for the M-M^+ half-cells are evaluated relative to the standard hydrogen electrode defined by Equation 2.119, i.e.,

$$2H^+ + 2e^- = H_2$$

which is arbitarily assigned a value of zero at all temperatures for a 1 *m* solution of H^+ ions ($a_{H+} = 1$) at pH = 0 and $p_{H_2} = 1$ atm. It follows that any metal exhibiting a positive electrode potential can be precipitated from solution by hydrogen gas. However, it is also possible to precipitate metals that show small negative potentials with respect to hydrogen. The Nernst equation for the hydrogen electrode can be written as:

$$E = E^{\ominus} + \frac{RT}{2F} \ln a_{H^+}^2 - \frac{RT}{2F} \tag{6.20}$$

and at 298K:

$$E = -0.0591\text{pH} - 0.0296 \log p_{H_2} \tag{6.21}$$

This is the equation that defines the position of the dashed line designated (a) in Figures 6.1 to 6.3. Thus, the equilibrium potential for the hydrogen half-cell reaction decreases (i.e., E becomes increasingly negative) as the pressure of the gas is increased and as the hydrogen ion activity is decreased or the pH of the solution is raised. The equilibrium potentials for the metal-ion half-cell reactions also decrease as the metal ion activities are lowered (*cf.* Equation 6.14). Figure 6.4a illustrates the change of potential with pH for the hydrogen electrode and with ion activity, relative to an infinitely dilute molal standard state, for the metal electrodes at 25° C.

Metal ions can be precipitated from solution by hydrogen under conditions described by the top portion of the diagram, above the hydrogen lines; but it is evident that the residual metal ion concentration in the solution rises with increasing electronegativity of the metal when equilibrium is attained with H_2 gas. Raising the gas pressure to 100 atm has only a small effect on the residual ion concentrations. Increasing the temperature to 200° C (Figure 6.4b) has a greater effect, but the reduction of strongly electronegative elements like Al and Zn is not feasible even at 200° C under 100 atm hydrogen pressure.

There is a further complication. As the pH of an acid solution is raised, a stage is eventually reached for each metal where an oxide or hydroxide starts to precipitate. The critical pH values are indicated by vertical bars on each of the metal lines in Figure 6.4a. It is evident that Co and Ni are the

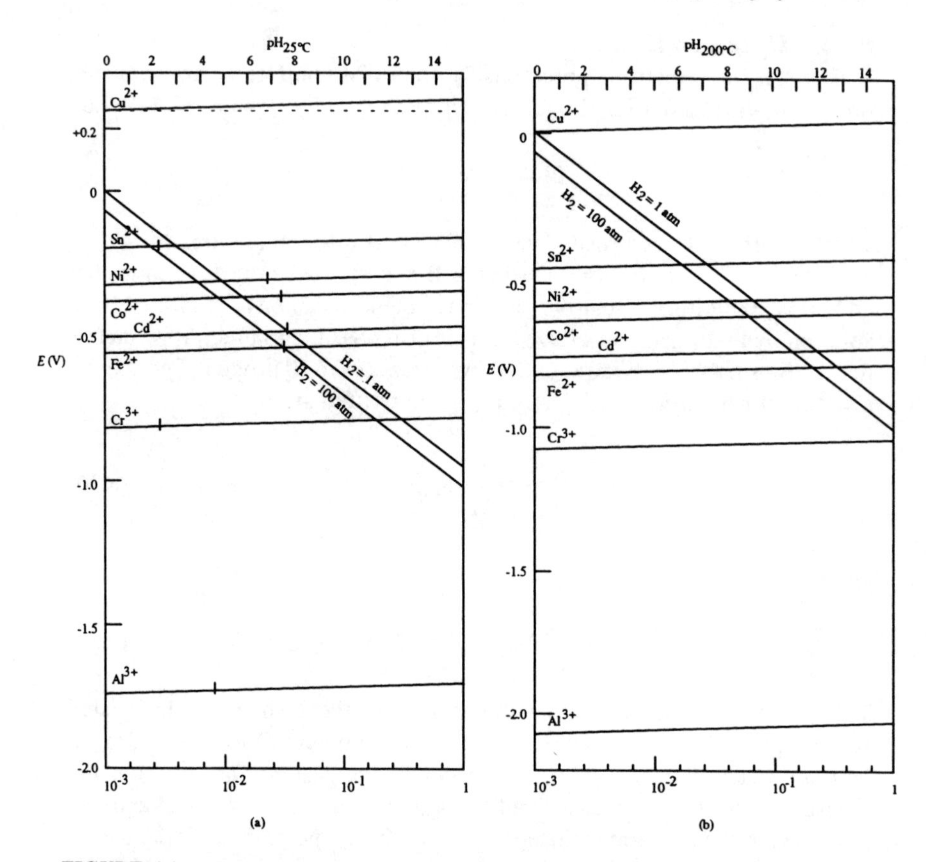

FIGURE 6.4. Electrode potentials as a function of pH for hydrogen and as a function of ion activity for various metal ions in sulfate solutions, (a) at 25° C and (b) at 200° C. (From Richardson, F. D., *Extraction and Refining,* Institute Metallurgists Review Course, November, 1970. With permission.)

only electronegative elements that can be precipitated from an acid leach solution by hydrogen, and the amounts of these two metals that can be recovered from solution are not very significant. This difficulty can be avoided by leaching in an alkali, and both Co and Ni can be recovered effectively by hydrogen reduction from ammoniacal carbonate solutions. When both metals are present in the solution, most of the Ni can be recovered by hydrogen reduction before the Co ions start to be reduced to the metal.

The high dissociation energy for the H_2 molecule results in very slow reduction by hydrogen at ambient temperature and pressure. Thus, the reaction is usually conducted at elevated temperature and pressure in an autoclave. Agitation is applied to increase the liquid contact with the gas and

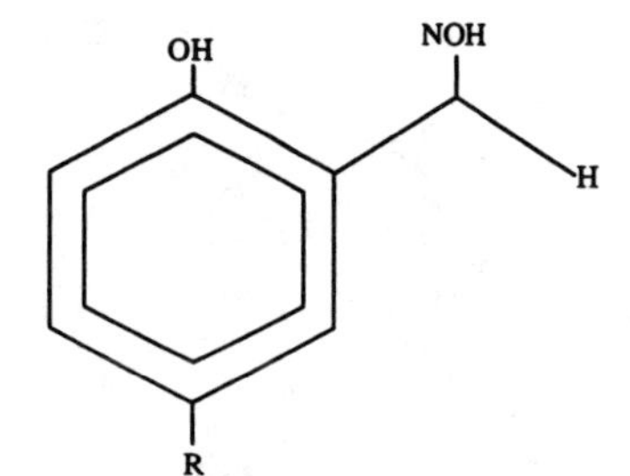

FIGURE 6.5. Typical structure of a hydroxeme reagent used for solvent extraction of ions from acidic solutions. R = C_9H_{19} or $C_{12}H_{25}$.

seed grains of the metal are added to remove the nucleation barrier to precipitation. Pure metal powders, suitable for powder metallurgy applications, can be obtained in this way.

Hydrogen gas can also be used to precipitate the less electronegative impurities from solutions containing the more electronegative metals, but the relatively high cost of the gas and the explosion hazard (particularly when the gas is pressurized) makes this route not very popular.

6.3.3. Solvent Extraction

Liquid-liquid exchange of ions by solvent extraction is an efficient means of both purifying and concentrating an ionic species to produce a solution that may be suitable as a direct feed to an electrowinning cell (see Chapter 7). The aqueous leach liquor is mixed with an immiscible liquid that is usually a relatively cheap organic species, such as kerosene or benzene. The organic phase contains a chemical substance capable of forming a compound with the ions required in the concentrate. The reagent should be selective in the extraction of the metal ions, and the ions should be easily recovered from the organic phase to regenerate the compound for recycling. Both the reagent and the complex it forms with the metal ions should also be very soluble in the organic phase and insoluble in the aqueous phase. A wide variety of extractant compounds is available. They are selected for use with reference to the particular ions to be extracted, the other impurity ions present in the liquor, the pH of the solution and the type of acid or alkaline lixiviant used for leaching the ore.

A typical structure of an LIX hydroxime reagent used for extraction of ions from acid solutions is shown in Figure 6.5. The exchange mechanism can be described as:

$$M^{2+}_{(aq)} + 2\,RH_{(org)} = R_2M_{(org)} + 2H^{+}_{(aq)} \qquad (6.22)$$

Equation 6.22 shows that the H^+ ions are exchanged with the metal ions and the equilibrium position is governed by the hydrogen ion concentration. The acidity of the aqueous solution increases during extraction (loading)

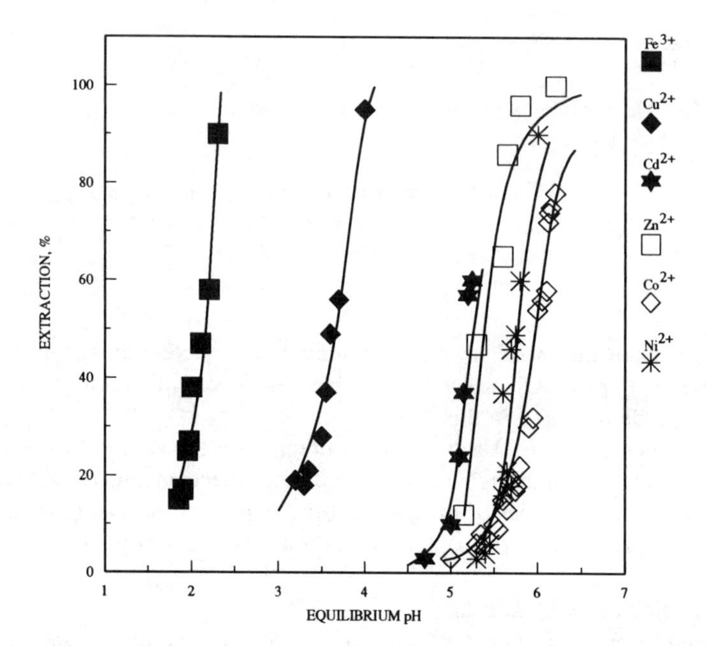

FIGURE 6.6. Extraction of ions into napthenic acid. (From Fletcher, A. W., *Metals and Materials,* 3, 9, 1969. With permission.)

into the organic phase, and care may be needed either to limit the acidity of the solution or to restrict the concentration of the ions in the leach liquor to control the decrease in pH of the solution. Each extractant is most effective above a specific pH value. For example, the type of aldoxime reagent illustrated in Figure 6.5, with an exchangeable H^+ ion, operates better at a pH greater than about 2.0. However, a more acidic environment is acceptable if a hydrocarbon radical such as CH_3 is substituted for the exchangeable hydrogen to form a ketone reagent. The loaded organic phase is separated from the aqueous solution and the metal ions are then released from the reagent by mixing with an acid solution with a lower pH to increase a_{H^+} and reverse the direction of Reaction 6.22.

The conditions required for loading and stripping can be illustrated by considering Figure 6.6. Copper can be extracted completely from the leach liquor by naphthenic acid at a pH of about 4.2, leaving Cd, Co, Ni and Zn in the solution. The Cu can be stripped from the organic phase at a pH of about 2.5. Some of the impurity ions are also extracted if the pH is raised above 5.0, but all the Cu is not extracted if the pH falls below 4.0 during loading. Any Fe^{3+} ions in the solution are also extracted by the naphthenic reagent, but remain in the organic phase if it is stripped at pH 2.5. Progressive build-up of the Fe^{3+} ions during recycling of the organic phase would then reduce the amount of Cu that could be extracted by each liter

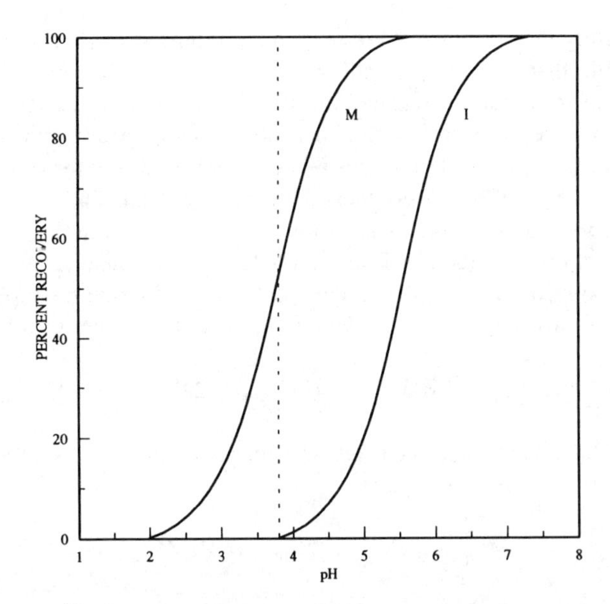

FIGURE 6.7. Schematic illustration of the effect of displacement of elements along the pH axis on the purity of a metal recovered from solution by solvent extraction. M and I indicate the recovery curves for the metal and an impurity element, respectively, in the same solution.

of reagent; thus, the iron and any similar elements that may be extracted preferentially must be removed before the leach solution contacts the organic phase.

It is more common to find that the extraction curve for a metal in a reagent is less clearly separated from the corresponding curves for the impurities, as shown schematically in Figure 6.7. In this case, the pH must not exceed the value indicated by the broken line if the impurity, I, is to be eliminated completely from the organic phase. However, less than half of the metal, M, can then be extracted, even when the pH is not allowed to fall as the loading reaction proceeds. Metal recovery is increased by passing the leach solution, which has reached equilibrium with the organic phase, to one or more additional tanks in which it is mixed with fresh extractant. Alternatively, when the recovery curves for two or three valuable metals are close together, it may be financially advantageous to extract all these elements simultaneously. The elements can then be separated by stripping at different values of pH or in different acids. This technique is used for the separation of the less common metals, such as Zr and Hf, which are difficult to part by other methods in aqueous solution.

Reagents are also available that allow the selective withdrawal of one element from a solution in the presence of another valuable element normally recovered simultaneously by other extraction methods. For example,

Co is slightly more electronegative than Ni and most of its compounds are more stable than the corresponding Ni compounds. However, Co can be extracted from a sulfate leach solution with a phosphonic acid chemical such as Cyanex 272, leaving the Ni in the aqueous phase.[106] Cu, Mn and Zn would also be extracted by this reagent, however, and must be removed before the pregnant liquor contacts the leach solution. The Ni ions can be recovered with other reagents after the Co has been extracted.

The reaction described by Equation 6.22 involves the release of H^+ ions in acidic solution; but the exchange mechanism can also be applied to alkaline solutions with ammonia, for example, as the exchangeable ion:

$$M(NH_3)^{2+}_{4(aq)} + 2RH_{(org)} = R_2M_{(org)} + 2NH^+_{4(aq)} + 2NH_{3(aq)} \quad (6.23)$$

Other extractants involve the transfer of organic pairs, as with the CLX 50 reagent for chloride solutions:

$$M^{2+}_{(aq)} + 2Cl^-_{(aq)} + 2R_{(org)} = R_2MCl_{2(org)} \quad (6.24)$$

The reagent is loaded at low NH_4^+ concentrations in the first, and at high Cl^- concentrations in the second of these examples. The reverse conditions are applied for stripping. The solution pH is not changed by either of these loading or stripping reactions.

Solvent extraction is most commonly performed in agitated mixers in which the aqueous and organic phases are intimately mixed by impellers. Two or three mixers may be used in series to improve the metal ion recovery. The liquid overflows into a gravity settler in which time is allowed for the organic phase to float to the top and permit separation of the two liquids by weirs located at the settler discharge point. The chemical compounds used as extractants are expensive, and care is taken to ensure that separation is as complete as possible to minimize losses in both the loading and stripping stages. Residual amounts remaining in the leach liquor could adversely affect the reuse of the solution and, likewise, could affect the metal recovery if left in the solution obtained after stripping.

The volume of the leach liquor and the organic phase may be roughly equal in the loading stage. However, if a strong acid or alkaline solution (as appropriate) is used for stripping, the volume of extract is low when compared with the volume of leach liquor and concentration of the metal ions in solution is increased accordingly. With appropriate selection of the extractant, pH and volume ratio of the aqueous and organic phases, it is possible to achieve an enrichment of about 50:1 in the concentration of the required metal ions while, at the same time, the ratio of that metal to specific impurity ions in the stripping solution is as high as 5000:1. Usually, these solutions are sufficiently pure and concentrated to pass directly to an electrowinning plant. Sometimes, however, the metal is recovered directly by precipitation from either the organic phase or the stripping solution. It is difficult to recover Co, for example, by acid stripping from a Kelox 100 decanol kerosene reagent, but the metal is readily recovered by hydrogen reduction from the organic phase at elevated temperature and pressure in an autoclave.[107]

Until comparatively recent times, the use of solvent extraction was restricted primarily to the recovery of Cu and U from low-grade ores, and this is still the major volume application. However, it is now also applied to the recovery of a wide range of the less common metals.

6.3.4. Ion Exchange

Some naturally occurring minerals, such as aluminosilicate clays (and zeolites which are synthesized from them), cellulose and coal, possess a residual electronic charge. When these substances are immersed in a solution, ions of opposite charge to that on the solid are attracted to and held by the solid. The attached ions can then be released (elutriated) by washing the solid with a liquid containing ions of the same sign as those on the original solid. This is the principle employed in the water softener.

A synthetic resin of a cross-linked polymer such as polystyrene is most commonly used for metal extraction, and acids are used to attach functional groups of the type Cl, NO_3, CO_3OH and COOH, the first two acting as exchangeable anions and the latter two as H^+ cations. The attached ion is positively charged for extraction of a cation:

$$R^-X^+ + M^+ = R^-M^+ + X^+ \tag{6.25}$$

and is negatively charged for the removal of anions:

$$R^+X^- + M^- = R^+M^- + X^- \tag{6.26}$$

where R signifies the resin and X is the attached ion. For example, complex uranium ions are recovered by reactions such as:

$$UO_2(SO_4)_3^{4-} + 4RNO_3 = R_4UO_2(SO_4)_3 + 4NO_3^- \tag{6.27}$$

In practice, the resin is usually in the form of very porous granules to give a large surface area per unit volume. A vertical column is packed with the granules through which the leach liquor flows until the resin is saturated and the metal ions begin to appear in the outlet liquid. The liquor is then replaced by a solution containing the ions displaced from the resin to elutriate the metal ions. The kinetics of the exchange reaction are improved as the particle diameter of the resin is decreased to increase the reactive surface area, but the flow resistance through the bed also increases with decreasing particle diameter, thereby raising the pressure drop across the bed. The best compromise is usually found with a granule size of about 1 mm diameter. More rapid exchange and elutriation can be achieved when the resin is in the form of fibers and these are now used in the form of woven filters and as endless belts.[108]

It is more difficult to obtain selective extraction of a particular metal ion by ion exchange than it is by solvent extraction. In general, it is easier to extract large ions and ions with a high charge than smaller ions with a low charge. However, some degree of selectivity can be introduced by varying the strength of the bonding of the functional group. For example, Cu can be extracted with a weak cation exchanger, but Zn is only removed from solution by a strong cation exchanger. Selectivity can also be obtained by elutriation of elements in succession, using solutions of progressively increasing strength.

Solvent extraction is usually a lower cost route than ion exchange for solutions containing more than about 1 g l^{-1} recoverable ions, whereas ion exchange is the preferred route for recovery of elements such as Au or U from solutions containing only about 0.1 g l^{-1} metal ions. Ion exchange is also very useful for the removal of heavy metals from effluents before discharge to the environment, since large volumes of liquid can be purified from very low residual concentrations before the resin becomes saturated.

FURTHER READING

Jackson, E., *Hydrometallurgical Extraction and Reduction,* Ellis Horwood, Chichester, England, 1986.

Murr, L. E., Toma, A. E. and Brierly, J. A., Eds., *Metallurgical Applications of Bacterial Leaching and Related Microbiological Phenomena,* Academic Press, New York, 1978.

Osseo-Asare, K. and Miller, J. D., *Hydrometallurgy,* TMS, Warrendale, PA, 1982.

Ritcey, G. M. and Ashbrook, A. W., *Solvent Extraction: Principles and Applications to Process Metallurgy,* Elsevier, Amsterdam, 1984.

Streat, M. and Naden, D., Eds., *Ion Exchange and Sorption Processes in Hydrometallurgy,* Critical Reports on Applied Chemistry, Soc. Chem. Ind., Vol. 19, 1987, New York, Wiley Interscience, New York, chap. 6.

Chapter 7

ELECTROLYTIC EXTRACTION AND REFINING

7.1. OVERVIEW

Electrolysis can be used for the extraction (electrowinning) of metals from either aqueous solutions or molten salts and for the purification (electrorefining) of crude metal obtained by other extraction routes. The principles involved are readily appreciated from consideration of a galvanic cell, which consists of two plates or rods of dissimilar metals partially immersed in an electrolyte and connected by an external circuit. When Cu and Zn are selected for the electrodes, both metals are soluble in a sulfuric acid electrolyte. The half-cell reactions are:

$$Cu^{2+} + 2e^- = Cu, \qquad E^\circ = +0.34\ V \tag{7.1}$$
$$Zn^{2+} + 2e^- = Zn, \qquad E^\circ = -0.76\ V \tag{7.2}$$

Since Zn is electronegative with respect to Cu, electrons flow from the Zn to the Cu when the two metals are connected through the external circuit. The loss of electrons from Zn causes the metal to dissolve and Zn^{2+} ions are released from the electrode. Reaction 7.2 therefore proceeds in the reverse direction. The electrons arriving at the Cu electrode are discharged by the conversion of Cu^{2+} ions into Cu atoms, which are deposited on the Cu electrode, so reaction 7.1 proceeds simultaneously in the forward direction. The overall cell reaction is:

$$Zn + CuSO_4 = Cu + ZnSO_4 \tag{7.3}$$

Therefore, the Zn electrode acquires a negative charge and becomes the anode, while the Cu is the positively charged cathode. By convention, the current is considered to flow in the opposite direction to the electrons (i.e., from cathode to anode).

When the electrolyte contains ions of both metals at unit activity and the electrodes are connected through a large external resistance, so that the reactions occur slowly and reversibly, the maximum voltage that can be generated across the cell is the sum of the standard electrode potentials for the two half-cells:

$$E = E^\circ_{Cu} - E^\circ_{Zn} = +0.34\ (-0.76) = +1.10\ \mathrm{V} \tag{7.4}$$

If, now, a DC potential of slightly greater than 1.10 V is imposed between the electrodes to make the Cu electrode more negative than the Zn then, theoretically, the reactions should reverse to cause Cu to dissolve at the negatively charged pole (anode) and the Zn to be deposited on the positively charged cathode. In practice, as explained later, a larger potential difference would be required to drive the reaction in the reverse direction and other factors may then intervene.

This sequence of a naturally occurring forward reaction and a driven reverse reaction corresponds to the discharging and charging cycles of a rechargeable battery. The discharging stage will be recognized as the basis of metal recovery from solution by cementation, which was described in Section 6.3.1. The recharging stage is basically similar to the reactions that occur during electrowinning. If, instead of dissimilar metals, the electrodes comprise a pure and an impure plate of the same metal, the standard electrode potential for the cell is virtually zero, depending on the extent to which the activity of the metal deviates from unity in the impure electrode. By imposing a very small DC potential to make the impure electrode more negative then, in theory, the metal should dissolve from the impure plate and transfer through the electrolyte to deposit on the pure plate. Again, in practice, a larger potential must be applied between the electrodes to drive the reaction. Elements insoluble in the electrolyte remain as solids, while ions of elements that are more electronegative than the metal forming the electrodes are retained in solution. This is the basis of electrorefining.

7.2. ELECTROWINNING

All metals can be extracted from their ores by electrowinning. The tonnage actually produced is very low in comparison with the quantity obtained by pyrometallurgical techniques, but the proportion of the total output produced by this route varies over a very wide range from one metal to another. At one extreme, iron and steel with the required purity and range of properties can be manufactured at much lower cost by pyrometallurgy. The output per square meter of ground space from electrolytic cells is very low in comparison with a modern blast furnace and BOS converter plant, and a very large area would be required to produce the same tonnage of metal by electrolysis. The total energy and other operating costs would also be very much higher. At the opposite extreme, because of the high stability of its oxide, aluminum can be extracted more economically by electrolysis than by any other route, and this is the only method used at present for the production of this metal. Magnesium, which also forms a very stable oxide, is also extracted primarily by electrolysis. And, as explained in the previous chapter, Zn is mainly obtained by leaching from the ore. Reduction of this metal from the leach solution is more economical by electrolysis than by any other method because of the strong electronegativity of the element. Electrorefining is often required to produce nonferrous metals with the required purity for very demanding applications, and the direct production

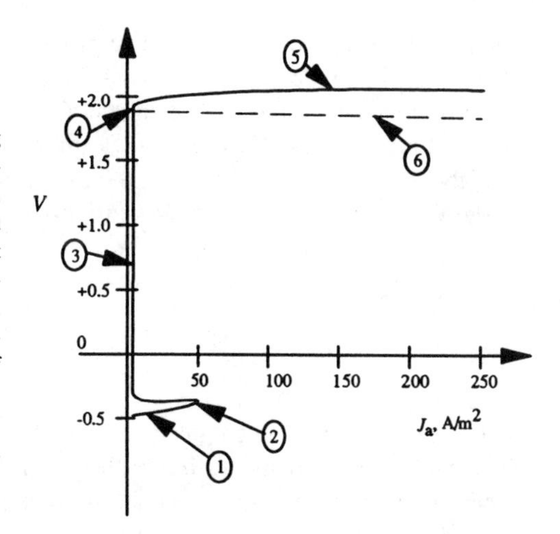

FIGURE 7.1. Schematic drawing of the potentiostatic anodic polarization curve for lead in sulfuric acid. The numbers indicate (1) active lead dissolution, (2) passivation current density, (3) lead passivated, (4) precipitate changes into PbO_2, (5) oxygen evolution, and (6) potential above which no discharge of PbO_2 film can occur. (From Winand, R. and Fontena, A., *Trans. Inst. Min. Metal.,* 92, C27, 1983. With permission.)

of the pure metal by electrowinning from leach solutions may then prove to be the most economical route, particularly from very low-grade ores.

7.2.1. Aqueous Electrowinning

7.2.1.1. Electrowinning of Copper

Metals recovered by aqueous electrowinning are most commonly reduced from sulfuric acid solutions. Metallic copper can be obtained from a sulfate leach liquor by electrodeposition of the metal on thin starter sheets of Cu as the cathodes with Pb plates as the anodes, and this is a suitable example to illustrate the complexities of the process. The type of reactions described by Equations 7.1 and 7.2 are not suitable for conducting the process, since the anode would be consumed and the electrolyte would become progressively enriched in ions of the anode metal as the Cu deposited on the cathode. An electrochemically inert anode material that will not dissolve in the electrolyte is required, and Pb can actually satisfy this requirement in sulfate solutions, even though it is electronegative with respect to Cu.

Figure 7.1 shows the dissolution (corrosion) behavior of Pb in such solutions as a function of the applied voltage and the current density, J_a, a measure of the rate of solution.[109] The metal dissolves at an increasing rate as the applied voltage is raised from -0.5 to about -0.3V (point 2 on the diagram). If the voltage is raised slightly above this level, however, the surface of the anode becomes covered with insoluble $PbSO_4$, protecting the underlying metal and preventing dissolution. The metal is then said to be **passivated.** The $PbSO_4$ layer is replaced by a series of basic sulfates as the applied potential is increased progressively from point 2 until the surface becomes covered with PbO_2, which is an electronic conductor, at about 1.65 V (point 4). Thus, the Pb electrode is almost completely inert at applied voltages between points 2 and 4. In practice, the anode is usually a Pb-Sb (4 to 6% Sb), a Pb-Ag (1.0% Ag) or a Pb-Ca-Sn alloy to increase the strength and reduce the risk of mechanical distortion of the plate, but this does not affect markedly the passivation behavior.

If the anode does not dissolve, the electrons consumed by the deposition of the Cu must be provided by some other anodic reaction. They can be supplied by the ionization of water:

$$H_2O = 2H^+ + \tfrac{1}{2}O_2 + 2e^-, \; E^{\ominus} = 1.23 \text{ V} \tag{7.5}$$

with the release of oxygen gas at the anode. The H^+ ions associate with the SO_4^{2-} anions that diffuse through the electrolyte toward the anode and regenerate the acid content, so the solution can be recycled for further use in leaching or solvent extraction after the metal ions have been recovered. However, the presence of H^+ ions in the solution introduces the possibility of an additional cathodic reaction:

$$2H^+ + 2e^- = H_2, \qquad E^{\ominus} = 0 \text{ V} \tag{7.6}$$

with the release of H_2 gas at the cathode. Since the hydrogen release reaction competes with the metal deposition for the available electrons, the mass of metal deposited by the passage of a given amount of electrical energy decreases with increasing amounts of hydrogen evolved. However, when Cu is electrowon, the standard electrode potential is more positive for the Cu half-cell than for hydrogen gas release, so Cu^{2+} ions are discharged preferentially from $CuSO_4$ solutions. The risk of significant hydrogen release only arises when the electronegative elements are being recovered by electrolysis.

Both of these reactions involve change in the H^+ ion concentration and hence the electrode potentials are dependent on the pH of the solution. When the water is present at unit activity and p_{O_2} is 1 atm in Equation 7.5, the equilibrium potential is given as a function of pH by the Nernst equation:

$$\begin{aligned} E &= 1.23 - \frac{RT}{2F}\ln a_{H^+}^2 \\ &= 1.23 - 0.0591\text{pH} \end{aligned} \tag{7.7}$$

and, similarly, for Equation 7.6:

$$E = 0 - 0.0591\text{pH} \tag{7.8}$$

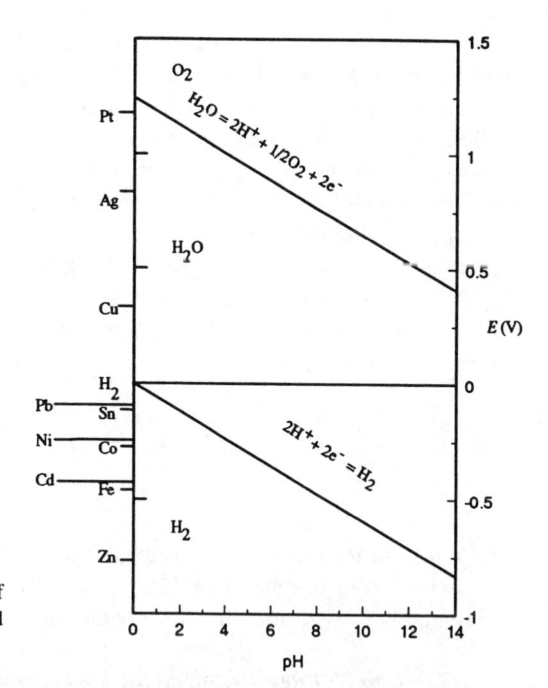

FIGURE 7.2. The stability of water as a function of the applied voltage and pH of the solution.

Equations 7.7 and 7.8 describe the position of the lines labeled (b) and (a), respectively, in Figures 6.1 to 6.3. They are replotted in Figure 7.2, and the standard electrode potentials for some of the more common metals are indicated on the ordinate. The positions of the lines shown on the diagram apply only when all the participants in the two reactions are present at unit activity. Figure 6.4 shows, for example, that the hydrogen line is moved downward on the diagram if the pressure of the hydrogen gas is increased above 1 atm and, conversely, the line is moved upward if the hydrogen is released at a lower pressure. Similarly, the line is moved downward if the activity of the H^+ ions is lowered below 1.0 by dilution with other ions in the solution.

As shown in Section 2.3.4, the equilibrium between a metal and its ions in solution is independent of the pH of the solution and the M-M^{2+} boundary is horizontal on a Pourbaix diagram. This can be seen in Figures 6.1 to 6.3. Hence, the equilibrium voltage that must be exceeded in order to deposit a metal at the cathode and liberate oxygen gas at the anode decreases as the pH of the solution is raised. The corresponding risk that H_2 gas will be liberated at the cathode when electronegative elements are deposited is also decreased as the pH is raised. This is evident, again, from Figures 6.1 to 6.3. However, these figures also show that there is a limit to which the pH can be raised for any particular metal, beyond which there is an increasing risk of precipitation of an oxide, hydroxide or other salt of the metal. Also, water at pH 7.0 is a very poor electronic conductor and the resistivity of an acid electrolyte rises as the pH is increased. The pH of the solution does not remain constant but decreases continuously as H^+ ions are released by Equation 7.5 accompanying the deposition reaction.

The concentration and hence the activity of the metal ions in solution decreases as the metal is plated out on the cathode. This increases the equilibrium voltage that must be exceeded to drive the reaction, but the effect is small in comparison with the effect of a change in the pH on the hydrogen evolution reaction. For example, the standard electrode potential for pure Cu in equilibrium with a 1 *m* Cu^{2+} solution is +0.34 V. If the activity of the Cu ions is lowered to 0.1 by metal deposition then, according to the Nernst equation, the electrode potential is changed to:

$$
\begin{aligned}
E &= E^{\ominus} + \frac{RT}{2F} \ln 0.1 \\
&= +0.34 - 0.0591 = +0.28 \text{ V} \qquad (7.9)
\end{aligned}
$$

The equilibrium potential continues to fall with further depletion of the concentration of the ions in solution but, in practice, the electrolyte is recirculated before all the ions have been plated out.

7.2.1.2. Cell Voltage Required for Deposition

The voltage required to drive the deposition reaction has been evaluated thus far only in terms of the equilibrium potential, as calculated from the Nernst equation. The actual voltage must be higher in order to overcome other resistances inherent in the process. The most obvious of these are the ohmic resistances arising from the flow of electricity along the bus bars, across the contact surfaces between the bus bars and the anode and cathode plates, and through the electrolyte between the plates. These resistances can be minimized by correct design and operation of the cell, with very small distances between the anodes and cathodes and by control of the conductivity of the electrolyte.

An additional potential drop across the cell can arise from **polarization.** When ions are being discharged by the passage of an imposed current, the cell is no longer behaving in a thermodynamically reversible mode. Consequently, the electrode potentials for the half-cell reactions are not equal to the values listed in Table 2.1, which apply only to reversible conditions. The change in the electrode potential is caused by polarization and the magnitude of the change is represented by the symbol η, which is called the **overpotential.** There are two principal components—concentration and activation polarization—which need to be considered.

The first of these terms can be regarded simply as a potential drop arising from variations in the ion concentration across the gap between the anode and the cathode. The electrolyte is circulated between the electrodes to reduce the concentration gradients and the ions are distributed within the bulk solution by mass transport. However, the deposition of metal on the cathode depletes the ion concentration at the liquid-solid surface and results in a rapid change in composition with distance near to the interface, similar to the concentration gradient across the boundary layer in liquid-liquid and similar reactions. Transport is by diffusion across this relatively static layer. Consequently, the concentration of the metal ions can fall to a very low level at the interface with the cathode, and the applied voltage must be increased above the equilibrium value for the bulk solution to overcome the concentration polarization.

Activation polarization is a result of the activation energy required for an atom in the solid at the anode to transform into an ion in the bulk solution and for the reverse transition at the cathode. In terms of the reaction rate theory (cf. Section 3.1.2 and Figure 3.1), the activated complex formed during the transition can be regarded as a condition of partial solvation of an ion in close proximity to the electrode surface. When an electrode and an electrolyte are in equilibrium, the free energies of the metal as an atom in the solid and as an ion in the solution are equal. The activation energy for the transition is the same in either direction (i.e., from liquid to solid and from solid to liquid), so the rates of dissolution and deposition at the electrode are equal. However, the electrode potential becomes polarized, with the overpotential η V, when the equilibrium is disturbed and atoms are transferred preferentially to (cathodic) or from (anodic) the electrode. The value of η is dependent on several factors, including the compositions of the electrode and the electrolyte and the reactions occurring at the electrode.

Activation polarization at the anode increases ΔG and raises the energy of the metal in the anode by $zF\eta$, since $\Delta G = -zFE$. The energy of the activated complex is also increased by $azF\eta$, where a is a symmetry factor defining the position of the maximum in the reaction coordinate, Figure 7.3. The reverse conditions, with a different value of η, apply at the cathode.

It is readily shown[110] that the net current at the electrode as a result of this polarization is given by:

$$I_{\text{net}} = I_0 \left[\exp\frac{(1-a)zF\eta}{RT} - \exp\frac{(-azF\eta}{RT} \right] \tag{7.10}$$

where I_0 is the difference between the currents entering and leaving the electrode, or the exchange current density. This is the **Bulmer-Volmer equation.** Since all the terms in the equation are constant for a given electrode process at constant temperature and composition, the equation is more usefully expressed as:

$$\eta = b \log I_0 \pm b \log I \tag{7.11}$$

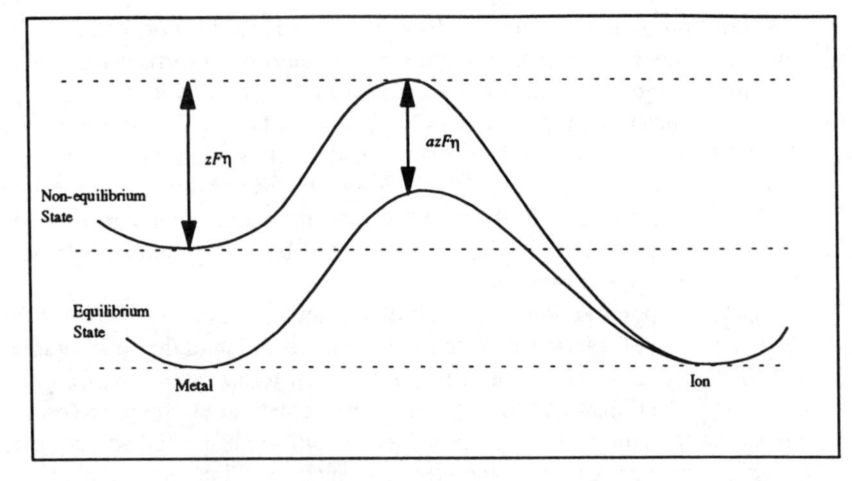

FIGURE 7.3. The effect of activation polarization on the activation energy for an electrolysis reaction.

where

$$b = 2.303RT/(1 - a)zF$$

This is known as the **Tafel equation.** The constant, b, which is the slope of an η vs. log I plot, is called the Tafel slope for the process. The positive sign in Equation 7.11 applies for the anodic overpotential at the anode, while the negative sign applies to the activation polarization at the cathode.

The anodic reaction involves the release of oxygen as a gas when the anode is inert, but a nucleation barrier must be overcome before the gas bubbles are formed and, in similar manner to the nucleation of CO bubbles in molten iron, the bubbles are not released into the atmosphere until the driving force is increased above the equilibrium value. The additional energy required for oxygen release is the anode overvoltage. It is the activation polarization for the release of gaseous oxygen, which is usually much larger than for metal dissolution or deposition, and is an appreciable proportion of the total voltage required. Various techniques can be employed to lower the overvoltage. These include plating the anodes with Pt or other precious metals that catalyze gas evolution (which is expensive), adding a sulfite compound that oxidizes to sulfate ions with the release of electrons:[111]

$$H_2SO_3 + H_2O = SO_4^{2-} + 4H^+ + 2e^-; E^{\circ} = 0.17 \text{ V} \qquad (7.12)$$

or by using hydrogen diffusion anodes to replace the water decomposition reaction by hydrogen oxidation.[112]

Theoretically, in terms of the Nernst equation, an applied voltage of $(+1.23 - 0.34) = +0.89$ V is required to deposit Cu from a 1 molal solution. The actual voltage that must be applied is the sum of the Nernst voltage, the anode and the cathode overpotentials, and the energy required to overcome the ohmic and the electrolyte resistances:

$$V = E + \eta_{(\mathrm{anode})} + \eta_{(\mathrm{cathode})} + IR \tag{7.13}$$

A cell voltage of 2.0 to 2.5 V must be applied for the electrodeposition of Cu. Thus, the efficiency of energy useage is low. This is demonstrated in Worked Example 14.

7.2.1.3. Elecrowinning of Other Metals from Aqueous Solutions

Figure 7.2 shows that the standard electrode potentials for Ag, Cu and Pt lie above the hydrogen line and these metals can be deposited from solution without risk of H_2 evolution. Ag and Pt, together with Au and the other electropositive precious metals, are insoluble or only sparingly soluble in sulfate solutions. Any traces of Ag in solution, for example, are readily precipitated by a small concentration of chloride ions in the electrolyte. These elements are retained in the ore, therefore, and are not coprecipitated with the other metals during electrowinning.

The other common metals are all electronegative with reference to the standard hydrogen electrode. Hydrogen gas release would consume the available electrons in preference to metal deposition if equilibrium conditions prevailed in the electrowinning of these elements from aqueous solutions. Fortunately, equilibrium is not attained in practice because a hydrogen nucleation overvoltage is also required for the cathodic release of H_2 gas. This overvoltage rises with increasing solution temperature and with increasing current density. The applied potential required for H_2 release also varies markedly from one metal to another. Pb, Sn and Zn cathodes are very poor catalysts for the gas reaction and Zn, with $E^{\circ} = -0.76$ V for the half-cell reaction, can be recovered from solution by electrowinning without excessive H_2 formation. However, other metals (such as Co, Fe, Ni and Sb) that are less electronegative than Zn are more difficult to recover because they have higher activation polarization values and function as better catalysts for the gas formation.

The probability of hydrogen evolution is decreased as the pH of the solution is increased (*cf.* Figure 6.3 for Ni) and the acidity of the solution is held at the lowest level at which the metal ions remain stable in the solution for the electrowinning of the electronegative elements. Even then, about 15% of the applied energy is lost through H_2 release when Ni is deposited from $NiSO_4$ solutions with the lowest practical pH value. Manganese, with $E^{\circ} = -1.1$ V, is the most electronegative element recoverable by aqueous electrolysis, albeit with a relatively low efficiency, from solutions with a pH of about 8.0—just low enough to avoid precipitation of $Mn(OH)_2$. The more strongly electronegative elements, including Al, the alkalies and the alkaline earths, cannot be electrowon even from strongly alkaline aqueous solutions.

The hydrogen overvoltage for any metal is lowered if a rough surface is formed on the cathode, since this provides crevices in which the gas bubbles can nucleate. Very small amounts (often in the parts per million range) of organic substances such as gelatin, glue and thiourea are often added to the electrolyte to prevent this from occurring. These substances are adsorbed preferentially onto any protuberances growing outward from the cathode surface and retard further growth at those points, allowing time for deposition to occur over the rest of the cathode area and resulting in the development of a smoother surface. Excessive quantities of these additives interfere with deposition over the whole surface, however, and raise the potential that must be applied to drive the reactions. Thus, very careful control is required to maintain their concentrations in the optimum range in the electrolyte. An alternative practice, which is growing in popularity, is to electropolish the cathode surface by periodically reversing the applied potential for a few seconds. This is an application of a technique that is used to produce a smooth deposit during electroplating, where an AC current is applied and partially rectified by the reactions in the plating bath.

Since Cu is electropositive with respect to all the common impurities soluble in sulfate solutions in the operative pH range, it is evident that the impurities in the solution will not be codeposited with Cu on the cathode if the applied potential is just sufficient to drive the Cu deposition reaction. A reasonably pure deposit of this metal, containing impurities only in the parts per million range, can be obtained directly by electrowinning. At the opposite extreme, all the less electronegative impurities are preferentially extracted before the Zn if they are present in a $ZnSO_4$ electrolyte, so the solution must be purified before electrolysis in order to obtain a pure metal deposit. It is particularly important that all impurities are removed that could catalyze the H_2 release reaction and lower the cathodic overpotential. Mn is the only impurity tolerated in Zn-rich electrolytes, but care must be taken to ensure that the oxygen overpotential is then not high enough to cause MnO_2 deposition on the anode. It is not necessary to remove all traces of impurities having similar electronegativities to the metal being recovered and not affecting the H_2 overpotential, since the steady-state potential for the deposition of an impurity decreases as its concentration in the solution is lowered.

7.2.1.4. Current Efficiency

According to Faraday's First Law of Electrolysis, the passage of 1 Faraday (96,487 C) of energy through the cell should deposit 1 gram-equivalent of any element dissolved in the electrolyte if the cell is operating at

100% efficiency. The gram-equivalent of an element is equal to its atomic weight divided by the valency which its ions exhibit in the solution. Thus, Cu ions are divalent and the atomic weight is 63.57, so the gram-equivalent is 31.79.

Faraday's Second Law defines the maximum weight, W, of an element that can be electrodeposited in a time, t, seconds:

$$\frac{dW}{M} = dm = \frac{I}{nF}dt \tag{7.14}$$

where M is the molecular weight, m is the number of moles, I is the current (ampere, A) and n is the valence of the ion. Hence, if the cell operates at 100% efficiency, the passage of 1 A for 1 h will deposit $[(60 \times 60)/(n \times 96487)]$ mol metal. The theoretical quantities that can be deposited by this amount of energy are, therefore:

Metal	**Co**	**Cu**	**Fe**	**Ni**	**Pb**	**Zn**
(grams)	1.100	1.184	0.695	1.095	3.864	1.219

Alternatively, 1 gram-equivalent of a metal can be deposited if a current of 1 A is passed for 96,487 s, or 26 h 48 min. Obviously, therefore, the rate of deposition should increase as the current density (i.e., A m^{-2} electrode) is raised.

Less metal than the theoretical amount is deposited in practice if some of the energy is lost through unwanted side reactions such as the release of hydrogen gas. A particularly troublesome energy loss is caused by the presence of Fe ions in solution. Iron is electronegative with respect to Cu, Pb, Sn and Ni and is largely or entirely retained in solution during the electrowinning of these metals. However, the iron can be oxidized to Fe^{3+} at the anode and reduced to Fe^{2+} at the cathode, consuming electrons in the process and reducing the amount of metal deposited. A limit of about 0.06 mol l^{-1} Fe^{2+} ions in solution is set in Cu electrowinning for this reason.

The current efficiency, CE, is usually defined as the amount of metal actually deposited multiplied by 100, divided by the amount that should have been deposited according to Faraday's Laws. In aqueous electrowinning, CE is usually in the range 80 to 95%, but it can fall to below 70% if there is much Fe in solution and even lower if hydrogen gas is also released during the recovery of the metal.

7.2.1.5. Aqueous Electrowinning Practice

An electrowinning tank may contain about 50 anodes and cathodes, each with an area of about 1 m^2, arranged in series. The voltage drop across the tank is thus equal to that required to drive the reaction between one anode and cathode pair. A number of tanks are connected in parallel, the bus bar that supplies the cathodes in one tank being connected also to the anodes in the adjacent tank (Figure 1.7h) so that the total voltage drop across the assembly is equal to the DC voltage supplied by the generator. So, for a given DC voltage supply, less tanks can be connected for electrowinning of Ni or Zn (*E* about 3.5 V) than for Cu (*E* about 2.5 V). The electrolyte flows from tank to tank, usually by gravity feed.

The electrolyte is often heated to about 50° C to lower its resistivity. The current density is usually in the range of 100 to 500 A m^{-2}. According to Faraday's laws, the amount of metal deposited should increase linearly with increasing current density, but the voltage required to overcome activation and concentration polarization in the electrolyte also rises with increasing rate of deposition, raising the total energy consumption for each unit weight of metal recovered from the solution. The roughness of the cathode surface also tends to increase as the current density is raised and, consequently, there is a greater risk of hydrogen evolution. As a result, the cost of production falls initially as the current density is raised, but it may pass through a minimum and start to rise again if it is raised beyond a critical value.

As indicated earlier, the anodes are usually prepared from Pb alloys, but these may not be sufficiently inert in strongly acidic solutions obtained, for example, from the solvent extraction of Cu. The pH of the solution must then be increased by dilution. More commonly, surface-oxidized sheets of Ti or a Ti alloy, or sheets coated with precious metal oxides or with titanium carbide (TiC), are used as anodes when the pH of the solution is very low. TiC is a good conductor and both Ti and TiC passivate under anodic polarization. The cathode starter sheet can be a thin plate of the metal to be deposited, or a cold rolled plate of another metal may be used to give greater rigidity. Thus, Cu may be deposited on a Ti sheet precoated with Cu by electrodeposition, while Al sheets are usually used for Zn electrowinning. The cathodes are replaced with fresh starter sheets when they have grown to a thickness of a few centimeters and the anode-cathode gap is becoming too small for adequate electrolyte circulation.

The spacing between the anodes and cathodes is usually 20 to 50 mm and decreases as the metal is deposited on the cathode. In order to limit the voltage drop across the electrolyte, the smallest initial practicable spacing used is that which allow sufficient thickening of the cathode by metal deposition without the risk of short-circuiting due to unevenness or distortion of the surfaces and will allow adequate circulation of the electrolyte between the plates. The spacing must be increased, however, when it is necessary to separate each anode and cathode by a porous membrane or diaphragm to prevent transfer of solids or some impurities to the cathode surface. A diaphragm may also be used, as in the electrowinning of Ni, to separate an acidic anolyte solution from a less acidic catholyte and lower the risk of hydrogen evolution.

Sulfate solutions are most commonly used for aqueous electrowinning; but some elements like Co, Ni and Zn can also be recovered readily from chloride solutions.[113] Copper is more difficult to extract in this way since it can form chloride anions in the solution. It is also difficult to avoid both dendrite growth on the cathode and chlorine gas liberation at the anode.

7.2.2. Electrowinning From Molten Halides

Many metals too electronegative for electrowinning from aqueous solutions can be prepared by electolysis from molten salts. Virtually all the present-day output of aluminum is produced in this way and, in tonnage terms, this is the principal application of the technique. It is also the major method for the production of Mg and the alkali metals.

The fused salt should have a high electrical conductivity and a low melting point to restrict the total energy consumption. It should be capable of dissolving the compound to be reduced, but have a low solubility for the reduced species and should have either a significantly higher or lower density than the metal to aid separation. It must be stable and not produce ions that could be discharged at a lower potential than is required for the production of the metal. The molten salt should also have a low vapor pressure to prevent change of composition of the melt and environmental pollution by volatilization. In some cases, as with Mg production, these conditions can be achieved in a simple chloride melt containing the metal chloride. In other cases, as with Al which is electrolyzed from the oxide, a cocktail of halides is required with a composition approximating a multicomponent eutectic point to give a suitably low working temperature range.

Since the metals produced from fused salts all form very stable oxides and are strongly electronegative, the ores must be purified to remove the less stable impurities prior to electrolysis. However, the benefits of prepurification would be lost if the metal product became contaminated during reduction, so the components added to form the salt melt and the materials used as electrodes and to line the reaction vessel must also be as free as possible from readily reducible impurities. The partial pressure of oxygen is very low beneath the surface of the melt, so carbon is suitable as a refractory if it is sufficiently pure. Petroleum coke is commonly used for this purpose, but it is expensive and adds significantly to the cost of extraction. The metal would readily reoxidize if it was exposed to air while still molten; thus, provision must also be made to prevent reoxidation.

7.2.2.1 Aluminum Production

Aluminum is produced by electrolysis of bauxite that has first been purified by the Bayer pressure leaching process (see Section 6.2.1). The Bayerite, $Al(OH)_3$, product obtained from the leach circuit would cause explosive release of steam if it was added to the molten salt, so it is first dehydrated in a rotary kiln to produce a pure anhydrous alumina (i.e., corundum, Al_2O_3) feedstock. The preparation cost of the ore can represent as much as 40% of the total cost of Al production.

The alumina is dissolved in a molten mixture of Na and Al fluorides and reduced to the metal in a Hall-Heroult furnace, Figure 1.7a. The stoichiometric composition of the compound, cryolite (Na_3AlF_6), melts at 1010° C, but the melting point is lowered by addition of AlF_3 and LiF, both of which also lower the solubility of Al and the surface tension, density and viscosity in the melt. Small amounts of CaO are introduced unintentionally as an impurity in the alumina. This forms CaF_2 which accumulates over a period of time to reach a steady-state concentration of about 5% in the salt, any excess thereafter being reduced into the metal. CaF_2 also lowers the melting temperature of the halide mixture and the solubility of Al, but it increases the surface tension, viscosity and density and lowers the solubility of alumina in the salt.

A ternary eutectic is formed, which melts at about 920° C, in a salt containing a weight ratio, NaF/AlF_3, of 1.1 in the presence of 5% CaF_2. The solubility of alumina increases with rising temperature, but the melting point rises rapidly with more than 20% alumina in solution. In practice, the operating temperature is usually in the range 950 to 980° C with up to 10% Al_2O_3 in the electrolyte. This relatively low superheat results in incomplete fusion of the salt mixture and formation of a solid crust above the melt. The crust acts as both a thermal barrier that lowers heat loss to the atmosphere and a transport barrier that restricts oxygen ingress to the melt. The metal product has a higher density than the electrolyte and collects as a molten layer below the salt, providing further protection against reoxidation. The density difference is small, however, and any disturbance at the metal-halide interface can adversely affect the metal separation.

The carbon lining serves as the cathode. Carbon anodes are suspended above the cell and project through the crust into the molten salt. Up to 30 prebaked carbon blocks extend into each furnace. Some plants still use one large Soderberg electrode, formed continuously by feeding carbon paste inside a steel shell suspended above the melt. The paste is baked into a solid by the heat within the cell.

The electrolyte can be regarded as ionized by reactions such as:

$$2AlF_3 = Al^{3+} + AlF_6^{3-} \tag{7.15}$$

$$NaF = Na^+ + F^- \tag{7.16}$$

$$Al_2O_3 = Al^{3+} + AlO_3^{3-} \tag{7.17}$$

The cathodic reaction is:

$$Al^{3+} + 3e^- = Al \tag{7.18}$$

Oxygen gas is released at the anode, so the anodic reaction can be visualized as:

$$4AlO_3^{3-} = 3O_2 + 2Al_2O_3 + 12e^- \tag{7.19}$$

The trivalency of the aluminum ion results in a reduced yield of metal per Faraday of electricity (i.e., $\frac{1}{3}$ mol F^{-1} or 9 g F^{-1}) passed through the cell, in comparison with the yield in the electrowinning of a divalent metal. This partially accounts for the high energy consumption in the production of Al.

The standard electrode potentials listed in Table 2.1 are valid only at 25° C, so the equilibrium potential that must be exceeded to drive the reaction at the operating temperature must be calculated from the standard free energy of formation of alumina at that temperature. The overall reaction per mole of oxygen gas is:

$$\tfrac{2}{3}\,Al_2O_3 = \tfrac{4}{3}Al + O_2 \tag{7.20}$$

and the standard free energy for this reaction at, say, 950 °C (1223 K) is +858,900 J. Four electrons are transferred from anode to cathode for each mole of oxygen gas released; so, using Equation 2.116, one obtains:

$$E^{\ominus} = -\frac{\Delta G^{\ominus}}{zF} = -\frac{858{,}900}{4 \times 96{,}487} = -2.23\ \text{V} \tag{7.21}$$

But this is not the complete story. The oxygen gas generated by the reaction combines with the carbon in the anode to form CO and CO_2; this lowers the total energy that must be supplied. If all the oxygen is converted to CO_2, the actual reaction should be written as:

$$\tfrac{2}{3}Al_2O_3 + C = \tfrac{4}{3}Al + CO_2 \tag{7.22}$$

for which the standard free energy change is +462,600 J at 1223 K, and the reversible electrode potential, $E^{\ominus}$, for the reduction reaction is lowered to −1.20 V. The gas released is actually a mixture of CO and CO_2, so the true value of $E^{\ominus}$ is somewhat greater than −1.20 V but is certainly lower than −2.23 V.

This calculation assumes that all the condensed species are present at unit activity and that the gas is released at a pressure of 1 atm. The molten Al and the carbon anode are both of high purity, but the activity of the alumina in the electrolyte is less than 1.0 if the melt is not initially saturated with the oxide and its activity falls if the concentration is not replenished continuously as the metal is produced. Any deviation of the activity of the alumina below unity makes the electrode potential more negative. The change is relatively small, however, and is only a small component of the actual potential of −4.5 to −5.0 V applied in practice to drive the reaction at a satisfactory rate. The additional voltage is required, as in aqueous electrolysis, to overcome the ohmic resistance in the electrolyte, the electrodes, bus bars and contacts, and the polarization resistances. A battery of cells is connected in series to create a voltage drop equal to the DC power supply.

Modern cells can operate with a current of up to 300 kA and with a current efficiency of 90 to 95%, the reduced efficiency resulting mainly from reoxidation of Al in solution in the electrolyte by the CO_2 released at the anode. Thus, one cell can produce about 2.5 t metal per day. This is a high productivity in comparison with that for an aqueous electrowinning cell, but is very small in comparison with the output from most of the pyrorefining processes.

The overall energy efficiency is less than 50% and the most efficient cells consume about 12.8 MW energy per ton metal produced. The excess energy is not wasted completely, however, for it is partially released as heat within the furnace. Heat is required to raise the temperature of the charge to the operating temperature. An alternative heating source would be required if the necessary energy was not supplied by the electrical inefficiency.

When the high cost and low efficiency of electricity generation from fossil fuels is taken into account, it is not surprising to find that aluminum production tends to be concentrated in countries such as Canada, Indonesia, Norway and Venezuela, where there is a surplus of hydroelectric power remote from large conurbations and hence the power is available at low cost. More recently, aluminum extraction plants have been constructed in oil-producing countries including Abu Dhabi, Bahrain, Dubai and Iran. The annual rainfall in these countries is not sufficient to supply the needs for domestic and agricultural use, and coupling aluminum production with water desalination results in a financially viable operation. The gas released from the oil wells is used to produce electricity in excess of local needs and the surplus is applied to the production of Al. The waste heat from electricity production and from the smelter is used for the desalination of seawater. In 1992, the profitability of Al production increased in those countries where the power was produced from oil well gas, but profits decreased for the other major Al producers.

Until comparatively recent times, the Al produced from a newly relined cell had a purity of only about 99.6%. The purity increased progressively as the solutes in the lining were removed by dissolution into the salt and transfer into the metal. The metal was tapped periodically from the furnace and graded according to the impurity content of each batch. The purity reached a maximum when all the soluble impurities had been removed from the surface layer of the lining. Eventually, it began to decline again as very small amounts of sodium (which had dissolved in the bath) penetrated through the carbon lining to the outer steel shell and acted as a transport medium to transfer some Fe to the bath. Now, with high purity refractories and close control of the electrolyte composition, it is possible to produce consistently by electrowinning metal with a purity of 99.999%[5] that requires no further refining for the most demanding applications.

7.2.2.2. ***Production of Other Metals***

Magnesium is obtained by electrolysis of $MgCl_2$ dissolved in molten NaCl, with the addition of sufficient $CaCl_2$ to lower the melting temperature and allow the cell to be operated at 700 to 750° C. If anhydrous $MgCl_2$ is charged to the cell, the reaction produces Cl_2 gas:

$$MgCl_2 = Mg + Cl_2 \tag{7.23}$$

which does not attack the carbon anode and, since Mg and Fe are virtually immiscible, the melt can be contained in a steel tank that also serves as the cathode. Molten Mg is less dense than the electrolyte, so it floats to the top where it must be protected both from reoxidation by the atmosphere and, by vertical curtains, from recombining with the Cl_2 gas released at the anode.

It is not possible to produce pure $MgCl_2$ by dehydration of the salt. Most of the combined water can be evaporated, but increasing amounts of Mg are hydrolyzed to MgO when the moisture content falls below about 2 wt%. The anode consumption is raised markedly by the formation of CO_2 gas if the partially hydrolyzed chloride is fed to the melt and some MgO is lost as a sludge that must be removed periodically from the bottom of the tank. A pure anhydrous feedstock is normally preferred and this is prepared by chlorination of MgO in the presence of carbon at about 1000° C, producing CO_2 gas and molten $MgCl_2$ that can be fed directly to the electrowinning cell. The theoretical reversible potential required for Mg electrolysis is −2.45 V, but an actual potential of 5 to 8 V is required to drive the reaction and the overall energy consumption is 16 to 18 kWh kg^{-1} Mg.

Sodium is produced similarly by electrolysis of a NaCl-$CaCl_2$ eutectic salt mixture. Both Na and Ca are produced at the cathode. The metals are separated by cooling to just above the melting point of Na, where the Ca separates as an immiscible solid phase. Precautions must be taken to exclude oxygen from the cell and prevent spontaneous reoxidation of the metal. Sodium can also be produced at a lower temperature from fused NaOH. Other metals, such as Cr, Nb, Ta, Ti, V, W and Zr, can also be extracted by fused salt electrolysis, but this is not economically attractive at present.

Extensive efforts have been made in recent years to decrease the energy consumed by fused salt electrolysis and the energy efficiency has been

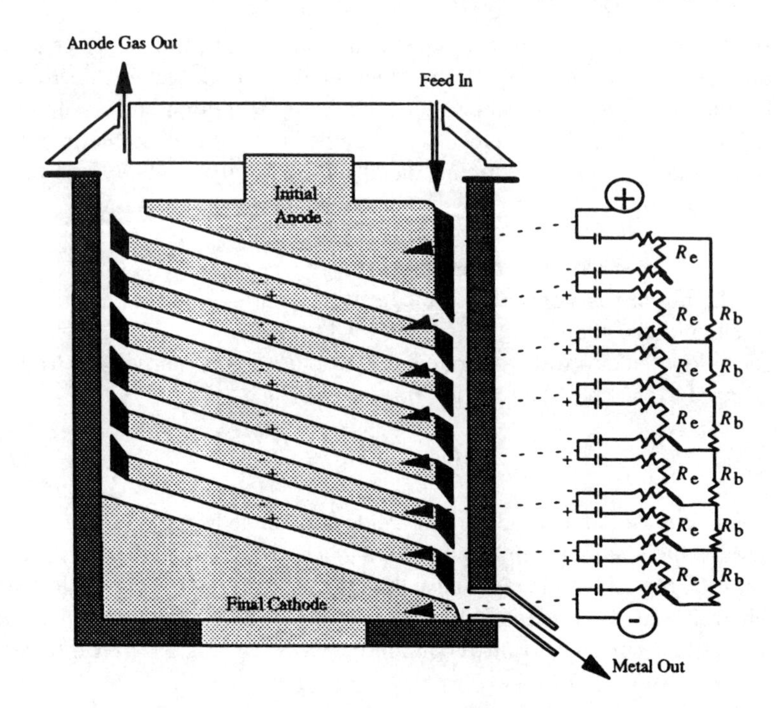

FIGURE 7.4. Schematic diagram of a bipolar cell. (From Griotheim, K. and Welch, B., *JOM*, Nov. 12, 1989, With permission.)

markedly improved, but it is still less than 50% and there seems to be only faint hope for further significant improvement with the existing cell design. Attempts are being made, therefore, to develop alternative cell designs and the bipolar arrangement, illustrated in Figure 7.4, appears to offer promise of more efficient performance. This cell contains a number of graphite slabs, each of which functions as an anode on one surface and as a cathode on the opposite face, spaced between the terminal anode and cathode at the ends of the cell. For efficient operation, the electrolyte resistance, R_e in Figure 7.4, must be markedly lower than the by-pass resistance, R_b, around the ends of the electrode plates. It is claimed that the total energy consumption can be reduced to less than 12 kW kg^{-1} for Al production from a chloride salt melt[114] and to 9.5 kW kg^{-1} for Mg production[115] using a bipolar cell.

7.3. ELECTROREFINING

An electrorefining cell is similar in design to those used for electrowinning, but the inert anodes are replaced with slabs, a few centimeters thick, of the impure metal to be refined. The anodes are replaced when they have dissolved and become so thin that there is a risk they will disintegrate.

The cathodes are, again, thin starter sheets of the metal to be deposited or sheets of some stronger material that have been precoated with that metal. The electrolyte is normally an acidic solution in which the anode metal can dissolve but in which, ideally, the impurities are insoluble.

The reversible cell potential is zero if the activity of the metal is the same (e.g., 1.0) in the anode and in the cathode. The anode is not pure but, with the exception of some Ni anodes, the impurities seldom exceed 1.0% and are often present at much lower concentration, thus, the activity of the host metal is very close to unity in the anode and the reversible potential is very small. As with electrowinning, a larger potential is required to overcome the other resistances in the circuit. Since the electrons arriving at the anode are consumed by the formation of metal ions, there is no requirement for oxygen gas release as in electrowinning. The largest component of the additional potential required in this case, therefore, is usually the electrolyte resistance, and the solutions are often heated to 50 to 70° C to increase the conductivity. The voltage that must be applied ranges from 0.2 to 0.25 V for Cu to 1.5 to 2.0 V for the refining of relatively impure Ni.

Electrorefining is particularly useful for the removal of impurities such as Bi that have a very detrimental effect on the properties of the metal and cannot be removed adequately from metals like Cu and Pb during pyrorefining but are insoluble in the electrolyte. It is also useful for the removal from smelted metals of impurities having similar or higher reducibilities with carbon. Gold, silver and other precious metals in an ore may be extracted into a metal produced by pyrometallurgy. For example, the Au and Ag contents of Cu anodes range from 8 to 73 ppm and from 90 to 7000 ppm, respectively.[116] These valuable elements are either insoluble in the electrolyte or, if they do dissolve, are readily precipitated by cementation since they are electropositive with respect to all the common engineering metals. They accumulate in a slime or sludge in the bottom of the electrolysis tank, together with other rare elements such as Se and Te, from where they are recovered as values that can markedly lower the cost of electrorefining.

Metals are primarily refined with sulfuric acid electrolytes, but other solutions are also used. Pb is insoluble in H_2SO_4, so this element is refined from a fluosilicate, $PbSiF_6$, solution containing hydrofluosilic acid, H_2SiF_6.

In addition to Ag and Au, copper anodes may contain Ni, Pb and oxygen in concentrations equal to or greater than 0.4%.[116] Complex oxides containing As, Bi, Ni, Pb and Sb may also be present, in addition to selenide and telluride compounds, in Cu with a high oxygen content.[117] Some of these dissolve partially or completely in an H_2SO_4 electrolyte, but most of the compounds may be found in variable amounts in the anode sludge. Small quantities of As, Ni and Sb are transferred to the cathode only when the anode has a very low oxygen content. Since the soluble impurities are present in only small amounts and they are electronegative with respect to

Cu, they remain in solution in the electrolyte. Thus, the solution is circulated through the electrorefining tanks and then to an external circuit in which the impurities are removed to prevent them from building up to concentration (activity) levels at which they might start to codeposit with the Cu.

When Ni is electrorefined from sulfate solutions, the anodes may contain 2% or more of Cu, together with variable amounts of Co and Fe as the main impurities requiring removal. These elements are above, or close to, Ni in the electromotive series (Table 2.1) and could coprecipitate on the cathode. This is prevented by inserting a permeable membrane or diaphragm between each anode and cathode. The elements dissolve from the anode in the anolyte section and the solution flows to an external circuit, where the Cu is precipitated either as the metal by cementation with solid Ni or as Cu_2S by the addition of sulfur. Co and Fe are precipitated by oxidation with Cl_2 and air, respectively. The purified and filtered solution is returned to the catholyte section where the Ni is electrodeposited. A higher hydrostatic pressure is maintained in the catholyte than that in the anolyte to ensure that the electrolyte flows through the diaphragm from the cathode to the anode sections and prevents the impurities reaching the cathode. The presence of the diaphragm and the consequent increase in the spacing between the anode and the cathode largely accounts for the higher voltage required for the electrorefining of Ni. The electrolyte resistivity is also higher than for the refining of Cu since a higher pH of about 4 must be maintained in the solution to prevent hydrogen evolution.

A diaphragm cell is also used for the production of pure Ni from anodes cast from Ni_3S_2 matte. The anodic reaction is then:

$$Ni_3S_2 = 3Ni^{2+} + 2S + 6e^- \quad (7.24)$$

for which $E^{\ominus} = -0.35$ V and the applied voltage must be increased accordingly. The solution is again externally purified, but the amount of slime formed in the anolyte section is increased markedly by the precipitated sulfur, so the anodes are enclosed in canvas bags to collect the slime. It would be an attractive process if pure Cu could also be produced directly from copper matte, thus avoiding the release of SO_2 in the pyrometallurgical converting operation. Unfortunately, Cu_2S passes through numerous stages of decreasing S/Cu ratio during electrolytic reduction, and the porosity of the sulfide layer is insufficient to sustain the reaction without a marked increase in the applied potential.

The slimes released from the anode impurities normally accumulate in the bottom of the tank, but they can also form a coating over the surface of the anodes. Porous deposits form and detach from the surface as the underlying metal is dissolved. However, nonporous deposits of compounds such as Cu_2O and Cu_2S can passivate the surface of the anodes and bring the dissolution process to a halt. The operating conditions must be carefully controlled to prevent the formation of adherent layers. The skeleton sulfur residue remaining on the surface during the electrolysis of Ni matte anodes similarly restricts the diffusion of ions to and from the dissolving surface, and the required voltage for continued dissolution increases progressively as the anode is dissolved.

7.4. ECONOMIC ASPECTS

The purity required for some applications of nonferrous metals can only be achieved economically by electrolysis, although the production rates are very low in comparison with most of the other methods of metal manufacture and the operating costs are increased by the relatively high energy consumption and the cost of electric power. Whereas the metal produced from the pyrometallurgical processes can be cast directly while still molten into artifacts or into semifinished shapes, the cathodes provided by electrolysis must first be remelted to refine the crystal structure and degassed to remove dissolved hydrogen, adding further to the total cost of the metal.

On the other hand, the processes are environmentally friendly. The electrolytes can be purified and recirculated either to a leaching circuit in aqueous electrowinning or to the electrolysis tank in electrorefining. Chlorine is liberated from fused salt electrolysis of metal chlorides, but the gas is consumed and replaced by a similar volume of CO_2 in the preparation of the chloride feedstock. No chlorine can escape to the atmosphere if the equipment is gastight, although a cost penalty is incurred in lining the equipment with chlorine-resistant materials and in providing gastight seals. CO_2 is also released in the Hall-Heroult type process for the electrolysis of metal oxides but, in both cases, the total amount of oxygen converted to CO_2 is derived only from the metallic mineral when the energy is supplied by hydroelectricity. Acid fume may contaminate the atmosphere in the vicinity of aqueous electrolysis tanks and must be removed, but no other pollutant gases are released into the atmosphere. The main environmental requirement is the neutralization of spent liquor and the conversion of noxious solutes into insoluble compounds before final discharge.

Hence, the cost disadvantages of electrolysis are diminishing as increasingly stringent environmental constraints are imposed on metal extraction processes, and it is probable that an increasing proportion of the production, particularly of the less common metals, will be obtained by this route.

FURTHER READING

Potter, E. C., *Electrochemistry,* Cleaver Hume Press, 1961.

Romankiw, L. T. and Turner, D. R., *Electrodeposition Technology, Theory and Practice,* Electrochemical Society, Pennington, NJ, 1987.

Winnard, R., *Application of Polarization Measurements in the Control of Metal Deposition,* Elsevier, New York, 1984.

Chapter 8

ENVIRONMENTAL ISSUES

8.1. INTRODUCTION

Environmental issues evoked little general interest until comparatively recent times. Legislation to restrict pollution was limited in scope and was often not strictly enforced. Attitudes have now changed markedly and preservation of the environment is at the forefront of public concern. Increasingly more stringent restrictions are being imposed and applied more rigorously on activities that could render damage to the ecology, and these have important consequences for the manufacturing industries. Unfortunately, the pendulum of public concern has swung too far, and decisions are sometimes being biased by emotive arguments, unsupported by proven fact, that can be counterproductive to environmental protection.

The extractive metallurgical industry has been a major contributor to ecological degradation. Damage was caused both in the initial production of the metal from the ore and in the disposal of the metal at the end of its useful life. The works could often be identified at a distance by the dense plumes of smoke emerging from the chimneys and, at closer range, by a coating of grit falling from the atmosphere and coating everything in the vicinity. Vegetation was sparse or completely destroyed by the fumes. The locations of mineral extraction sites were all too frequently characterized by spoil heaps marring the landscape. These conditions are still apparent in some parts of the world, though the situation in most countries has improved dramatically during the last 30 or 40 years.

The industry is now a relatively minor contributor to ecological degradation in comparison, for example, with the many problems arising from overpopulation and domestic consumption. These changes have been brought about during a period in which metal production has increased very rapidly. It is estimated that the output of metals during the 1980s was more than five times greater than the total quantity produced from when mankind first started to use metals in the Bronze Age up to the beginning of the 20th century. The ecological damage now caused by the industry is often important only on a local scale, where it can be more easily controlled.

The improvement has not resulted entirely from altruistic motivation of the manufacturer. Energy is a major factor in the cost of metal production. The increases in the price of oil in the 1970s and again in the early 1980s were quickly reflected in the price of coal and gas fuels. Action was taken to contain the effect on the cost of production, resulting in a rapid decrease in the energy consumption by the extraction industries. Marked improvements in energy efficiency have led, in turn, to marked reductions in the amounts of CO_2 and SO_2 released into the atmosphere per tonne of metal product. Efforts to reduce fuel consumption and increase productivity predate the sudden increases in the prices of fuels, which merely stimulated the activity. Improvements in the overall efficiency resulting from a better understanding of the controlling parameters of existing processes and the development of new processes, such as flash smelting, made a major contribution. Industry has also responded to laws limiting the emission of noxious and toxic substances. Fume and particulate discharges have now been largely or entirely eliminated. There are many examples of spoil heaps that have been covered with soil and merged into the landscape to create a better environment and of open-cast mining sites that have been fully restored for agricultural use. Regrettably, these improvements have not yet been effected in some Eastern and Third World countries that still allow severe environmental pollution to continue without control. The public image of the industry is sometimes distorted by reports of the damage caused by obsolete processes and practices still operated in those areas.

The extractive industry in some parts of the world, notably North America and Western Europe, is now faced with the danger that restrictions are being imposed that cannot readily be achieved. Actions required to limit ecological damage and that do not also increase the efficiency of the process increase the cost of the product and can alter the balance between financially viable and nonviable operations. Extractive operations could be forced to cease in those countries that impose the most stringent requirements, with the consequent loss of employment and adverse effect on the balance of trade, and be transferred to other countries where ecological damage is less strictly controlled.

Some legislation on pollution is counterproductive. Wet batteries are a major source of Pb scrap metal, but the U.S. and some other countries impose limits on Pb release into the atmosphere which cannot be achieved economically by the secondary (scrap) metal melter. Thus, the remelters are forced to close down and old batteries are incorporated into landfills, thereby creating a greater hazard than arises from the small amounts of fume released on remelting. Meanwhile, more Pb ore must be extracted from the earth and processed to replace the metal stock that otherwise would have been provided by recycling the batteries. Numerous other examples can be cited.

The technologist employed in the metal-producing industry has a responsibility to ensure that the various operations are conducted in the most environmentally acceptable manner. However, he or she also has a responsibility to persuade legislators to make decisions based on proven information or well-supported hypotheses and after consideration of all the probable consequences (such as the loss of employment and Pb battery disposal) and not in response to unsubstantiated emotional arguments.

8.2. POLLUTION SOURCES AND CONTROL

Waste products from metal production processes may be released as solids, liquids, gases and as mixtures of these (e.g., fumes, slimes and slurries). Some discharges, such as uncontaminated cooling water, do not cause chemical damage to the environment and can be released without treatment. However, if they are discharged while still hot, they can cause thermal pollution that can lead to changes in the ecological environment (such as the growth of algae). Most of these wastes constitute an actual or potential hazard that should be eliminated as far as is practicable before release.

8.2.1. Solids

Mining and beneficiation treatments produce solid waste in quantities that vary over a wide range, from relatively small amounts with high-grade iron ores to hundreds of tonnes of waste for each tonne of nonferrous metal extracted from very low-grade ores. The total quantity of waste produced annually has increased markedly as a result of the rise in the amount of metal produced, the exploitation of lower grade, nonferrous metal ore deposits and improvements in the beneficiation methods that produce a richer concentrate and reject more of the gangue materials at this stage. There is often little or no commercial outlet for clay-like constituents. Crushed rock can be used as aggregate for concrete, road foundations and for other civil engineering works, but the mines and quarries are often remote from large conurbations and transportation costs may restrict this utilization. Some may be disposed of as backfilling, but spoil heaps are usually an unavoidable consequence of deep mining operations.

Comminution of the rock may expose minerals neither wanted nor recovered by flotation, etc. but which can be leached subsequently by surface waters. Exposed metal sulfides in underground mines may also be leached when the workings are abandoned if they subsequently become flooded by water percolating through the surrounding strata. Legislation should ensure that the mine owner has a responsibility to test and, where necessary, treat the water in the mine and the runoff from spoil heaps until the risk of pollutant release has been eliminated.

Slag and/or dross is produced in almost all pyrometallurgical smelting and refining operations. The amount of slag generated per tonne of metal has decreased markedly with the use of higher grade iron ores and richer nonferrous concentrates and with better understanding of the optimum conditions that must be maintained to secure maximum retention of the impurities in the slag. Whereas, 40 years ago, European blast furnaces produced about 1 tonne of slag for each tonne of molten Fe, the modern furnace smelting high-grade ores generates only about 250 to 350 kg t^{-1}. In some cases, refining has been separated into discrete stages, as in the ladle refining of steel, with slags formulated to remove specific impurities at each stage and with a lower total slag volume than when attempting to achieve the same degree of metal purification with a single slag. When a slag does not contain harmful impurities, which might revert to the metal, it is common practice to recirculate the slag obtained from the later stages of metal production back to an earlier stage to recover the metal values. Sometimes, it is worth treating the slag obtained from the production of one metal to recover another valuable element that has partitioned preferentially into the slag. Drosses are also recirculated to an earlier stage in the process and may be a valuable source of other elements. Usually, only small amounts of metal remain as insoluble compounds when the slag or dross is finally rejected as waste.

Slag can be sold for a variety of applications, ranging from aggregates in concrete and road and railroad ballast to foamed products used for high-temperature insulation. The composition of the molten slag may have to be controlled when it is to be used subsequently for these applications in order to prevent the formation of compounds such as calcium orthosilicate ($2CaO{\cdot}SiO_2$) that undergo a crystallographic transformation with a large volume increase during cooling. The volume change causes the slag to disintegrate into a powder.

When iron was extracted from low-grade ores with a high phosphorus content, the calcium phosphate concentration in the steelmaking slags was high and the slag could be sold as an agricultural fertilizer. This is not pertinent to present-day practice, but crushed basic slag with a high lime content is still marketed as a conditioner for acid soils. In most cases, slags do not constitute any environmental hazard.

Solid residues from leaching circuits should be washed thoroughly and neutralized to remove or immobilize all traces of the lixiviant. The residues are normally dewatered to reduce the bulk before disposal. Very fine residues are often buried or coagulated to prevent dispersal as dust.

Airborne particulate material may be released into the atmosphere whenever ores, fluxes or fuels are blended or discharged from hoppers or are transported. High wind velocities may cause dust losses from stockpiles, but this can be avoided by spraying the surface of the pile with water or a polymer coating agent. The risk of particulate loss increases as the average particle size is diminished and, partially for this reason, the final stages of comminution (milling and grinding) are often conducted in wet circuits. The proportion lost into the atmosphere is diminished sharply after the fines have been sintered, pelletized or briquetted.

Particulate matter, ranging in size up to about 50 μm but mainly less than 1.0 μm in diameter (fume), is transported by the hot gases leaving most of the pyrometallurgical processes. The composition of the particulates depends upon the metal being treated. For example, iron oxides are the major components from steelmaking furnaces, with smaller amounts of the oxides of Al, Ca, Mn, P and Si. Oxides of other elements such as Cu, Mo, Pb and Zn may also be present when scrap steel is remelted, while the fumes from Cu processing may contain the oxides of As, Cu, Pb and Zn, together with particles of the fluxes and fuels.[118] Fumes may also be emitted from the surface oxidation of a molten metal when it is tapped from a furnace and poured into a ladle. Escape of these species into the atmosphere can be prevented by collecting the effluent gases under suction hoods and passing them through wet or dry cyclones, nylon or fabric filter bags, venturi scrubbers, electrostatic precipitators, etc. where the dust and fume are removed before the gas is vented into the atmosphere. The solids recovered are often sufficiently rich in metal values to justify consolidation by sintering or other means and return to the extraction circuit. However, continuous recirculation may be unacceptable due to the accumulation of hazardous volatile impurities such as As, Bi and Sb in the dust, and it may then be preferable to divert the residue to a hydrometallurgical circuit, or it may be diverted to alternative circuits for the recovery of other metals.

8.2.2. Gases

Carbon dioxide and water vapor are commonly regarded as the two largest contributors to the "greenhouse" effect of global warming, with lesser contributions from other species such as hydrocarbons and nitrous oxides, NO_x. Acid rain, defined as rainfall with pH less than 5.0, is caused by SO_2, SO_3 and NO_x combining with water in the atmosphere. All these gases may be released at some stage during metal extraction. In addition, CO may be released by incomplete combustion, fluorine and gaseous fluorides are released in Al electrowinning and other species including bromine and chlorine may be released from some processes.

8.2.2.1. Carbon Dioxide

Carbon dioxide absorbs energy strongly in the infrared wavelengths and this is why it is often regarded as a major contributor to global warming. However, there are many dissenting arguments. The variation over the last 150 years of the CO_2 concentration in the atmosphere and the air temperature, averaged from a number of compilations, is illustrated in Figure 8.1. Actual mean temperatures may vary by more than ±0.2 °C from year to year. A general trend is apparent, but it is evident that variation in the CO_2 content is not solely responsible for the temperature changes. Indeed, the temperature change in recent years is statistically not separable from natural fluctuations.[119] Reports of temperature increases are not a new phenomenon. For instance, a paper published in the 1850s by J. Glaisher, a superintendent at the British Royal Observatory, reported an increase in the average temperatures during the previous 100 years. There followed a much cooler period toward the end of the 19th century.

A caveat must be entered here. The earlier measurements, which indicated variations in CO_2 concentrations between 200 and 300 ppm over the last 15,000 years, were obtained from analysis of air bubbles trapped in

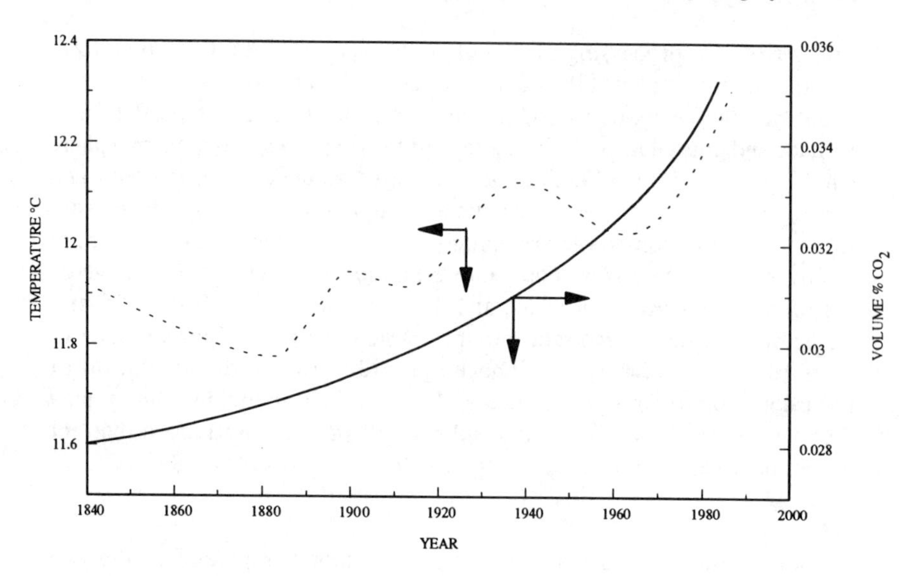

FIGURE 8.1. Smoothed variations of the carbon dioxide content and average temperature of the earth's atmosphere.

polar ice caps, whereas, more recent data were derived from air measurements at a small number of fixed sites. In both cases, it is commonly assumed that the local concentrations truly represent the variation in the average concentration in the whole of the earth's atmosphere. Likewise, the method, accuracy and locations of the temperature measurements have varied markedly over the time period indicated on the diagram and the error band must be wide.

Data obtained over the last 20 years should be more reliable; but there has been a major switch from coal to oil and gas as the fuels for heat and electricity generation during this period, and this has caused a marked decrease in the amount of SO_2 liberated into the atmosphere from combustion of the fuel. It is now recognized that SO_2 causes condensation of atmospheric moisture and the formation of clouds that reflect the infrared solar radiation back into the stratosphere. Hence, the increase in the average temperature of the earth, recorded over the last 20 years, could be dependent more on the decrease in the SO_2 content than on the reported increase of 0.003% in the CO_2 concentration. The actual variation in the recent temperature trend is too large for it to be used as a basis for accurate predictions of the effect of CO_2 on global warming.

There is a danger that legislation on CO_2 release, which could seriously affect the future of the metal extraction industries, is being enacted on the basis of a "worst case scenario" to allow for the more extreme predictions. There is evidence of the problems that can be created by incorrect interpretation of data. In the 1960s, the Club of Rome projected the rate of consumption of various commodities against the reserves proven at that time and forecast that the world's resources of oil and the ores of some of the common metals would be exhausted by the present time. This led to the formation of the Oil Cartel and the rapid escalation in the price of oil, followed by a recession that severely damaged the economies of the poorer countries and was only ended when vast new reserves of oil were discovered in the North Sea, Alaska, Venezuela and elsewhere. Despite increased rates of metal production, there are sufficient proven reserves of the ores of all metals to satisfy demand over the next few decades and new reserves are still being discovered. Similarly, predictions made on the basis of limited data suggested that the world was entering another Ice Age only a few years before reports of global warming captured the headlines.

Metal extraction and refining generates less than 1% of the carbon gases released by the combustion of fossil fuels. Only about 5% of the total amount of CO_2 released into the atmosphere is, in turn, produced by fuel combustion, the major contribution emanating from living and decaying vegetation.[120] Thus, cessation of all metal extraction activities would have negligible effect on global carbon emissions. Fossil fuels are a nonrenewable resource, however, and there is a strong argument, apart from profitability, for the development of energy-efficient processes to conserve these natural resources. Significant progress has been made in this direction during the last few decades. Progress is still being made, but the difference between the actual fuel consumption and the theoretical minimum, evaluated for example from the free energy of dissociation of the mineral compounds, has decreased markedly. Thus, the potential for further decrease in energy use by the best practices is small in comparison with the economies that have already been obtained. A few examples can be used to demonstrate the changes that have been effected.

Since iron and steel constitute 93% of the total production of all metals and the output is obtained from pyrometallurgical processes, it is evident that this industry consumes a major portion of the fossil fuels used for extraction. Forty years ago, a typical blast furnace consumed 750 to 1000 kg of coke for each tonne of molten iron produced. The most efficient furnaces today are using only 450 kg t^{-1} of coke and other forms of carbon. This improvement has resulted from numerous changes, including:

1. The use of larger furnaces with less heat loss per ton of metal
2. Smelting higher grade ores with a consequent decrease in the amount of gangue that has to be melted
3. Markedly lower slag volumes resulting from both the decrease in the gangue content and a better understanding of the slag chemistry that has enabled smaller amounts of slag to achieve the same amount of solute removal

4. Improved burden preparation (e.g., sintering and pelletizing) and distribution to avoid channeling within the furnace
5. Improved control of the balance between direct and indirect reduction of the oxides
6. Higher hot blast temperatures, oxygen enrichment of the blast and high top pressure operation

Larger coke ovens are now being built, thus diminishing the heat losses. The sensible heat is being recovered from the coke as it is discharged from the ovens by dry quenching with recirculated air and using the recovered heat to generate electricity. These changes have markedly reduced the total amount of energy expended in the manufacture of metallurgical coke, and further savings in coke consumption are being achieved by the injection of gaseous and solid fuels into the furnace to reduce the demand for coke. Improved methods of monitoring and control (see Chapter 9) have also contributed to the increased efficiency of the blast furnace and other processes.

The open-hearth steelmaking furnace, which consumed large amounts of energy to heat the metal, has now almost entirely been replaced by the autogenous BOS converter process and by the more energy-efficient electric arc furnace. Significant savings are being achieved by the progressive substitution of continuous casting of the molten steel into square, round and flat sections to replace the traditional ingot casting route. Whereas the yield of semifinished products is only 78% via the ingot route, 89% of the molten metal is converted into useful material by continuous casting. Thus, a smaller quantity of ore must be mined, crushed, sintered, smelted and refined to produce the same tonnage of end products by continuous casting, with a marked reduction in the energy consumed per tonne of finished steel.

The British Steel Corporation, plc, recorded a decrease in energy consumption, excluding the energy required for electricity generation, from 28 GJ t^{-1} cast steel in 1961 to 20 GJ t^{-1} in 1988. Looking to the future, a further decrease in energy demand, with complete elimination of the requirement for metallurgical coke, should arise with the commercial development of the continuous smelting reduction processes described in Section 4.4.2.

The nonferrous metals industry has also reduced energy consumption and CO_2 emissions very significantly, both by improvements in the efficiency of traditional processes and by the development of new processes. The fuel required for heating in the reverberatory furnace is saved when sulfide ores are flash smelted. Semicontinuous practices such as the Mitsubishi process for Cu production (Section 4.5.1) and the ISASMELT process for Pb extraction (Section 4.5.3) reduce the fuel requirements by using oxygen-enriched air and conserve heat by launder transfer of the molten metal from one stage to the next.

Electrowinning of metals that form very stable compounds is energy intensive, but improvements in energy efficiency have been achieved and further gains should result from the development of alternative techniques such as the bipolar cell (see Figure 7.4). Carbon gas release has also been lowered by the change from coal as the energy source for the generation of the electric power required for electrolysis. Only about 0.25 kg CO_2 is produced per kilowatt-hour when natural gas is used compared to about 0.45 kg kWh^{-1} from coal combustion, the actual saving depending on the calorific value of the fuel and the efficiency of power generation. CO_2 production is limited to the amount formed from the oxygen in the ore when hydroelectric power is used for electrowinning.

The gross energy consumption by the primary metals industry in the U.S. decreased by 53% between 1968 and 1985, and the CO_2 emissions decreased by 57%, the higher value for emissions arising primarily from the switch from coal to oil and gas for electricity generation.[119] However, the energy consumption by the steel industry in the U.S. was still high at that time, compared to the best practice in other countries, and further improvements have now been obtained merely by the adaptation of proven technology. Similar reductions in the amount of CO_2 released per ton of metal produced have been achieved throughout the world and further improvements will occur as, for example, obsolete processes and practices are replaced by those that are more energy efficient.

8.2.2.2. Other Gases

Sulfur is released as SO_2 and SO_3 when sulfide minerals are heated for roasting and smelting. Smaller amounts of these gases are produced during some refining treatments and when fossil fuels (particularly coal) are combusted. Although the damage attributed to acid rain has sometimes been found to arise from other causes (e.g., drought damage to the Black Forest in Germany), there is little doubt that sulfur gases do cause damage to buildings and to the ecology, and the emissions of these gases should be severely restricted. The concentrations of the gases that can be tolerated in the atmosphere before damage is caused have not yet been defined, however, and the limits imposed can only be regarded as arbitrary.

Flash smelting of sulfide ores with oxygen-enriched air produces a smaller volume of exhaust gas with a higher SO_2 content than is obtained from the older processes. Recovery of most of the sulfur as sulfuric acid is then much easier. The residual SO_2 remaining in the gas after acid formation and the SO_2 present in the exhaust gas from other processes can be removed with lime. Conventionally, the gas is passed through a lime slurry in an absorption tower:

$$Ca(OH)_2 + SO_2 = CaSO_3 \cdot \tfrac{1}{2}H_2O + \tfrac{1}{2}H_2O \qquad (8.1)$$

and the product is then converted to gypsum, which is sold for board manufacture, in an oxidizing tower:

$$2CaSO_3 \cdot \tfrac{1}{2}H_2O + O_2 + \tfrac{3}{2}H_2O = 2CaSO_4 \cdot 2H_2O \qquad (8.2)$$

The SO_2 content can be reduced to a few parts per million in this way. A magnesia slurry can be used instead of lime, or the gas can be treated with ammonia to yield an $(NH_4)_2SO_4$ fertilizer. Chlorine and other halide gases can be removed in a similar way. Metals can be extracted from sulfide ores without SO_2 release by reaction with lime in the presence of carbon or by using a sulfating roast prior to leaching (see Section 4.2) and by precipitation of elemental sulfur during leaching (Section 6.2).

Nitrous oxides are produced during the combustion of fuels. The amount formed is usually restricted by controlling the amount of oxygen supplied for combustion, by limiting the combustion temperature and by decreasing the residence time of the gases in the high-temperature zone. There is limited scope for the application of these techniques in metal production, so formation of the compounds is limited primarily by reducing the fuel consumption and the total volume of gas and by passing the exhaust gas through water sprays. Nitrous oxides can be removed completely by reacting the gas with ammonia over an activated Pt catalyst in the temperature range 250 to 400 °C, but all species that could contaminate the catalyst must first be removed from the gas and the treatment is expensive.

Hydrocarbons with high molecular weights dissociate readily on heating, but methane is not easily oxidized at temperatures below about 900 °C. Some CH_4 may be present in the exhaust gas when hydrocarbons are used as the reducing agent at low temperatures, as in the direct reduction of iron oxides. The hydrocarbons are usually reformed, however, at a higher temperature before feeding to the reduction unit. Hydrocarbons can also escape from poorly sealed doors on coke ovens and during the filling of the ovens if the charging machine is not properly shrouded. H_2S and ammonia may also be released into the atmosphere when the hot coke is quenched with water as it is discharged from the oven. Dry quenching with air avoids this gas release.

Ammonia is usually removed from coke oven and other gases by scrubbing with, or bubbling through, a dilute solution of sulfuric acid. Several other noxious gases can escape from coke ovens and the associated by-product plant; thus, the decreasing demand for blast furnace coke is environmentally beneficial not just because of decreased energy demand.

The capital and operating costs of gas cleaning equipment are high. They increase with the number of stages required to remove all the hazardous species and with the decrease in the residual amounts of these species allowed to be discharged into the environment. Air pollution is monitored with devices that measure chemical or physical phenomena such as chemiluminescence. The sensitivity is often limited and the analysis is restricted to the composition of the atmosphere in the vicinity of the sampling point, so the information provided may not be very significant. Legislation limiting emissions is meaningless, therefore, if the maximum concentrations permitted cannot be measured unambiguously.

8.2.3. Liquids

Aqueous effluents are produced from numerous stages of metal production, ranging from the water used in milling, grinding and flotation circuits and in the transportation of slurries to the spray cooling of exhaust gases to condense and coagulate volatile species. Large amounts of water are used for cooling furnace shells, tuyere nozzles and gas injectors. The most heavily contaminated effluents are usually produced in leaching and electrolysis circuits. Unintentional pollution may also arise from natural mechanisms such as the contamination of water in old mine workings and water draining from spoil heaps, as mentioned earlier in this chapter.

Water used solely for cooling in closed systems should not be contaminated, but all other liquid effluents may contain solids in suspension and hazardous substances in solution that must be removed before final discharge into the environment. Some metals, such as Fe and Mg, are not regarded as toxic; but the concentration of other elements including Be, Cr, Cu, Hg, Pb, V and Zn must often be lowered to less than 0.1 mg l^{-1}, and As, Cd, Sb and Se must be less than 0.01 mg l^{-1} in the effluents.

Solids in suspension are normally removed by gravity sedimentation in settling tanks or classifiers. Lime, magnesia or caustic soda may be added at this stage to adjust the pH of acidic solutions. Flocculants, (including alum, lime and ferric sulfate) may be added to coagulate colloidal suspensions, but polymers may be used for this purpose when the recovered solids are recycled for metal extraction. Very finely divided toxic particles may require removal by dissolved air flotation under pressure (see Section 1.3.2) in a separate tank. Dry methods using filters, centrifuges, electrostatic precipitators, etc. are now being favored for the removal of particulate matter from exhaust gases, since this avoids the problem of solids removal from water that arise when wet scrubbing techniques are used.

Metallic elements in solution can usually be precipitated by adjusting the pH to a suitable range, as indicated by the appropriate Pourbaix diagram. Thus, Cu, Cd, Mn, Ni and Zn can be precipitated as insoluble compounds by the addition of caustic soda, NaOH, sodium carbonate, or with a lime or magnesia slurry; eg.,

$$M^{2+} + MgO + H_2O = M(OH)_2 + Mg^{2+} \tag{8.3}$$

The pH of the solution is raised by these additions, but it is buffered at a maximum pH of 10.5 at 25° C by the precipitation of $Mg(OH)_2$ when magnesia is used.[121] A higher pH is reached in the presence of an excess of the other reagents, and the solution must then be neutralized before discharge with acid additions after the metals have been removed.

Metallic elements may also be precipitated as sulfides, which usually have a lower solubility than the corresponding hydroxides. In some cases, the element must first be reduced to a lower valency with SO_2, $FeSO_4$ or sodium metasulfide before precipitation. Arsenic can be precipitated as ferric (or copper-ferric) arsenate by adding gypsum when iron and copper oxides are present in the solution. Microorganisms can be used for the biosorption of some toxic metals and ion exchange can remove small amounts of metals from large volumes of effluent before the exchange membrane requires regeneration.

Biodegradation via aerobic oxidation of cyanides, phenols, thiosulfates and other toxic compounds can be accomplished by holding the effluent in ponds containing the bacteria. Cyanides can also be destroyed with NaOH and sodium hypochlorite to release CO_2 and N_2, or oxidized in the presence of SO_2 with Cu as the catalyst (INCO process):

$$CN^- + SO_2 + O_2 + H_2O = CNO^- + H_2SO_4 \tag{8.4}$$

to produce the insoluble and much less toxic CNO^- radical.

The capital costs for treatment of liquid effluents are usually low in comparison with gas cleaning, but operating costs are often high and can markedly increase the cost of the products. Problems may arise in the measurement of the contamination in the discharged effluents. Samples are normally taken at fixed time intervals, and a sudden surge in the level of pollution may be missed or not detected until considerable quantities have been released. There is a need for simple and reliable techniques for continuous monitoring.

8.3. SCRAP METAL RECYCLING

New ore bodies are still being formed in the earth's mantle by igneous intrusions and as sedimentary deposits by natural weathering. However, the rate of formation is extremely slow in comparison with the rate at which the ores are being extracted, so metallic minerals are regarded as non-renewable resources. The recycling of metal from artifacts whose serviceable life has expired thus conserves the natural resources. The landscape is also protected by recycling, for less ore has to be mined to satisfy the demand for metals. The energy consumed in recycling is very low when compared with primary metals extraction, varying from 30 to 35% of the quantity required for steel manufacture to only about 5% of that required to produce Al from the ore, since the stages of mining, comminution, concentration and reduction are eliminated.

The reuse of metals is not a new concept. Since it is easier to remelt metal than to produce a new supply from the ore, reuse has been practiced continuously since the Bronze Age. It is probable that a greater proportion of old metal was regenerated before the bulk methods for the production of metals were introduced than is the case today. The companies that remelt and refine obsolete scrap are often referred to as "secondary metal producers" in order to distinguish them from those who produce primary metal directly from the ore; but the distinction is not always clear-cut because old scrap is often included as part of the charge when new metal is refined.

8.3.1. Sources and Preparation of Scrap Metal

Scrap metal is classified by source of origin into three categories. Home or "in house" scrap is uncontaminated with other metals. It arises within the producer's plant from runners, top and bottom discards from ingots and partially filled ingot molds when ingots are cast, and from croppings and rejects when the metal is shaped by forging, rolling, extrusion, etc. The composition of this metal is known accurately from the cast analyses. New, prompt or process scrap comes from further shaping by mechanical working, stamping, machining and other operations by the customer in the manufacture of usable artifacts. The composition is known and it is uncontaminated if the scrap metal is segregated at each stage according to its origin. Both of these categories are available for more or less immediate recycling. The third category, old or obsolete scrap, is obtained from artifacts that have completed their useful life and this may be 30 or more years after the metal was first produced. The approximate composition may be readily recognized from the original use (e.g., Al beverage cans, railroad lines or ship plate), but it is more commonly contaminated with several different metals and nonmetals that have been assembled into a structure (as in an automobile) and the composition is uncertain.

The demand for new metal produced from the ore is only reduced by the recycling of obsolete scrap. Home and new scrap do not make any contribution to this saving. In fact, the total energy consumed in the production of 1 t of artifacts is increased in proportion to the amount of metal rejected and recycled during the manufacturing processes. This is illustrated in Figure 8.2. If y tonnes of new scrap are generated by the product manufacturer then the metal producer must supply the manufacturer with $(1 + y)$ tonnes to produce 1 t of finished goods. Similarly, $(1 + x)$ tonnes of metal must be produced if x tonnes of home scrap are generated for each tonne supplied. If f is the fraction of the output obtained from primary

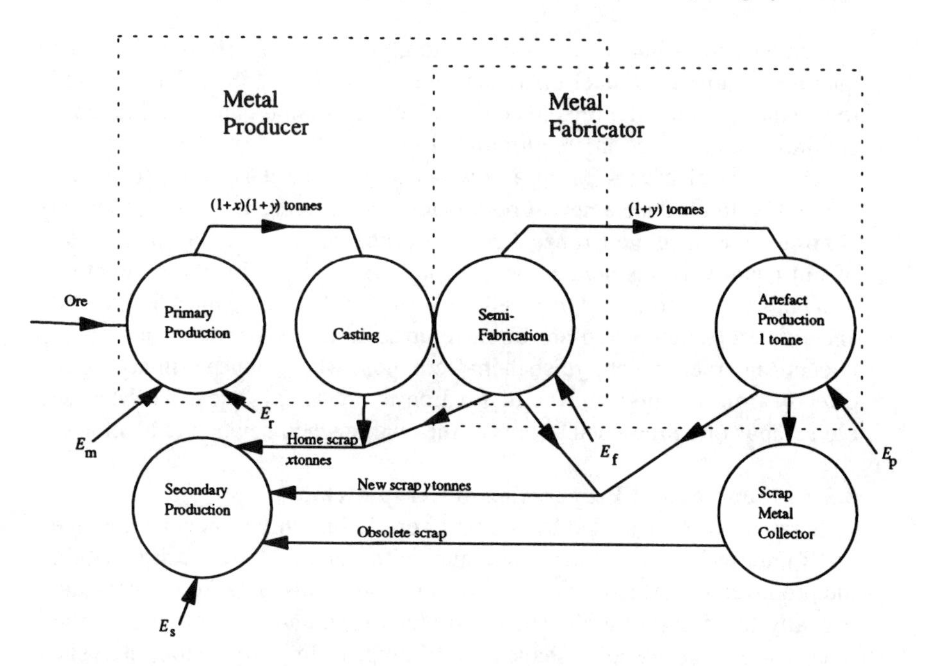

FIGURE 8.2. Metal consumption and scrap metal production in the manufacture of 1 tonne of artifacts.

production and $(1 - f)$ is the fraction from home and new scrap, the gross energy requirement, GER, per tonne of products is given by:[122]

$$\text{GER} = f(1 + x)(1 + y)[E_m + E_r] + (1 - f)(1 + x)(1 + y)E_s + (1 + y)E_f + E_p \tag{8.5}$$

where E_m, E_r, E_s, E_f and E_p represent the energy inputs for ore mining and preparation, smelting and refining, secondary metal production, fabrication, and product manufacture, respectively. Sometimes, as in BOS steelmaking, the energy required for melting the recirculated scrap is supplied by the excess exothermic heat generated by refining the primary metal. In other cases, the energy, E_s, must be supplied externally and, in all cases, an additional amount of energy, yE_f, is consumed in fabrication. Thus, there is an incentive to minimize the amount of home and new scrap produced in order to conserve energy.

Obsolete scrap, on the other hand, requires less energy to process than is needed for primary metal production, and every ton of metal fed into the fabrication line from this source directly replaces the same amount of primary metal production. The decrease in the energy demand is usually reflected, in turn, in a decrease in CO_2 emissions. This is not always the case, however. When the energy required for primary metal production is obtained from surplus hydroelectric power and secondary processing is conducted nearer to the point of collection, but remote from a source of hydroelectric power (as is often the case with Al recycling), the fossil fuel consumption and CO_2 release into the atmosphere is actually increased by recycling.

Obsolete scrap is useful only if it can be collected and processed to yield a product at a cost less than or equal to the cost from primary metal production. Otherwise, it is a waste material and a cost is incurred in its disposal. In many cases, it is not recovered as a single metal but as an alloy composition that has been assembled with other metals, alloys and nonmetallics into a usable article. The first stage of recovery may thus involve dismantling into the separate components, cutting into pieces of suitable size and identification of the composition of the components by a variety of techniques, including visual, magnetic, spark, spectrographic, chemical and X-ray examination—all of which add to the cost. Different metals may be separated from each other and from nonmetallics by exploiting differences in their physical properties using levitation, magnetic, heavy media, electrostatic, eddy current, cyclone and similar separation processes. Each of the separated batches are then compressed into bales or bundles and graded according to the approximate composition and average thickness of the scrap, since thinner gages oxidize more rapidly than thicker sections on remelting.

The problems that arise if the different metals are not completely separated can be illustrated by consideration of the recycling of automobiles and tin cans, both of which contain steel as the major component. Cu and Sn are not readily removed during the refining of a scrap steel melt; and if the concentration of these elements, alone or in combination, exceeds critical levels, the steel is prone to cracking (hot shortness) during mechanical working at elevated temperatures. Thus, the concentration of these elements must be lowered as much as possible by physical or chemical separation before charging to the melting furnace.

The automobile contains up to 1 t metal as almost pure Fe in the body shell and as alloys in which Fe is the predominant element in the engine, suspension and transmission. Cu is present in the form of solenoids, motor windings and wire looms. The engine block may be cast in Al. Zn die castings are used as fittings, and the body shell is protected from corrosion by a coating of Zn, either alone or in combination with Al or Ni. The plastic trim and upholstery would burn and release toxic fumes if they were included in the furnace charge and the window glass would increase the slag bulk. Unalloyed and low-alloy steel with a ferritic microstructure becomes brittle at subzero temperatures; thus, the popular method today for disposing of old automobiles is to cool the chassis with liquid nitrogen to below the ductile-brittle transformation temperature and break up the metal into small pieces in a shredder. The fragmented Fe is then readily parted from other materials with a magnetic separator. A heavy media separator can be used to part the other metallic components from the glass, plastic and other refuse. Typically, about 70% of the weight of the automobile is recirculated as clean steel scrap and 3 to 5% as nonferrous metals. The steel scrap commands a lower price and the other metals are lost when the entire automobile is compressed without fragmentation.

Tin plated steel is used extensively for beverage and food containers. The quantity of Sn in each can has declined markedly in recent years as the coating thickness has been decreased and as welded seams have replaced soldered joints. Although Sn is a relatively high-priced metal, it is not profitable to treat the containers solely as a means of recovering the Sn; but the element is undesirable in the steel remelt. It can be removed from the surface of the containers by alkaline leaching with caustic soda under oxidizing conditions, but the Pourbaix diagram for the Sn-water system[123] shows that a pH of about 15 (i.e., almost pure hydroxide) must be maintained in the solution to obtain a soluble stannate complex. Alternatively, the Sn can be removed by electrolysis of shredded scrap in a caustic soda solution or by chlorination. If the cans are heated to high temperature in a domestic waste incinerator, however, the Sn coating reacts with the Fe to form a series of Fe-Sn compounds in which the activity of the Sn is lowered and removal by these means is then not feasible. The scrap can be recirculated if the Sn content is diluted by melting with directly reduced iron, DRI, or with other scrap of very low Sn content. Over 12 billion steel beverage cans are recycled annually in the European Community.

8.3.2. Recirculation Rates

Obsolete scrap could be processed to produce metal of any required purity, but financial restraints usually limit the amount of refining that it is worth while to undertake. More stages are often required to produce a pure metal from an alloy than are needed to refine the primary metal. The operations are usually conducted on a smaller and, therefore, more costly scale than for primary production. The scrap may still be contaminated with plastic coatings or with oils (on swarf and turnings) and may contain volatile elements not normally present when the metal is produced directly from the ore, thereby increasing the costs required to satisfy environmental legislation. Dismantling and sorting scrap is labor intensive. Wage costs have risen rapidly during the last few decades, while the real costs of primary metal have fallen, so the financial return from recycling has tended to decrease.

The proportion of primary metal production actually recirculated as obsolete scrap varies markedly from one metal to another. Scrap arising from metals with a high intrinsic value (such as Au, Ag and Pt) is recovered intensively throughout the fabrication stage and this, together with remelted artifacts, is reconstituted to form a major part of the total production of these metals.

Lead is readily reclaimed, primarily from disused batteries that represent the major use of the metal, but also in the form of Pb sheet, pipe and sheathing that contain relatively low concentrations of other elements. Pb-Sn solder is also readily reconstituted from scrap, dross, etc. by melting, drossing and adjustment of the composition via addition of Pb or Sn, although the purity of the remelt solder is lower than that prepared from the primary metals. About 50% Pb is now recycled and this is a higher proportion than for any of the other common metals, although about 60% was recycled during the 1970s. The current decrease is perhaps indicative of the extent to which increasingly stringent regulations on hazardous emissions have resulted in a reduced incentive to collect obsolete scrap and increased the amount of primary production.

Iron and steel have the next highest recirculation rate and 38% of the world steel production is currently derived from scrap metal. Large quantities of steel are used in the form of thick sections such as girders and plates, which only require dismantling and cutting to suitable size for recirculation. The shredder has proved to be an effective means of separating the different materials in an automobile, and scrap metal from this source is readily accepted for remelting. Protests are now being voiced, however, against the incineration of the fluffy organic and inorganic wastes produced by the shredding plants, and these plants have been forced to cease operation in some localities. However, as with the incineration of domestic waste, negligible damage is caused to the environment if the incinerators are operated at the correct temperature. Considerable environmental damage is created when old automobiles are abandoned and set on fire in the countryside.

The recirculation rates for the other common metals are lower than for Pb and Fe. The portion of the total output provided by the secondary metal melters is rarely greater than 20% (Al, Cu) and is often much less. This is due primarily to the use of these elements as alloy additions to other metals and as coatings. The fraction of the total output of a metal obtained from recycled material, however, is also influenced by the rate of increase in the annual output of that metal and by the average life cycle of the products in which it is incorporated. Thus, the fraction is lowered when the output of a metal is now very much higher than it was at the time when the scrap artifact was produced. In general, high recirculation rates apply when the demand for the metal is static, or is growing only slowly; when the metal is used in a relatively pure form or as easily identified alloys (e.g., brass, bronze, stainless steel); and when the scrap can be easily recovered and reconstituted as metal or alloy with an acceptable purity.

Recovery could be improved if manufactured products contained a smaller range of materials that could be disassembled more readily into their component parts at the end of their useful life. Designers are now beginning to take a cradle-to-grave approach and modifying designs to enhance the recyclability. It is claimed that the BMW Series 3 automobile is the first to be built along these lines and that it is 80% recyclable. The traditional tin-coated steel beverage can has an aluminum end for easy tear-opening, but an all-steel can with a tear opening has now been perfected. There are encouraging indications that this approach is being adopted for other products. However, there would be no environmental gain if, for example, the modified design resulted in an increase in the energy consumption in the fabrication or by the artifact during its useful life that exceeded the energy savings resulting from the enhanced recycling rate.

Recovery rates can also be increased by offering a financial incentive. A small premium is paid in the U.S. for aluminum beverage cans returned to collection points. This has also stimulated the use of the metal. Al cans are dominant in the U.S. and over 50% are recycled; whereas, in the European Community, where no incentive is offered, only 20% of Al cans are recovered and steel containers still dominate the market.

8.3.3. Scrap Melting and Refining

Secondary metals are most commonly produced by pyrometallurgical techniques and the equipment, procedures and gas cleaning arrangements are often similar to those used in the refining of primary metal. The major differences are the extent to which the metal may be oxidized during melting and the number and concentration of the other elements needing removal.

When scrap metal is charged into an empty furnace and heated for melting, it is not protected from oxidation by the atmosphere until a molten slag has formed and risen on the pool of molten metal to submerge the remaining solids. Extensive oxidation can occur during this period, the rate increasing with an increase in the exposed surface area per unit volume of scrap. It is common practice to melt only a portion of the charge and to add the remainder when the slag cover has formed, thereby restricting the amount of oxidation. Light (thin gage) scrap is compressed into bales to reduce the exposed surface area.

Metals that form the less stable oxides can be reduced back to the metallic form during refining. Strongly reducing conditions would be required, however, to reduce all of the oxide and some of the metal is usually left in the slag. Consequently, the concentration of impurities that form even less stable oxides (such as Cu and Sn in Fe) is increased slightly in the remaining metal. The content of these elements thus shows a small, cumulative increase every time the metal is recycled, even when no extraneous source in which they are present is included in the charge, and eventually exceeds the permissible threshold. The concentration of these impurities must then be diluted by addition of virgin metal or the remelted material is downgraded to less demanding applications.

The more stable oxides formed during the melting of metals such as Al cannot be easily reduced, and as much as 20% of the charge may be lost as Al droplets trapped in an Al_2O_3 dross when light aluminum scrap is melted. The metallic Al can be recovered from the dross by fluxing with chloride additions, but this creates an environmental problem in the disposal of the slag. Alternative treatments are being developed. For example, the BOC Group process uses a plasma furnace to fuse the dross in the presence of lime to form a calcium aluminate melt that floats on top of the molten Al.

Alloy elements that form more stable oxides than the metal are readily removed by oxidation from the melt, but a small residual concentration remains in the metal, the amount increasing as the difference between the stabilities of the oxides of the metal and the alloy element decreases. Thus, Cr forms oxides somewhat more stable than FeO but, at low concentrations of Cr, a_{Cr} is lower than a_{Fe} in an Fe melt in equilibrium with a refining slag and the Cr cannot be eliminated completely before the Fe begins to oxidize rapidly. The residual concentration remaining when refining is completed, together with any Ni and Mo in the scrap (that form less stable oxides and are not removed), increase the hardenability of the steel and may cause cracking during cooling after welding. The quality and the price of steel scrap is downgraded accordingly as the total concentration of (Cu + Cr + Mo + Ni + Sn) in the scrap is increased.

Scrap metal may contain large amounts of valuable elements with similar oxide stability to that of the principal metal. It may be more profitable to simply remelt the scrap with minimal or no refining, but with alloy additions to replace any loss through oxidation and regenerate the original alloy composition. This is usually acceptable, for example, with brass, bronze and stainless steel, but the unavoidable increase in the content of trace impurities, whose concentration must be rigorously controlled at the parts per million level for very demanding applications, often precludes this way of recycling alloys such as the Ni-based compositions used in the high-temperature regions of a jet engine. In such cases and also when the alloy elements cannot be removed by oxidation (as with Al), it is often more economic to melt the scrap metal with additions to produce an alloy suitable for use, say, in castings more tolerant of the impurities.

Scrap metal may contain elements that are partially or completely volatilized at the refining temperature, such as Zn in brass and on galvanized steel and Pb in free-cutting alloys. These elements are usually not present, or appear in only small amounts, in the ore from which the primary metal is derived. A more elaborate gas cleaning plant is then required for the treatment of the furnace exhaust gas than is needed for primary metal production. The dust and fume collected from the gas may be sufficiently rich in metal values to justify its inclusion in the charge for the primary or secondary production of other metals. It is sometimes difficult to satisfy environmental legislation for the disposal of less rich residues, and a variety of techniques are being explored for metal extraction from the discharges.

8.4. POSTSCRIPT

The metal-producing industries have made major improvements, both in the amounts of pollutants released into the environment and in more efficient use of nonrenewable resources. Progress continues to be made, but pressure needs to be applied to the few recalcitrant operators to ensure that these achievements are not defiled in the public image by the occasional bad example. There is need for more careful and comprehensive evaluation of the balance between the potential benefits that may be achieved and the probability of other types of damage being caused by consequential changes to practice before new environmental restrictions are enacted that might affect the performance of the industry.

It is improbable that mankind will voluntarily revert to a primitive existence. The Industrial Age will continue and the demand for metals will not diminish. This should be recognized when new legislation is introduced. For example, imposition of a carbon tax in one or more countries with the objective of reducing emissions of carbon gases, or tighter control on industrial emissions, could make metal production uneconomic in those regions and force the manufacturers to cease operations. However, there would be no gain to the ecology if the operations were simply moved to countries that imposed less stringent restrictions. In fact, global pollution would be increased by the action if an efficient operation in one country was replaced by a less efficient practice in another country.

Scrap metal recirculation is a very effective means of conserving both the countryside and nonrenewable resources. Encouragement should be given to schemes intended to increase the amount recirculated at least to the levels attained a few decades ago. Efforts should also be intensified to ensure that artifacts are designed with the total life cycle in mind and are readily recyclable at the end of their useful life, without sacrificing the quality of the product. More attention should also be given to the total energy costs throughout the life cycle of an artifact. For example, metal-matrix composites may allow a smaller mass of material to be used to satisfy, say, a given strength requirement. However, the cost of metal recovery from the composite may exceed the value of the metal produced and the composite becomes a waste material at the end of its useful life. The energy lost when the metal and the reinforcement in the composite are discarded may exceed the energy gain from the use of the composite.

The effect of a financial incentive is well illustrated by the increase in aluminum can recycling in the U.S. and there is a need for more schemes of this type. On the other hand, the secondary metal producer is operating in a commercial market and cannot sell the products if a high cost of collection and sorting increases the price of the metal above that of the primary metal producers. Government-sponsored schemes could be very beneficial here, and it is probable that the ecology would be better protected by diverting some of the funds currently applied to enforce environmental legislation into stimulation of material recycling.

FURTHER READING

Henstock, M. E., *Design for Recycling,* Institute of Materials, London, 1988.

Kaplan, R. S. and Ness, H., *Recycling of Metals, Conservation and Recycling,* Pergamon Press, Oxford, 1987.

Taylor, P. R., Sohn, H. Y., and Jarrett, N., Eds., *Recycle and Secondary Recovery of Metals,* TMS, Warrendale, PA, 1985.

Van Linden, J. H. L., Stewart, D. L., and Sahai, Y., Eds., *Recycling of Metals and Engineering Materials,* TMS, Warrendale, PA, 1990.

Chapter 9

COMPUTER CONTROL OF METAL PRODUCTION

The first Industrial Revolution started in Great Britain in the middle of the 18th century and spread slowly to the rest of the world. Iron was the main engineering metal at that time and supplies were limited by the shortage of charcoal. The revolution resulted largely from the marked increase in the production rate when coke was substituted for charcoal in the blast furnace and with the development of Cort's process for the production of wrought iron. Both of these developments lowered the cost and increased the quantity of metal available. A coincident rapid increase in the population of industrialized countries led to an increased demand for metals and encouraged the developments.

Only five metallic elements were in common use at that time. Steel was a relatively expensive and not very reliable material before the invention of Huntsman's crucible process, and aluminum was first produced for the Paris Exhibition in 1854. The scale and complexity of the extractive metallurgical industries have grown steadily over the last 2 centuries, with occasional spurts as new processes have been developed to the stage of commercial production; but the major growth has only occurred during the last few decades, over 200 years after the start of the revolution.

We are now living in the early stages of the second Industrial Revolution, which is being brought about by the application of the computer. The availability of the computer offers the prospect of fully automated control of manufacturing processes. The rapid developments in computing power and the reduction in the real cost of computers is already resulting in changes in the manufacturing industries no less significant than those started over 200 years ago. Eventually, computer optimization of the processes will secure maximum output of a metal from a production unit at minimum cost, while process control by computer will largely replace human effort and ensure that molten metal of consistent quality is produced within the required composition and temperature range. Significant progress has been made already, and many computer packages have been produced. There is still a long way to go before complete control of all processes can be achieved, but the rate of progress will be much more rapid than in the first Industrial Revolution, and manual control will be progressively simplified or replaced by computer outputs over the next few decades.

9.1. SENSORS

Complete control of an extraction process to achieve a desired end point requires knowledge of the equilibrium state that would be attained under the imposed conditions and how the equilibrium would be changed by altering the conditions. Knowledge is also required of the kinetic factors that determine the rate of approach to equilibrium. The former can be calculated from knowledge of the free energies for the appropriate reactions and the activities of the reactants and products at the operating temperature. The kinetics are less readily quantified. Many of the reactions are terminated before equilibrium is attained. For example, the oxygen supply is halted during Cu converting and in oxidative refining when most of the impurities have been eliminated but before the metal starts to oxidize rapidly. Furthermore, many of the processes are not operated under what could be termed ''steady-state conditions.'' New material is charged and treated metal is withdrawn continuously or periodically in processes like the blast furnace. The composition of the metal and the slag are changing continuously with time as solid materials melt and elements partition between the slag and the metal in batch-type processes.

The numerical values of a large number of variables may change with time as the reactions proceed. It is desirable that these changes—or at least the most important ones—are measured and incorporated into the equilibrium and kinetic calculations in order to control the process and attain the required changes in the metal composition in the shortest possible time. Measuring devices, or **sensors,** can provide this information.

9.1.1. Traditional Measurement Techniques

Until comparatively recent times, control was achieved on the basis of the accumulated knowledge and skills of the operatives that were acquired from past experience and aided by relatively few quantitative measurements. The weights of materials entering and leaving a process were easily measured and the size distribution could be determined by sieving. Gas flows could also be measured with reasonable accuracy, but composition and temperature measurements were more difficult.

Temperature was originally assessed from the color of the molten metal and could often be estimated by eye to within ±20 °C. When refractory sheaths were first introduced about 50 years ago, to enable thermocouples to be immersed through the slag layer into the molten steel, the visual temperature estimate was sometimes more reliable than the measured value.

The chemical analysis of a few elements dissolved in a metal, such as C and S, could be determined fairly rapidly by combustion techniques, but the assessment of the concentration of the majority of elements by acid dissolution, separation and titration, electrodeposition or precipitation techniques was time consuming and might take an hour or more to complete. Analysis of the concentration of impurities present in only very small amounts often took considerably longer. This was acceptable for the analysis of input materials, but was of limited value for the control of batch-type processes in which the reaction stages were completed in only a few hours. It was useless when the reactions were completed in about 20 min, as in the original Bessemer process for steelmaking. Slag analyses were even more time consuming, so the slag condition was estimated by observing the surface appearance of a sample allowed to solidify as a flat pancake and the color of a fractured section of the sample. The viscosity was estimated from the distance a slag sample flowed around a spiral mold before it solidified.

These examples are indicative of the very limited feedback of information that could be used in process control until comparatively recent times. The small percentage of rogue casts that failed to meet the required specifications is a tribute to the acquired skills of the operatives and a salutary warning that a complex computer logic is required if it is expected to produce better results than are obtained by human perception.

9.1.2. Modern Techniques

A wider range of robust, accurate and reliable sensors is now available to measure the changes that occur in the process variables and simplify process control. For example, the height to which a reaction vessel is filled by a given quantity of metal changes progressively with time as the vessel lining is eroded. The height of the molten metal can be measured accurately with a laser or ultrasonic beam, and the distance that a gas injection lance must be lowered to a position at some fixed distance above the melt surface can thus be determined. If the lance is subsequently immersed in the slag layer, the depth of immersion can be ascertained with an audiometer from the noise level emitted by the blast. The height of the stock line in a shaft-type process, such as the blast furnace, can also be determined by ultrasonic and laser techniques.

Perhaps the greatest change is in the speed and accuracy with which chemical analyses can be determined. Measurement of the composition of the gas leaving a furnace can often provide useful information on the extent of the reactions taking place within the furnace. Gas sampling probes that filter solids from the gas can now be installed in very dirty environments to feed the gas to analytical cells that give continuous recordings of the changes in the gas composition. The SO_2 content of the exhaust gas indicates the rate of sulfur removal (dS/dt) in matte converting, and a rapid fall in the SO_2 content indicates the stage when air or oxygen injection should cease in order to avoid overoxidation of the metal. Similarly, the CO-CO_2 content of the exhaust gas shows the rate of carbon removal in the BOS refining of steel, and a decrease in the concentration of carbon gases shows when very low carbon contents have been achieved, Figure 9.1. Probes moved around just above the stockline, or immersed to known depths in the stack of the blast furnace, supply information on the

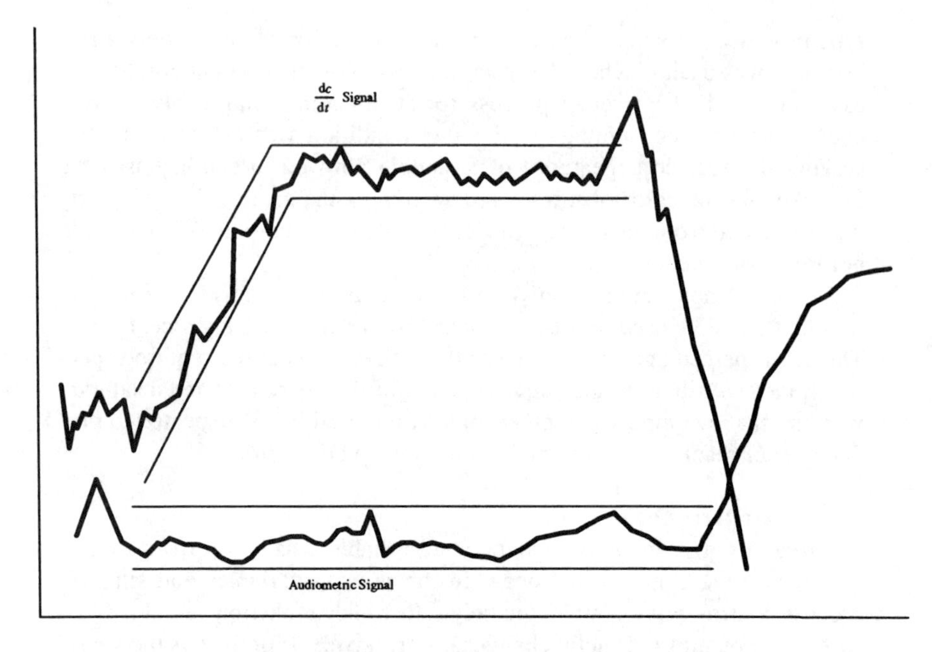

FIGURE 9.1. Rate of carbon removal assessed from exhaust gas analysis and audiometric signal indicating the height of the oxygen lance above the bath in a BOS heat.

uniformity of gas distribution within the solid charge and show immediately when corrective action is required to alter the charge distribution to prevent channeling of the gas.

A complete chemical analysis of molten metal, including the concentrations of the trace impurities, can be provided by physical techniques such as X-ray or optical emission spectroscopy and reported back within a few minutes from when the sample was withdrawn from the furnace. This allows time for corrective action to be taken while the reactions are still in progress, or it provides a more accurate basis for the assessment of the further time required to reach the target composition.

More simple techniques can sometimes be used to obtain an approximate indication of the composition that is sufficiently accurate for use at a particular stage in a process. For example, the melting point of Fe is depressed by carbon. If other elements are present in only low concentrations, the carbon content of the steel can be assessed, with reference to the Fe-C phase diagram, by thermal analysis to determine the liquidus temperature of a sample of metal. Sub-lances have been developed to enable a sample of metal to be withdrawn from the bath during decarburization without interruption of the oxygen blast. Due to the difficulty of nucleating CO bubbles, the carbon in the melt is not usually in equilibrium with the oxygen, but it can be related to the actual FeO content of the slag through empirical relationships derived from observations for any particular refining practice, such as those shown in Figure 5.9. Hence, this measurement gives an indication of the oxidizing power of the slag. Since a_O in the metal is dependent on a_{FeO} in the slag, the metal oxygen content can also be calculated to within a few parts per million, and the amount of deoxidizers required to remove the oxygen can be determined from the measurement of the carbon content.

The amount of oxygen in solution in a molten metal can be determined directly by measuring the emf generated across a solid electrolyte sensor. One face of the electrolyte is placed in contact with the molten metal. The opposite face is in contact with a reference electrode, which consists of a mixture of some other metal with its oxide. The potential drop across the electrolyte is given by:

$$E = \frac{RT}{zF} \ln \left[\frac{p_{O_{2(\text{metal})}}}{p_{O_{2(\text{reference electrode})}}} \right] \tag{9.1}$$

The partial pressure of oxygen in equilibrium with the reference electrode at the melt temperature is readily determined from the standard free energy of formation of the reference metal oxide at the reaction temperature and use of Equation 2.95, i.e.,

$$\Delta G^{\circ} = RT \ln p_{O_2}$$

since the metal and its oxide are pure and are therefore present at unit activity. The oxygen in solution in the melt is then obtained from the free energy change for the solution of gaseous oxygen in the metal:

$$\tfrac{1}{2}O_{2(g)} = a_{O(\text{in solution})} = \%O \times h_{O(\text{in solution})} \tag{9.2}$$

where h_O is the activity coefficient for oxygen in the metal with reference to the dilute, weight percent, standard state.

Equation 9.1 is valid if the solid electrolyte shows only ionic and no electronic transport within the relevant range of oxygen partial pressures at the working temperature. Very few substances that remain as solids in the appropriate temperature range satisfy this requirement. Magnesia-stabilized zirconia is satisfactory and is commonly used as the electrolyte, while a mixture of molybdenum and its oxide can be used for the reference electrode[124] (see Worked Example 6). A change of 8 mV in the output from this cell corresponds to a change of 55 ppm in the dissolved oxygen content of molten steel at 1670 °C,[125] so the sensitivity is quite high. A similar arrangement can be used to measure a_{FeO} in molten slags.

Analysis of the composition of the aqueous solutions used in hydro- and electrometallurgy is relatively straightforward since a sample can be withdrawn at any convenient time without interruption of the process. Sampling is more difficult with pyrometallurgical operations. Ideally, the concentrations of all elements in the molten metal and slag would be monitored continuously throughout the heat, but no technique has yet been devised to accomplish this. In some cases, an analysis can only be obtained at infrequent intervals, when a quantity of metal or slag is withdrawn (tapped) from the furnace. In other cases, it is possible to gain access to the bath, but the flow of heating, oxidizing or reducing gas has to be disrupted to allow access and withdrawal of a sample. The total time taken to complete the reactions is thus extended each time a sample is taken, so very few samples for analysis are normally obtained during the course of a heat.

A major problem is encountered in the determination of the initial molten metal composition when the furnace charge is compounded partially or completely from obsolete or old scrap. Much effort is expended in sorting and analyzing the scrap, but the precise composition of the scrap metal charge is rarely known. Melting is sometimes not completed until shortly before the end of the refining stage, so a sample of molten metal taken before refining is started does not give a true measure of the charge composition. In such cases, the initial composition is estimated and the refining path is chosen based on previous experience with similar charges. A sample taken when melting is judged to be complete then provides an indication of the corrective action required to obtain the required end point. There is increasing tendency, particularly with scrap metal heats, to aim to achieve a composition close to that required while the metal is held in the furnace and to make final adjustments in the ladle, based on the composition of a sample taken while the metal is being tapped from the furnace.

9.2. COMPUTER CONTROL

A computer can be used to provide data, either for completely automatic control of a process or as an aid to human control if the values of all the variables affecting the process can be expressed as a consistent set of mathematical equations. These equations are assembled into a model that simulates the process and that should be able to extrapolate forward to the condition that the process will achieve after a given lapse of time.

Mathematical models are frequently used without recognizing them as such. For example, the ideal and regular solution concepts are really models that allow the changes in the thermodynamic properties to be quantified when two or more species are mixed to form a liquid or solid solution. The computation of phase diagrams is performed with the aid of mathematical equations that define the changes in the free energy of the various phases with variation in the temperature and composition. The operating line in the Rist diagram for the blast furnace (Figure 4.8) can be expressed as an equation that aids control of the process to minimize carbon consumption, while the phosphate capacity (Figure 5.6) is a model of the partitioning of phosphorus between the metal and the slag as a function of temperature and composition.

Static models base predictions on a set of data representing the system at a fixed moment in time. These are used, for example, to perform a heat and mass balance to select from the available stock the ingredients of ore, metal, flux, etc. that make up the charge for a pyrometallurgical operation. They quantify the amounts of each of these materials needed to produce the required quantity and composition of metal at the end of the process. The input data in this case includes the compositions and the available amounts of the stock materials, the constitution of charges used previously and the final compositions obtained with each of those charges.

Dynamic models modify and update the predictions by incorporating data fed in continuously or intermittently by sensors or other means. These models are used for real-time process control.

A control model often incorporates an interdependent subset of models, each of which describes some particular feature of the process. An estimator program then combines information supplied by the model with the sensor inputs to calculate the current state of the process, while an optimizer program uses the current state and other data from the model to select the optimum path for the remaining process time.[126] A review article[127] has been published that provides a useful description of the principles involved and an indication of the types of models available. Descriptions of new models appear regularly in the technical journals and advanced treatise describing the state of the art in modeling for a particular process have also been published (see, for example, Reference 128).

9.2.1. Fundamental Models

The ideal model aims to completely describe the system in terms of the physical and chemical laws that govern the progress of the various reactions which can occur. These are usually called fundamental or mechanistic models. If a valid model can be formulated on this basis, it can be used to predict the material outputs to be obtained from a wide range of input materials. It can also be used to forecast the extent to which the performance can be improved by changes to the inputs or to the operating procedures. However, several problems can arise in the implementation of this approach.

Some processes, such as comminution and multiple-hearth roasting, can be regarded as essentially steady-state in the sense that the operations continue with relatively little change for a long period of time after stable operating conditions have been achieved. These can be modeled fairly easily, but most pyrometallurgical processes deviate markedly from steady-state conditions. A complete description of such a process in terms of a fundamental model can be very complex. The computer time required to obtain a prediction tends to increase with increasing complexity of the model, so more powerful (and expensive) computers are then required to obtain a useful printout within an acceptable time period. More commonly, the mathematical expressions are simplified to reduce the computation time, with a corresponding decrease in the accuracy of the forecasts.

Frequently, it is found that an attempt to construct a model to give a complete description of a process is frustrated by an inadequate understanding of either the actual mechanisms involved, or the effect of some of the variables on the progression of the reactions. For example, mineral flotation is controlled by relatively few variables in comparison with, say, a smelting process, and the recovery rates for different minerals in an ore can be predicted with reasonable accuracy from knowledge of the size, composition and surface characteristics of the ore particles fed into the circuit. However, the middlings from the rougher cells are usually reground before passing to the cleaner or scavenger cells. The particle size and the surfaces are changed by regrinding, and mineral recovery in these cells cannot be predicted accurately from the characterization of the ore fed into the start of the circuit.

Relatively few variables are also pertinent to the sintering process, but the amount of liquid bond formed and the period of time it remains molten is dependent on the temperature profile through the bed. These, in turn, are dependent on the bed permeability and the rates of the coke combustion and the endothermic and exothermic reactions, which are interdependent variables. It also depends on the composition, amount and distribution of the gangue minerals in the ore, the rate at which they can react to produce liquids with a low melting point and the change with time in the composition of the liquid. It is difficult to predict these changes with sufficient accuracy from fundamental laws. Flotation and sintering are relatively simple processes in comparison, say, with a reaction across an interface, where the transport to, across and away from the interface and the chemical reaction are each dependent on a large number of variables.

Even when the fundamentals are adequately understood, evaluation of the derived equations is often limited by the scarcity or by the low reliability of the available data, and computer modeling of a process often identifies the topics for which more experimental measurements are needed. Consider any pyrometallurgical extraction or refining process. Thermodynamic data can be used to describe the equilibrium partitioning of the elements present between the molten metal and the slag or the dross. Values of the relevant standard free energies and the equilibrium constants for the reactions can be obtained from the various data banks. However, the heat capacity data used to calculate the temperature dependence of the free energies are often known to no greater accuracy than $\pm 2\%$. The error can be quite large when the free energy for a reaction is a relatively small quantity obtained as the difference between much larger numerical values for the standard free energies of other reactions (see, for instance, Worked Example 7). The thermodynamic activities of many solutes have been evaluated in binary and ternary alloys with a solvent metal. Several solutes may be present in the metal, and their activities in the multicomponent solution can be determined by combination of the appropriate interaction parameters for ternary alloys (using Equations 2.84 or 2.85). Interaction coefficients have been evaluated for all the common solutes in Fe and Cu, but data are sparse or nonexistent for other solvents. Where values have been determined, they are often known only at one temperature and for low concentrations of the solutes that may only be approached toward the end of the reaction stages.

Data on the thermodynamic properties of slags is usually limited to binary, ternary and, occasionally, quaternary melts; but actual slags may contain many more species including, of course, the solutes partitioned from the metal into the slag. Minor constituents, particularly if they are surface active, may have an influence on the properties that is far greater than might be expected in terms of their concentration by weight or volume. Thus, estimates of the equilibrium partitioning of solutes are subject to an error band that may be quite large.

Equilibrium calculations give no indication of the time required to complete the reactions. Profitable operation of a process generally requires that reactions be completed in the shortest possible time, but knowledge of the quantitative values of the parameters that control the kinetics of the reactions is often more limited than knowledge of the pertinent thermodynamic data. For example, the rate of partitioning of a solute across a slag-metal interface is dependent on many factors that affect the rate of transport of solutes up to and away from the interface. Three of the fundamental properties affecting the kinetics of these reactions are the density, surface tension and viscosity of the two liquid phases; but errors arise in the measurement of these properties, even under closely controlled laboratory conditions. The error limits are estimated as:[129]

Density	±2%
Surface tension	±10%
Viscosity	±25%

A large component of the error in the viscosity measurements arises from the calibration of the viscometers with reference liquids at room temperature and also from the assumption that the calibration constant thus obtained is applicable at the high temperatures relevant to the production processes. The temperature of the bulk metal may be measured accurately, but it is more difficult to measure the temperature of a relatively thin slag layer, and the uncertainty in the estimated slag viscosity may be greater than ±25% since the viscosity is temperature dependent.

Surface tension is also a temperature-dependent property, and the temperature at the slag-metal interface may differ from that for the bulk solution. Furthermore, the surface tension can fall to very low values as a result of the induced interfacial agitation when there is rapid mass transfer across an interface. Laboratory measurements made with a static interface are of little value under these conditions.

The rate of mass transport caused by natural and induced circulation within each of the liquid phases is dependent on a number of factors, including the geometry of the containing vessel, the temperature gradients within the bath and the way in which the turbulence is created. Thus, the turbulence created by gas bubbles is affected by the volume of gas released in unit time, the average bubble size, the depth below the surface at which the bubbles are released and the volume of the melt in which the bubbles are contained. It is evident, therefore, that reaction rates predicted from fundamental principles are dependent on the assumptions made in formulating the model and may vary from one vessel to another in which, ostensibly, the same procedure is being followed. Similar problems arise in the fundamental modeling of other phenomena (such as heat transfer), particularly when foaming or emulsified slags separate the heat source from the metal.

9.2.2. Empirical Models

A process model can be constructed purely on the basis of observations and measurements of the way in which the process normally operates, or from experimental measurements on a pilot-scale or full-size plant. The derived mathematical relationships are empirical expressions of the data, without reference to the basic laws applicable to the process. The process description is thus usually incomplete and restricted to quantification of those aspects that have been observed.

A development of this approach is the neural network. Data are fed into the computer describing the starting composition and the decisions taken during the processing of previous quantities of material. The neural network computer derives its own mathematical model to fit these data and can learn complex relationships from them. It assigns an arbitrary weighting of the importance to each of the data inputs. The models are designed to function in a similar way to the neurons (i.e., independent cells that conduct nerve impulses) in the human brain and to make decisions on the basis of the weighted inputs and the required outputs. The program is "trained" by comparing the computed decisions with the actual results. If the prediction is incorrect, the weightings are changed progressively by small increments until the correct decisions are obtained. In use, the weightings may be changed frequently as the reactions being evaluated progress toward the desired end point.

This type of model is best suited to maintaining steady-state operation of an existing procedure such as a continuous process in which relatively few variables are involved, or repetitive batch processes in which the inputs and the required outputs remain more or less constant. However, the true relationships between inputs to and outputs from the process are not clearly defined. Empirical models often bear little resemblance to the actual process, and major errors can arise if they are used to extrapolate to conditions outside the applicable range when the data used in the model were determined. The model is often backed up, therefore, with an expert system that monitors the actual conditions in the process and overrides the computer predictions if they appear to be unrealistic.

9.2.3. Dynamic Models

The majority of process control models in current use are based on a structure of equations derived in terms of the underlying fundamental laws and on using the relevant experimental data for evaluation. Simplifying assumptions are made where the data do not extend to, say, the composition or temperature range of interest. Thus, regular solution behavior may be assumed in order to extrapolate to the relevant range, even though it is evident from other considerations that the entropy of mixing is not ideal; or, the data may be supplemented by empirical relations derived from observations on the operating plant. However, it must be remembered that a mathematical model cannot predict values that are more accurate than the input data, and the uncertainty in the forecasts increases as the proportion of estimated data and empirical relationships is increased. Ideally, when a model has been constructed, it should be tested against actual experimental or plant data and fine-tuned by adjustment of the estimated numerical values of the equations until it accurately reproduces the actual results, in similar fashion to the tuning of a neural network.

Prediction of material outputs on the basis of the amounts and compositions of the material inputs becomes more complicated as the number of variables that must be taken into account is increased. So, simplified dynamic models are often constructed that do not completely describe the process, but in which the computer predictions are regularly updated by evaluation of data supplied by sensors measuring the real-time values of the parameters affecting the process. The computer updating calculations must be completed very quickly in order to achieve effective control when the total time taken for a package of material to pass through the process is comparatively short. This places a limit on the amount of on-line data that can be analyzed in a single program. It requires the application of parallel processing techniques, in which various components of the input data are evaluated simultaneously in a series of subroutines and the outputs from all of these are used to update the main program; or the fundamental relationships must be further simplified in order to incorporate the real-time data. The model is then less useful for predicting the effect of changes to the operating procedures or in the input materials that lie outside the range for which the program has been tuned. More inputs can be accommodated for processes having a relatively long time span, and the total inputs and outputs can exceed 5000 for a blast furnace.

The data supplied by the sensors must be interpreted correctly by the model and lead to meaningful predictions; otherwise, the data are useless.[127] The selection and positioning of sensors should be decided on the basis of the inputs regarded as the most important and that can be usefully evaluated for the real-time updating of the model predictions, in preference to attempts to use all the data available from sensors in use before the computer program was written.

Readings from a sensor may be incorrect due to malfunctions of the instrument, poor calibration, zero drift, mist or dust accumulation on optical windows, blockage of filters, accidental misalignment and a variety of other causes. An operative who is observing both the actual process and the readouts from a variety of sensors should quickly detect when one of the sensors is providing improbable information. The computer program must also examine trends in the data that are interrogated from each of the sensors, and sound an alarm and disregard the input from any sensor that is inconsistent with that supplied by the other sensors. It is sensible practice to ensure that at least all the more important sensors are duplicated and can be monitored, one against the other, when the readings appear to be dubious.

9.2.4. Examples of Control Systems

Although the objective of fully automatic control by computer of extractive processes is still over the horizon, a large number of computer packages have been produced that supply information to simplify the task of the human operative in the control of the process. The skills required by the operatives have thus changed from an ability to react to visual observations on the basis of cumulative experience to an ability to interpret and respond quickly to data supplied by the computer. In many cases, the operative is now located in a clean environment, surrounded by screens displaying computer data and graphics, remote from the reaction vessel. Throughput times are being decreased, and productivity and quality of the product are being increased progressively as more comprehensive computer programs are being introduced. Many of these programs have been devised and tuned to a specific production unit, but a wide variety of software packages are available that can be incorporated into more complex models to deal with subroutines such as chemical equilibrium, heat transfer and fluid dynamics.

Modeling of pyrometallurgical processes has been applied most extensively to the iron blast furnace and BOS steelmaking, and new developments in the control systems for these processes appear regularly in the technical literature. Models have also been derived for a wide range of other processes, including the sintering of iron ores,[130] reactions in the stack of the Pb blast furnace,[131] reduction of Cu sulfides[132] and volatilization of impurities[133] in flash smelting, converting of Ni,[134] mass transfer by gas injection in the furnace[135] and in ladle refining,[136] and the simulation of vacuum refining.[137]

The control system operated by a leading steel manufacturer for the production of steels via a BOS vessel, a ladle refining station and a continuous casting machine is indicative of the progress that has been achieved. The charge inputs and the amount of oxygen required for the BOS blow are calculated in terms of a heat and mass balance with a static model. Sensors on the teeming ladle at the continuous caster measure the metal temperature and the weight of metal remaining in the ladle, from which a dynamic model continuously updates the time required to empty the ladle. This indicates the time remaining for completion of the ladle refining to ensure that the metal will arrive at the continuous caster just in time to maintain the sequence casting. The ladle refining model determines the actions required to complete the adjustment of the metal composition, complete the deoxidation and adjust the temperature of the metal in the time available. Analysis of the CO and CO_2 contents of the exhaust gas from the BOS vessel indicates the rate of carbon removal (dC/dt), while the audiometric signal denotes the depth to which the lance is immersed and hence the rate of accumulation of FeO in the slag. The BOS model controls the lance position and the oxygen flow rate to maintain these two signals within predefined limits, shown as boundary lines in Figure 9.1, to ensure that the blow is completed at the time required for the metal to be tapped from the vessel and transferred to the ladle station to maintain the sequence.

Looking to the future, the computer will progressively reduce the decision input required from the operatives. Expert systems and fuzzy logic are already being introduced to allow the computer to mimic the deductions that would be made by the operator and to make decisions on the actions required. It is much more difficult to fit a mathematical model to an existing process than to derive a new process from basic principles. Modeling will be used to devise new processes and practices that are more efficient and economical in the use of time and materials, which are more readily controlled to produce a high-quality product and are environmentally friendly. Numerous attempts have been made in the past to devise truly continuous processes with, for example, a pyrometallurgical plant in which a slag incorporating the gangue is continuously discharged from one end of the furnace, while metal of a constant composition is discharged from the other end. These attempts have not been successful for a variety of reasons, including inadequate heat transfer, insufficient contact time for completion of the slag-metal reactions and incomplete separation of the metal from the slag. Problems of this type should be avoided if the reactor is designed efficiently on the basis of a computer model. There is a need for more highly skilled personnel to enter this field who understand the possibilities and limitations of existing processes and can construct models for the more efficient processes.

Appendix

WORKED EXAMPLES

The following examples indicate types of equilibrium calculations often required in the evaluation of extraction and refining operations. The first three examples illustrate the use of heat capacity data in the determination of heat balances. These are followed by a materials balance calculation (Example 4) and then by a nucleation problem (Example 5). Examples 6 through 12 demonstrate various applications of the equilibrium constant. Examples 13 and 14 are concerned with aspects of hydro- and electrometallurgy. The examples are all typical of the types of calculations required in practice but, in some cases, they have been simplified to aid understanding.

Tabulated data often list values of free energy or enthalpy to six significant figures, which implies an accuracy of better than ± 0.001%. The uncertainty is usually much larger than this. For practical purposes, it is not necessary to consider data to more than four significant figures, and this practice is adopted in the following calculations.

Example 1

A charge of 10 kg copper is placed in a high-frequency induction furnace at 25 °C. The useful heat input to the charge (i.e., total heat input less thermal losses) is 7.0×10^6 J. Calculate the temperature to which the Cu is heated by this amount of energy.

The first step is to calculate the heat which is required to just melt the Cu. The data required are:

$$C_{p(\mathrm{Cu,s})} = 22.64 + 6.28 \times 10^{-3}T \quad \mathrm{J\,K^{-1}\,mol^{-1}}$$
$$C_{p(\mathrm{Cu,l})} = 31.38\ \mathrm{J\,K^{-1}\,mol^{-1}}$$
$$L_f = 13{,}000\ \mathrm{J\,mol^{-4}}$$

where L_f is the latent heat of fusion of Cu at the melting point of 1083 °C (1356 K).

The atomic weight of Cu is 63.57. Hence, the charge comprises (10,000/63.57) = 157.3 mol Cu. The enthalpy required to heat this quantity of Cu to the melting point is:

$$157.3 \int_{298}^{1356} (22.64 + 6.28 \times 10^{-3}T)\mathrm{d}T\ \mathrm{J}$$

which is equal to:

$$157.3\ [22.64 + 3.14 \times 10^{-3}T^2]_{298}^{1356}\ \text{J.} = 4{,}632{,}000\ \text{J}$$

The latent heat required to melt 10 kg Cu is:

$$(13{,}000 \times 157.3) = 2{,}045{,}000\ \text{J}$$

Hence, the heat available to raise the temperature above the melting point is equal to:

$$\begin{aligned}&\text{Heat available} - \text{heat required to just melt the Cu}\\&= [(7.0 \times 10^6) - (4{,}632{,}000 + 2{,}045{,}000)] = 323{,}000\ \text{J}\end{aligned}$$

The heat required to raise the temperature of 10 kg molten Cu through 1 °C is (31.38 × 157.3) = 4936 J. The temperature attained is therefore [1356 + 323,000/4936)], which is equal to 1421 K or 1148 °C.

Example 2

A gas comprising 20% CO and 80% N_2 by volume is preheated to 800 °C and fed into a reverberatory furnace where it is completely combusted with the stoichiometric quantity of air. If the exhaust gas leaves the furnace at 1150 °C, calculate the maximum flame temperature and the quantity of heat supplied to the furnace per cubic meter of the inlet gas.

Heat capacity data are often tabulated in the form:

$$C_p = a + b \times 10^{-3}T + c \times 10^5T^{-2}\ \text{J K}^{-1}\ \text{mol}^{-1}$$

where *a*, *b* and *c* are constants. The values required here and the standard enthalpies at 298 K are:

	a	b	c	ΔH°_{298} J mol^{-1}
CO	28.43	4.10	−0.46	−110,500
CO_2	44.17	9.04	−8.54	−393,700
O_2	29.98	4.19	−1.67	0
N_2	27.88	4.27	0	0

The combustion reaction is:

$$2CO_{(g)} + O_2 = 2CO_2$$

and the heat of the reaction at 298 K is given by:

$$\begin{aligned}\Delta H^\circ &= [(2 \times -393{,}700) - (2 \times -110{,}500) - 0]\\&= -566{,}400\ \text{J}\end{aligned}$$

The change in the heat capacity for the reaction is obtained by summing the given data:

$$\Delta C_p = 1.50 + (5.69 \times 10^{-3}T) + (14.49 \times 10^5 T^{-2}) \text{ J.K}^{-1}$$

A completely general expression for the enthalpy as a function of temperature can be obtained by solving Equation 2.3 as an indefinite integral to find the value of the integration constant:

$$\Delta H^{\ominus}_{T} = \Delta H_0 + \int_0^T \Delta C_p \mathrm{d}T$$

Hence, at 298 K:

$$-566{,}400 = \Delta H_0 + 1.50T + (2.85 \times 10^{-3}T^2) + (14.49 \times 10^5 T^{-1}) \text{ J}$$

and

$$\Delta H^{\ominus}_{T} = -571{,}900 \text{ J}$$

The enthalpy at any temperature can now be found simply by inserting the required temperature in place of T in the above equation. The heat of reaction at the gas entry temperature of 1073 K (800 °C) is then given by:

$$\begin{aligned}\Delta H^{\ominus}_{1073} &= -571{,}900 + (1.5 \times 1073) + (2.85 \times 10^{-3} \times 1073^2) \\ &\quad + (14.49 \times 10^5 \times 1073^{-1}) \\ &= -565{,}700 \text{ J for the reaction of 10 mol gas containing} \\ &\quad \text{2 mol CO.}\end{aligned}$$

(This value could be obtained directly from the tabulated values for the products and the reactants at 1073 K, which can be found in the data tables given, for example, in References 26 and 27.)

This enthalpy heats the combustion products plus the nitrogen from the air and in the inlet gas to the maximum flame temperature. Assuming for simplicity that air contains 20% O_2 and 80% N_2 by volume, the outlet gas from the furnace consists of: 2 mol of CO_2 from the combustion of CO and 12 mol N_2 (8 from the inlet gas and 4 from the air).

Therefore,

$$\begin{aligned}565{,}700 &= \int_{1073}^{T_M} (2C_{P_{CO_2}} + 12\, C_{P_{N_2}})\, \mathrm{d}T \\ &= [422.90\, \mathrm{T} + (34.66 \times 10^{-3}T^2) \\ &\quad + (17.08 \times 10^5 T^{-1}]_{1073}^{T_M}\end{aligned}$$

and

$$1{,}061{,}000 = 422.90T_M + (34.66 \times 10^{-3}T_M^2) + (17.08 \times 10^5 T_M^{-1})$$

Since T_M is greater than the gas exit temperature (1073 K), the numerical value of the last term in the equation is less than 1000 and can be ignored, leaving a quadratic equation that can be solved to yield:

$$T_M = 2135\ \text{K} = 1862\ °\text{C}$$

The heat supplied to the furnace is the difference between the heat of combustion and the sensible heat lost in the outlet gas. The latter is given by:

$$\int_{1073}^{1423} (2C_{p_{CO_2}} + 12\ C_{p_{N_2}})\ dT = 177{,}900\ \text{J}$$

Therefore, the heat supplied to the furnace by 10 mol inlet gas is:

$$(565{,}700 - 177{,}900) = 387{,}800\ \text{J}$$

Since 1 mole of any gas occupies 22.4 l at NTP, the volume of 10 moles at the inlet temperature of 1073 K is:

$$[10 \times 22.4 \times (1073/273)] = 880\ \text{l} = 0.88\ \text{m}^3$$

Therefore, the heat supplied to the furnace is:

$$(387{,}800/0.88) = 440{,}700\ \text{J m}^{-3}\ \text{inlet gas}$$

Example 3

An iron ore is reduced in a shaft-type direct reduction kiln by a gas mixture of CO and H_2. The sensible heat supplied by the inlet gas is sufficient to raise the temperature of the ore to 827 °C in the final reduction zone in which FeO is reduced to metallic iron. Reduction of FeO by CO is exothermic, whereas reduction with H_2 is endothermic. Assuming that the ore has been completely reduced to FeO when it enters the final reduction zone, determine the ratio p_{H_2}/p_{CO} required in the inlet gas when (p_{CO} + p_{H_2} = 1), which will maintain a constant temperature in the reaction zone.

From data tables, the enthalpies at 1000 K for the reactions concerned are:

$$\begin{array}{lr} & \Delta H^{\ominus}_{1000\ K} \\ H_{2(g)} + \frac{1}{2}O_{2(g)} = H_2O_{(g)}: & -247{,}900\ J \\ CO_{(g)} + \frac{1}{2}O_{2(g)} = CO_{2(g)}: & -283{,}600\ J \\ Fe_{(s)} + \frac{1}{2}O_{2(g)} = FeO_{(s)}: & -267{,}400\ J \end{array}$$

Let X equal the partial pressure of hydrogen in the inlet gas. If no heat change occurs, then the heat balance is:

$$247{,}900X + 283{,}600\,(1 - X) = 267{,}400$$

Therefore,

$$X = p_{H_2} = 0.454$$
$$(1 - X) = p_{CO} = 0.546$$

and the ratio p_{H_2}/p_{CO} required in the inlet gas is 0.832.

Example 4

A concentrate containing 29.0% Cu, 31.5% Fe, 32.0% S and 6.5% SiO_2 is fed into a flash smelter. Air is blown at a rate of 200 $m^3\ min^{-1}$ at NTP to produce a Cu_2S-FeS matte containing 50% Cu. Iron is oxidized to FeO and silica sand is added to form a slag containing 60% FeO. All concentrations are expressed in wt%.

Neglecting Cu loss into the slag and assuming that 90% of the oxygen in the air is consumed by the reactions, calculate the weight of matte and slag produced per hour and the partial pressure of SO_2 in the off-gas.

The atomic weights of the elements are:

Cu 63.57
Fe 55.85
O 16
S 32

There are several ways to obtain a solution, and it will be recognized that a shorter procedure could be used than the step-by-step method outlined here.

The copper is conserved in the reaction (i.e., all the Cu in the concentrate reports to the matte). 1 kg of concentrate contains 290 g Cu and the matte contains 50% Cu. So,

$$[(0.29/0.50) \times 1000] = 580\ \text{g matte}$$

are produced from 1 kg concentrate.

Two mol [or (2 × 63.57 g)] Cu combine with 1 mol (or 32 g) S to form 1 mol Cu_2S. Hence, 290 g Cu combine with 73 g S in the matte to form 363 g Cu_2S. Since the matte only contains Cu_2S and FeS, the weight of FeS in the matte formed from 1 kg concentrate is (580 − 363) = 217 g FeS.

The molecular weight of FeS is (55.85 + 32 g) = 87.85 g. Therefore, the weight of Fe in the matte is:

$$[(217 \times 55.85)/87.85] = 138 \text{ g kg}^{-1} \text{ concentrate}$$

Furthermore, 1 kg concentrate contains 315 g Fe. If 138 g of this report to the matte, then 177 g Fe are transferred to the slag. This weight of Fe forms:

$$177 + [177 \times (16/55.85)] = 228 \text{ g FeO}$$

Since the slag contains 60 wt% FeO, the slag weight is 380 g per kilogram concentrate.

One kg concentrate contains 320 g or 10 mol S. The matte formed from this amount of concentrate contains (217 − 138) = 79 g or 2.47 mol S in the FeS and 73 g or 2.28 mol S in the Cu_2S. Since the sulfur is conserved in the system, a materials balance can be written in the form:

$$\Sigma\, n_S = n_{S\ (\text{concentrate})} = n_{Cu_2S} + n_{FeS} + n_{SO_2}$$

where n is the number of moles of each species. Substituting the known values, $n_{SO_2} = 5.25$.

Similarly, the amount of oxygen consumed by the reactions, per kilogram concentrate, is:

$$\Sigma n_{O_2} = \tfrac{1}{2} n_{FeO} + n_{SO_2} = 6.79$$

Air contains 21 vol% O_2, and 22.4 m^3 oxygen gas contains 1000 mol oxygen. If air is blown at a rate of 200 $m^3\ min^{-1}$ and 90% of the oxygen reacts with the concentrate, the amount of oxygen consumed per hour is:

$$[200 \times 60 \times 0.21 \times 0.9 \times 1000)/22/4] = 101{,}250 \text{ mol}$$

Therefore, the weight of concentrate processed is:

$$(101{,}250/6.79) = 14{,}912 \text{ kg h}^{-1}$$

and the weight of matte produced is:

$$[(14{,}912 \times 580)/1000] = 8649 \text{ kg h}^{-1}$$

The slag weight is:

$$(14{,}912 \times 0.380) = 5667 \text{ kg h}^{-1}$$

For each kilogram concentrate processed, 177 g or 3.17 mol Fe are oxidized into the slag and this consumes 1.59 mol O_2. In 1 h of operation, the total amount of oxygen consumed by Fe oxidation is:

$$[(1.59 \times 14{,}912 \times 22.4)/1000] = 531 \text{ m}^3\text{h}^{-1} \text{ oxygen}$$

and the gas volume is decreased by this amount. (The gas volume is not changed by the conversion of O_2 to SO_2. 5.25 mol SO_2 are generated by the oxidation of 1 kg concentrate, so the volume of SO_2 produced per hour is:

$$[(14{,}912 \times 5.25 \times 22.4)/1000] = 1754 \text{ m}^3 \text{ h}^{-1}$$

The partial pressure of SO_2 in the outlet gas is:

$$1754/[(200 \times 60) - 531] = 0.153$$

which is equal to 15.3 vol%.

Example 5

Particles of a solid catalyst are suspended in a molten metal. If a nucleus of solid metal forms heterogeneously as part of a sphere on the surface of the catalyst, determine the effectiveness of the catalyst for nucleation of solidifiation of the metal when the values of the surface energies are:

Surface	**J m^{-2}**
Liquid-solid, l-s	0.25
Liquid-catalyst, l-c	0.16
Solid-catalyst, s-c	0.10

The surface tensions are identified in Figure 3.5 and the balance of the forces is given by Equation 1.4, i.e.,

$$\gamma_{(l\text{-}c)} = \gamma_{(s\text{-}c)} + \gamma_{(s\text{-}l)} \cos \theta$$

Hence,

$$\cos \theta = [(0.25 - 0.16)/0.10] = 0.9$$

and

$$\theta = 25.8°$$

The activation energy for heterogeneous nucleation is given by Equation 3.40, which can be rearranged as:

$$\Delta G^* = \left[\frac{16\pi \, \gamma^3_{(s\text{-}1)}}{3\Delta G_V}\right]\left(\frac{2 - 3\cos\theta + \cos^3\theta}{4}\right)$$

The first term in brackets on the right-hand side of this equation is equal to the activation energy required for homogeneous nucleation (Equation 3.39). The second term in brackets is, therefore, a measure of the energy change caused by the presence of the catalyst. Substituting the value of θ into the latter term:

$$\frac{[2 - (3 \times 0.9) + 0.9^3]}{4} = 0.0073$$

Hence,

$$\Delta G^*_{(heterogeneous)} = 0.0073 \times \Delta G^*_{(homogeneous)}$$

That is, the catalyst reduces the activation energy required to nucleate the solid to 1/137 of the energy required for homogeneous nucleation.

Example 6

Calculate the partial pressure of oxygen in equilibrium with solid molybdenum and its solid oxide at 1600 °C (1873 K) when (a) both solids are pure and (b) when the activity of the oxide is 0.5.

The standard free energy for the formation of the oxide, MoO_2:

$$Mo_{(s)} + O_{2(g)} = MoO_{2(s)}$$

is given by:

$$\Delta G^{\circ} = -578{,}200 + 166.5T \text{ J } (298\text{–}2273 \text{ K})$$

The figures in brackets indicate the range of temperatures within which the free energy expression is valid. It is always important to check that the free energy equation for a reaction selected from the data tables is valid at the temperature of interest. At 1873 K,

$$\Delta G^{\circ} = -266{,}300 \text{ J}$$

The equilibrium constant for the reaction is given by:

$$\Delta G^{\circ} = -RT \ln K = -19.15T \log K$$

Hence,

$$\log K = [+266{,}300/(19.15 \times 1873)] = 7.43$$

and

$$K = \left[\frac{a_{MoO_2}}{a_{Mo} \cdot p_{O_2}}\right] = 2.66 \times 10^7$$

(a) When Mo and MoO_2 are pure, both solids are in their standard states and their activities are unity. Hence,

$$\begin{aligned} p_{O_2} &= 1/(2.66 \times 10^7) \\ &= 3.7 \times 10^{-8} \end{aligned}$$

(b) When the activity of the oxide is 0.5:

$$\begin{aligned} p_{O_2} &= 0.5/(2.66 \times 10^7) \\ &= 1.88 \times 10^{-8} \end{aligned}$$

Example 7

The outlet gast from a gaseous reduction process is refrigerated to condense out the water vapor and an additional quantity of hydrogen is added. The composition of the gas is then 40% CO, 10% CO_2, 30% H_2 and 20% N_2 by volume. The gas is heated to 1000 °C at a pressure of 1 atm, and the composition of the gas is changed by the water gas reaction (Equation 4.5) before it is fed back into the reduction furnace. Determine the equilibrium composition of the gas at 1000 °C.

The data required are:

$$\begin{aligned} C + O_2 &= CO_2\text{: } \Delta G^{\circ} = -394{,}800 - 0.83T \text{ J (298–2273 K)} \\ C + \tfrac{1}{2}O_2 &= CO\text{: } \Delta G^{\circ} = -112{,}900 - 86.51T \text{ J (298–2273 K)} \qquad \text{(a)} \\ H_2 + \tfrac{1}{2}O_2 &= H_2O\text{: } \Delta G^{\circ} = -247{,}000 + 55.85T \text{ J (298–2273 K)} \end{aligned}$$

Subtracting (b) and (c) from (a):

$$CO + H_2O = CO_2 + H_2\text{: } \Delta G^{\circ} = -34{,}500 + 29.83T \text{ J (298–2273 K)}$$

Assuming ideal behavior of the gases, the equilibrium constant for this reaction is:

$$K = \left(\frac{p_{CO_2} \cdot p_{H_2}}{p_{CO} \cdot p_{H_2O}}\right) = 0.721 \text{ at } 1273 \text{ K}$$

There are four unknown quantities in the equation, so three other relations between the partial pressures are required in order to solve the equation.

The total pressure of the gas is 1 atm, and there is 20 vol% inert nitrogen; so,

$$p_{CO} + p_{CO_2} + p_{H_2} + p_{H_2O} = 0.8 \qquad \text{(d)}$$

The number of moles (n) of each of the elements in the gas is unchanged by the reforming reaction, and these numbers can be evaluated from the initial gas composition. Thus,

$$n_C = n_{CO} + n_{CO_2} = 0.5$$
$$n_H = 2n_{H_2} + 2n_{H_2O} = 0.6$$
$$n_O = n_{CO} + 2n_{CO_2} + n_{H_2O} = 0.6$$

Hence,

$$n_C = \tfrac{5}{6} n_H \qquad n_H = n_O$$

Therefore,

$$(n_{CO} + n_{CO_2}) = 1.67\,(n_{H_2} + n_{H_2O})$$
$$(n_{CO} + 2n_{CO_2} + n_{H_2O}) = 2\,(n_{H_2} + n_{H_2O})$$

Since the partial pressure of a gas is proportional to the number of moles, i.e.,

$$p_i = \left[\frac{n_i}{n_{total}}\right] P_i$$

these relations can also be written in partial pressures:

$$(p_{CO} + p_{CO_2}) = 1.67\,(p_{H_2} + p_{H_2O}) \qquad \text{(e)}$$

$$(p_{CO} + 2p_{CO_2} + p_{H_2O}) = 2(p_{H_2} + p_{H_2O}) \qquad \text{(f)}$$

Equations (d), (e) and (f) can be used to substitute for three of the unknown quantities in the equilibrium constant, and K can be written in terms of any one of the gas species. For example, p_{H_2} can be replaced by p_{H_2O} by subtracting (e) from (d) to give:

$$2.67p_{H_2} + 2.67p_{H_2O} = 0.8$$

or

$$p_{H_2} = 0.3 - p_{H_2O} \tag{g}$$

Similarly, subtracting (f) and (g) from (d) yields:

$$p_{CO_2} = 0.1 - p_{H_2O} \tag{h}$$

and combining (d), (g) and (h) gives:

$$p_{CO} = 0.4 - p_{H_2O} \tag{j}$$

Substituting these relations into the equilibrium constant:

$$K = 0.721 = \frac{(0.1 - p_{H_2O})\,(0.3 - p_{H_2O})}{(0.4 - p_{H_2O}) \cdot p_{H_2O}}$$

or

$$0.28\, p_{H_2O}^2 - 0.69 p_{H_2O} - 0.03 = 0$$

This is a quadratic equation that can be solved by the formular method to yield:

$$p_{H_2O} = 0.04$$

Substituting this value into (g), (h) and (j) gives:

$$p_{H_2} = 0.26$$
$$p_{CO_2} = 0.06$$
$$p_{CO} = 0.44$$

In practice, the gas would normally be reformed by injecting a hydrocarbon species to react with and reduce or eliminate entirely the CO_2 content from the gas. The solution of the problem is then a little more complicated, but can be achieved by a similar method.

Example 8

Silver is removed from molten Pb by adding Zn in the Parkes process (Section 5.4.1.). When 3 wt% Zn is added to Pb containing 1 wt% Ag at 450 °C, the Ag-Zn dross that separates contains 25% Ag and 75% Zn. Determine the residual concentrations of Ag and Zn remaining in the Pb when equilibrium is established.

The activity coefficient for Ag, $\gamma_{Ag(Zn)}$, in the Ag-Zn dross is 0.10 at 450 °C. Ag conforms to Henry's Law in dilute solution in Pb, and $\gamma^0_{Ag(Pb)}$ is 5.7.

Converting the wt% concentrations to atom fractions, the initial composition of the Pb before Ag partitioning occurs is $0.019N_{Ag}$, $0.089N_{ZN}$ and $0.892N_{Pb}$. The equilibrium composition of the dross is $0.355N_{Ag}$ and $0.645N_{Zn}$.

The activity of Ag is the same in both phases at equilibrium (see Section 2.3.1.); i.e.,

$$a_{Ag(Zn)} = a_{Ag(Pb)}$$

and the equilibrium constant for the partitioning of Ag between the two phases is unity. Since $a = \gamma \cdot N$ (Equation 2.29), it follows that:

$$\frac{N_{Ag(Zn)}}{N_{Ag(Pb)}} = \frac{\gamma_{Ag(Pb)}}{\gamma_{Ag(Zn)}} = \frac{5.7}{0.1} = 57$$

Substituting the value given for $N_{Ag(Zn)}$:

$$N_{Ag(Pb)} = 0.355/57 = 0.006$$

Now,

$$n_{Ag(Zn)} + n_{Ag(Pb)} = 0.019$$

where n is the number of atoms of Ag is each of the phases in equilibrium with 1 mol Pb when partitioning has occurred. Furthermore,

$$N_{Ag(Pb)} \simeq (n_{Ag}/n_{Pb})_{Pb} \simeq n_{Ag(Pb)}$$

since n_{Ag} and n_{Zn} are very small quantities when equilibrium is established. Therefore,

$$n_{Ag(Zn)} = 0.019 - 0.006 = 0.013$$

The atom fraction of Zn atoms in the Ag-Zn phase is given by:

$$N_{Zn(Zn)} = \left(\frac{n_{Zn}}{n_{Ag} + n_{Zn}}\right)_{Zn}$$

Substituting the known values:

$$0.645 = [n_{Zn}/(0.013 + n_{Zn})]_{Zn}$$

whence $n_{Zn(Zn)}$ is 0.024 and

$$N_{Zn(Pb)} \simeq n_{Zn(Pb)} = 0.089 - 0.024 = 0.065$$

Converting back to wt%, the residual concentrations in the Pb are 0.32 wt% Ag and 2.03 wt% Zn. Thus, 68% of the Ag is removed from the Pb by this treatment. A further quantity can be removed by skimming off the dross and adding more Zn to form a second dross. In practice, the dross contains about 68 wt% Pb and only about 32% of the Ag-Zn phase.

Example 9

After the silver has been removed from the Pb in Example 8, the Zn remaining in the Pb can be removed by adding $PbCl_2$ to the melt (see Section 5.4.1). The activity coefficient of Zn in dilute solution in Pb ($\gamma_{Zn(Pb)}$) is 11.0 at 500 °C. Assuming that Pb containing a very small amount of Zn is ideal and that the $PbCl_2$-$ZnCl_2$ dross conforms to ideal solution behavior, calculate the residual Zn content remaining in the Pb when the dross contains 0.05 N_{PbCl_2} at 450 °C.

The free energy of the reaction:

$$PbCl_2 + Zn = ZnCl_2 + Pb$$

is given by:

$$\Delta G^{\circ} = -78{,}150 + 28.18\ T\ J\ (695\text{–}1005\ K)$$

and

$$K = 1.5 \times 10^4 \text{ at } 723\ K$$

The activity coefficient of Zn is given at 500 °C (773 K). Assuming that the solute conforms to regular solution behavior, the value at 723 K can be found from the equality:

$$RT \ln \gamma_{Zn(773\ K)} = RT \ln \gamma_{Zn(723\ K)}$$

Hence,

$$\gamma_{Zn(723\ K)} = 13.0$$

The equilibrium constant for the reaction can now be written as:

$$K = \frac{a_{ZnCl_2} \cdot a_{Pb}}{a_{PbCl_2} \cdot a_{Zn}}$$

$$= \frac{N_{ZnCl_2} \cdot (1 - N_{Zn})}{N_{PbCl_2} \cdot \gamma_{Zn} \cdot N_{Zn}}$$

Inserting the known values:

$$N_{Zn(Pb)} = 9.7 \times 10^{-5}, \text{ wt\% Zn} = 0.003$$

Example 10

Copper can be removed from solution in molten Pb by the addition of solid PbS to precipitate solid Cu_2S. Assuming that Cu conforms to regular solution behavior in molten Pb, described by the relation:

$$\overline{H}^{M}_{Cu} = 14{,}000N \text{ J mol}^{-1}$$

and that the sulfides are immiscible, calculate the residual concentration of Cu in Pb when excess PbS is added to the melt at 400 °C.

Data tables provide the following information:

$$2Cu_{(s)} + \tfrac{1}{2}S_{2(g)} = Cu_2S \qquad \text{(a)}$$
$$\Delta G^{\circ} = -131{,}800 + 30.81T \text{ J } (298\text{–}1356 \text{ K})$$
$$Pb_{(l)} + \tfrac{1}{2}S_{2(g)} = PbS \qquad \text{(b)}$$
$$\Delta G^{\circ} = -163{,}200 + 88.03T \text{ J } (600\text{–}1386 \text{ K})$$

Equation (a) applies only to solid Cu, but the Cu is molten in the Pb melt. The enthalpy of fusion of Cu, $L_{f(cu)}$, is 10,900 J mol^{-1} at the melting point of 1356 K. Assuming that L_f is independent of temperature and using Equations 2.78 and 2.79,

$$Cu_{(s)} = Cu_{(1)}\text{: } \Delta G^{\circ} = 10900 - 49.19T \text{ J} \qquad \text{(c)}$$

Subtracting (b) and (c) from (a),

$$2Cu_{(1)} + PbS_{(l)} = Pb_{(l)} + Cu_2S_{(s)}$$
$$\Delta G^{\circ} = +20{,}500 - 49.19T \text{ J}$$
$$= -12{,}600 \text{ J at } 673 \text{ K}$$

and

$$\log K = 0.98$$
$$= \log \frac{a_{Pb}}{a_{Cu}}$$

since the sulfides are immiscible and their activities can therefore be set at unity. Hence,

$$2 \log a_{Cu} = \log a_{Pb} - \log K \qquad \text{(d)}$$

Now, from the given relationship for regular solution behavior of Cu in Pb and Equation 2.53:

$$\begin{aligned} \log \gamma_{cu} &= 14{,}000/(19.15 \times 673) \\ &= 1.09 \text{ at } 673 \text{ K} \\ &= a_{Cu}/N_{Cu} \text{ from Equation 2.29} \end{aligned}$$

Furthermore, if N_{cu} is small, a_{Pb} can be equated to unity without serious error. Hence, substituting for a_{Cu} in Equation (d):

$$2\log N_{Cu} + 2.18 = -0.98$$

and

$$N_{Cu} = 0.026$$

Example 11

When a metal is vacuum melted and dissolved oxygen is removed, there is a danger that the oxides in the refractory container may be reduced and limit the minimum oxygen content that can be obtained. Assuming that equilibrium is established, determine the minimum wt% oxygen content that can be attained in pure iron which is exposed to a vacuum of 0.001 atm in a pure MgO crucible at 1600 °C.

The oxygen content is obtained directly as wt% if the calculation is made with reference to the dilute wt% standard state for oxygen in Fe. The free energy equations required are:

(a) $$Mg_{(g)} + \tfrac{1}{2}O_{2(g)} = MgO_{(s)}$$
$$\Delta G^{\circ} = -732{,}700 + 206.0T \text{ J } (1378\text{–}2000 \text{ K})$$

(b) $$\tfrac{1}{2}O_{2(g)} = O_{(\% \text{ in Fe})}$$
$$\Delta G^{\circ} = -117{,}200 - 2.9T \text{ J } (>1800 \text{ K})$$

Subtracting (b) from (a),

$$\begin{aligned} Mg_{(g)} + O_{(\% \text{ in Fe})} &= MgO \\ \Delta G^{\circ} &= -615{,}500 + 208.9T \text{ J} \\ &= -224{,}300 \text{ J at } 1873 \text{ K} \end{aligned}$$

and

$$\log K = \log\left(\frac{a_{MgO}}{p_{Mg} \cdot a_O}\right) = \left(\frac{224{,}300}{19.15 \times 1873}\right) = +6.253$$

Since MgO is pure, $a_{MgO} = 1$. p_{Mg} can be set equal to the vacuum pressure. The activity of oxygen in iron can be expressed as the product of the activity coefficient, h_O (relative to the dilute wt% standard state) and the wt% concentration. Using Equation 2.85,

$$\log h_O = e_0^0 \text{ wt\% O}$$

and $e_0^0 = -0.20$. Hence,

$$\log 0.001 - 0.20\% \text{ O} + \log \% \text{ O} = -6.253$$

Therefore, wt% O = 0.00055.

Example 12

A sample of metal taken during the refining of molten steel had the following analysis:

Solute	**C**	**Mn**	**P**	**S**
Wt %	0.12	0.10	0.025	0.015

The temperature of the metal was 1600 °C and a_{MnO} in the slag was 0.10. Assuming that the Mn oxidation reaction was in equilibrium, calculate the pressure of CO gas in equilibrium with the carbon dissolved in the Fe.

The equilibrium constants, relative to the wt% standard state, for the oxidation of Mn and C from Fe are given by Equations 5.7 and 5.9, viz:

$$Mn_{(\%\text{ in Fe})} + O_{(\%\text{ in Fe})} = MnO_{(l)}$$

$$\log K_{Mn} = \frac{11{,}770}{T} - 5.07$$

$$C_{(\%\text{ in Fe})} + O_{(\%\text{ in Fe})} = CO_{(g)}$$

$$\log K_C = \frac{1168}{T} + 2.07$$

The interaction coefficients required are:

Solute X	**C**	**Mn**	**P**	**S**
e_C^X	0.21	−0.001	0.042	0.044
e_{Mn}^X	0	−0.002	0	−0.032

Hence, from Equation 2.85,

$$\begin{aligned}\log h_C &= e_C^C \text{ wt\% C} + e_C^{Mn} \text{ wt\% Mn} + e_C^P \text{ wt\% P} + e_C^S \text{ wt\% S} \\ &= (0.21 \times 0.12) + (-0.001 \times 0.10) \\ &\quad + (0.042 \times 0.025) + (0.044 \times 0.015) \\ &= +0.0247\end{aligned}$$

and

$$h_C = 1.058$$

Similarly,

$$\begin{aligned}\log h_{Mn} &= (-0.002 \times 0.10) + O + O + (-0.032 \times 0.015) \\ &= -0.00068\end{aligned}$$

and

$$h_{Mn} = 0.998$$

$$K_{Mn} = 16.4 = \left(\frac{a_{MnO}}{a_O \cdot h_{Mn} \cdot \%Mn}\right) \text{ at 1873 K}$$

and a value of 0.061 is obtained for a_O by substituting the known values. Likewise, at 1873 K,

$$K_C = 494 = \left(\frac{p_{CO}}{a_O \cdot h_C \cdot \%C}\right)$$

and, substituting the values for a_O and h_O, one obtains $p_{CO} = 3.83$ atm.

Note that the activity coefficients for both C and Mn are approximately equal to unity in this calculation. This is frequently the case with very dilute solutions. The error then incurred by assuming that the activity of a solute is equal to its wt% concentration is not usually significant. This is not true, however, when one or more of the interaction coefficients has a large value or when the solute concentrations are higher.

Example 13

The oxygen blow for converting a copper matte is discontinued when the Cu contains 1.2 wt% O and the temperature of the metal is 1200 °C. Assuming that the pressure of SO_2 gas within the melt is 1 atm, calculate the equilibrium wt% concentration of S remaining in the metal.

The interaction coefficients required are:

$$e_S^S = -0.18;\ e_S^O = -0.38;\ e_O^O = -0.16;\ e_O^S = -0.21$$

Hence, using Equation 2.85,

$$\log h_O = (-0.16 \times 1.2) + (-0.21 \times \%S)$$
$$\log h_S = (-0.18 \times \%S) + (-0.38 \times 1.2)$$

Both equations contain the unknown quantity %S and make the solution more complicated. If it is assumed, however, that the equilibrium value of %S is very small, these terms can be ignored and the activity coefficients are then evaluated as:

$$h_O = 0.64,\ h_S = 0.35$$

The free energy of formation of SO_2 must be expressed in terms of the wt% standard state. The data required are:

$$\tfrac{1}{2}S_{2(g)} + O_{2(g)} = SO_{2(g)} \qquad \Delta G^{\circ} = -361{,}700 + 72.68T\ \text{J}$$
$$O_{2(g)} = 2\ O_{(\%\ \text{in Cu})} \qquad \Delta G^{\circ} = -170{,}800 + 37.19T\ \text{J}$$
$$\tfrac{1}{2}S_{2(g)} = S_{(\%\ \text{in Cu})} \qquad \Delta G^{\circ} = -119{,}700 + 25.25T\ \text{J}$$

Hence,

$$S_{(\%\ \text{in Cu})} + 2\ O_{(\%\ \text{in Cu})} = SO_{2(g)}$$
$$\Delta G^{\circ} = -71{,}200 + 10.24T\ \text{J}$$
$$= -56{,}100 \text{ at } 1473\ \text{K}$$

and

$$K = 97.5 = \left(\frac{p_{SO_2}}{h_O^2 \cdot \%O^2 \cdot h_S \cdot \%S}\right)$$

Substituting the known values obtains %S = 0.053.

If this value is now inserted into the equations for the interaction coefficients and the calculation is repeated, the change in the answer is not significant and the approximation made in the determination of the interaction coefficients is acceptable in this case.

Example 14

An electrowinning cell for the production of copper from $CuSO_4$ solution operates at 50 °C with an applied voltage of 2.0 V and a current efficiency of 87%. Determine the energy efficiency if the electrolyte contains 0.5*m* $CuSO_4$ and 1.6 mol H_2SO_4.

For the reaction:

$$Cu^{2+} + H_2O = Cu + 2\,H^+ + \tfrac{1}{2}O_2$$
$$\Delta G^{\ominus} = -167{,}500 \text{ J at } 323 \text{ K}$$

and, from Equation 2.116, one obtains:

$$E^{\ominus} = -\frac{\Delta G^{\ominus}}{2F} = +\left(\frac{167{,}500}{2 \times 96{,}484}\right) = 0.860 \text{ V}$$

The Nernst equation (2.122) for this reaction is written as:

$$E = E^{\ominus} + \frac{RT}{2F} \ln\left(\frac{a_{Cu} \cdot a_{H^+}^2 \cdot p_{O_2}^{1/2}}{a_{Cu^{2+}}^2 \cdot a_{H_2O}}\right)$$

The activities of ions in concentrated solutions are usually not known accurately as a function of composition and temperature. Assuming complete dissociation of $CuSO_4$ and H_2SO_4 and that the activities of Cu^{2+} and H^+ ions are equal to their molalities, then:

$$m_{H^+} = 2m_{H_2SO_4};\ m_{Cu^{2+}} = m_{CuSO_4}$$

Assuming also that $a_{H_2O} = 1$ and that the oxygen gas is evolved from the solution at 1 atm pressure, the theoretical voltage required for Cu deposition is given by:

$$E = 0.860 + \left(\frac{8.343 \times 323}{2 \times 96484}\right) \ln\left(\frac{(2 \times 1.6)^2}{0.5}\right)$$
$$= 0.860 + 0.042 = 0.902 \text{ V}$$

According to Faraday's Second Law (Equation 7.14), the passage of 96,486 A through the cell should deposit [(63.57/2) = 31.79] g Cu on the cathode (63.57 is the atomic weight of Cu). Hence, the deposition of, say, 1 t Cu in 1 h requires a current of:

$$\frac{96{,}484}{60 \times 60} \times \frac{10^6}{31.78} = 843 \text{ kA}$$

and the theoretical energy consumption is:

$$843 \times 0.902 = 761 \text{ kWh hr}^{-1}$$

The cell actually operates at 2.0 V with a current efficiency of 87%. Hence, the energy consumption is:

$$843 \times (100/87) \times 2.0 = 1938 \text{ kWh h}^{-1}$$

Thus, the energy efficiency is [(843/1938) × 100] = 43.5%

The use of molalities in place of activities of the ions produces an error of less than 5% in the calculated efficiency.

REFERENCES

1. **Sorensen, H.,** Materials resources, in *Materials in Modern Society,* 10th Riso International Symposium on Metallurgy and Materials, Roskilde, Denmark, 1989, 53.
2. **Chapman, P. F. and Roberts, F.,** *Metal Resources and Energy,* Butterworths, London, 1983, 56.
3. **Gochin, R. J. and Solari, J. A.,** *Trans. Inst. Min. Metall.* 92, C52, 1983.
4. **Bhappu, R. D.,** in *Frontier Technology in Mineral Processing,* Soc. of Mining Eng., AIME, New York, 1985, chap. 12.
5. **Grjotheim, K. and Welch, B. J.,** *Aluminium Smelter Technology,* 2nd ed., Aluminium Verlag, Dusseldorf, 1988.
6. **Bodsworth, C. and Bell, H. B.,** *Physical Chemistry of Iron and Steel Manufacture,* 2nd ed., Longman, London, 1972.
7. **Biswas, A. K. and Davenport, W. G.,** *Extractive Metallurgy of Copper,* 2nd ed., Pergamon, Oxford, 1980.
8. **Heubner, V., Nilman, F., Reinent, M., and Veverschaer, A., Eds.,** *Lead Handbook,* Metallgesellschaft, 1983.
9. **Wright, P. A.,** *Extractive Metallurgy of Tin,* Elsevier, Amsterdam, 1966.
10. **Morgan, S. W. K.,** *Zinc and Its Alloys and Compounds,* Ellis Horwood, Chichester, 1985.
11. **Emley, E. F.,** *Principles of Magnesium Technology,* Pergamon, New York, 1966.
12. **Burkin, A. R., Ed.,** *The Extractive Metallurgy of Nickel,* John Wiley & Sons, New York, 1987.
13. **Stensholt, E. O., Zackariesen, H., and Lund, J. H.,** Trans. Inst. Min. Metall., 95, C10, 1986.
14. **Chapman, P. F. and Roberts. F.,** *Metal Resources and Energy,* Butterworths, London, 1983, 201.
15. **Ek, C. S.,** in *Mineral Processing at a Crossroads,* NATO ASI series, Martinus Nijhoff, The Hague, 1986, 135.
16. **Nutting, J.,** *Metals and Materials,* July/August, 31, 1977.
17. **Jackobson, D. M. and Evans, D. S.,** *Mater. Soc.,* 9, 1985.
18. **Hexner, E.,** *The International Steel Cartel,* University of North Carolina Press, Chapel Hill, North Carolina, 1943.
19. **Chapman, P F. and Roberts, F.,** *Metal Resources and Energy,* Butterworths, London, 1983, 24.
20. **Everett, D. H.,** *An Introduction to the Study of Chemical Thermodynamics,* Longman, London, 1959.
21. **Gaskell, D. R.,** *Introduction to Chemical Thermodynamics,* McGraw-Hill, New York, 1981.
22. **Darken, L. S. and Gurry, R. W.,** *Physical Chemistry of Metals,* McGraw-Hill, New York, 1953.
23. **Gokcen, N. A.,** *Thermodynamics,* Techscience, Hawthorne, CA, 1975.
24. **Warn, J.,** *Concise Chemical Thermodynamics,* D. Van Nostrand, New York, 1969.
25. **Kubaschewski, O. and Alcock, C. B.,** *Metallurgical Thermochemistry,* 5th ed., Pergamon Press, Oxford, 1979.
26. **Stull, D. R. and Prophet, H., Eds.,** *JANAF Thermochemical Tables,* 2nd ed., U.S. Dept. of Commerce, Washington, D.C., 1971; Supplements in *J. Phys. Chem.;* Reference Data, 3, 311, 1974; 7, 793, 1978; 14, 1985.
27. **Barin, I.,** *Thermochemical Data of Pure Substances,* VCH Publishers, Weinheim, West Germany, 1989.
28. **Desai, P. D., Hawkins, D. T., Gleiser, M., Kelly, K. K., and Wagman, D. D.,** *Selected Values of Thermodynamic Properties of the Elements,* ASM, Metals Park, OH, 1973.

29. **Prankatz, L. B.,** *Thermodynamic Properties—of Oxides,* Bulletin 672, 1982; *Halides,* Bulletin 674, 1984; *Sulphides,* Bulletin 689, 1987, U.S. Dept. of Interior, Bureau of Mines.
30. **Elliott, J. F. and Gleiser, M.,** *Thermochemistry for Steelmaking,* vol. 1, 1960; Vol. 2 (with Ramnakrishna, V.), 1963, Addison-Wesley, Reading, MA.
31. **Bale, C. W. and Eriksson, G.,** Metallurgical thermochemical databases, *Can. Metall. Q.,* 29, 105, 1990.
32. **Hildebrand, J. H.,** *J. Am. Chem. Soc.,* 51, 66, 1929.
33. **Wagner, C.,** *Thermodynamics of Alloys,* Addison-Wesley, Reading, MA, 1952.
34. **Turkdogan, E. T.,** *Trans. Iron Steel Inst. Japan,* 1042, 23, 1983.
35. **Sigworth, G. K. and Elliott, J. F.,** *Met. Sci. J.,* 8, 298, 1974.
36. **Sigworth, G. K. and Engh, T. A.,** *Scand. J. Metall.,* 11, 143, 1982.
37. **Sigworth, G. K. and Elliott, J. F.,** *Can. Metall. Q.,* 13, 445, 1974; 15, 123, 1976.
38. **Sigworth, G. K., Elliott, J. F., Vaughn, G. and Geiger, G. H.,** *Can. Metall. Q.,* 19, 104, 1977.
39. **Bodsworth, C. and Appleton, A. S.,** *Problems in Applied Thermodynamics,* Longman, London, 1965, 144.
40. **Ellingham, H. T. T.,** J. *Soc. Chem. Ind.,* (London), 63, 125, 1944.
41. **Richardson, F. D. and Jeffes, J. H. E.,** *J. Iron & Steel Inst.,* 161, 229, 1949.
42. **Kellog, H. H. and Basu, S. K.,** *Trans. AIME,* 218, 70, 1960.
43. **Barry, T. I. and Davies, R. H.,** *MTDATA Handbook, COPLOT Module,* National Physical Lab, Teddington, 1987.
44. **Bale, C. W., Thompson, W. T. T., and Pelton, A. D.,** *Can. Metall. Q.,* 25, 107, 1986.
45. **Crank, J.,** *The Mathematics of Diffusion,* Clarendon Press, Oxford, 1957.
46. **Jost, W.,** *Diffusion in Solids, Liquids and Gases,* Academic Press, New York, 1952.
47. **Lee, H. G. and Rao, Y. K.,** *Ironmaking Steelmaking,* 12, 221, 1985.
48. **Lee, H. G. and Rao, Y. K.,** *Ironmaking Steelmaking,* 15, 238, 1988.
49. **Rao, Y. K. and Cho, W. D.,** *Ironmaking Steelmaking,* 17, 273, 1990.
50. **Fruehan R. J.,** *Met. Trans.,* 8B, 279, 1977.
51. **Szekely, J., Choudhary, Y. and El-Tawil, Y.,** *Met. Trans.,* 8B, 693, 1977.
52. **Tien, R. H. and Turkdogan, E. T.,** *Met. Trans.,* 8B, 305, 1977.
53. **Turkdogan, E. T.,** *Physical Chemistry of High Temperature Technology,* Academic Press, New York, 1980.
54. **Swisher, J. H. and Turkdogan, E. T.,** *Trans. Met. Soc. AIME,* 239, 427, 1967.
55. **Turkdogan, E. T.,** *Met. Trans.,* 9B, 168, 1978.
56. **Abraham, K. P., Davidson, M. W., and Richardson, F. D.,** *J. Iron Steel Inst.,* 196, 82, 1960.
57. **Kay, A. R. and Taylor, J.,** *J. Iron Steel Inst.,* 201, 68, 1963.
58. **Lumsden, J.,** in *Metallurgical Chemistry* Symposium, Brunel University and NPL, July 1971, Her Majesty's Stationery Office, London, 1972, 533.
59. **Rist, A. and Meysson, M.,** *Rev. de Metall.,* 61, 121, 1964.
60. **Fitzgerald, F.,** *Ironmaking Steelmaking,* 19, 98, 1992.
61. **Kundrat, D. V., Miwa, T., and Rist, A.,** *Met. Trans.,* 22B, 363, 1991.
62. **Campbell, D. A., Flierman, G., Malgarino, G. and Smith, R. B.,** in 2nd European Ironmaking Congress, Glasgow, Inst. Materials, 1991, 263.
63. **Parker, R. H., Mitchell, A. R., and Lai, F. H.,** *Trans. Inst. Mining Metall.,* 99, C93, 1990.
64. **Miller, J. and Stephenson, R. L., Eds.,** *Direct Reduced Iron,* AIME, Warrendale, PA, 1980.
65. **Direct Reduction of Iron Ore,** Commission of the European Communities, Brussels, English ed., The Metals Society, London, 1979.

66. **Nixon, I. G., Bodsworth, C., Taheri, S. K., and Mullett, S. K.,** *Ironmaking Steelmaking,* 12, 103, 1985.
67. **Meschter, P. J. and Grabke, H. J.,** *Met. Trans.,* 10B, 323, 1979.
68. **Proctor, M. J., Hawkins, R. J., and Smith, J. D.,** *Ironmaking Steelmaking,* 19, 194, 1992.
69. **Leckie, A. H., Millar, A., and Medley, J. E.,** *Ironmaking Steelmaking,* 9, 222, 1982.
70. **Stephens, F. A. and Williams, W. E. J.,** *Steel Technol. Int.,* 60, 1992.
71. **Delport, H. M. W.,** in *2nd European Ironmaking Congress,* Glasgow, Inst. Metals, 1991, 289.
72. **Smith, R. B.,** *Ironmaking Steelmaking,* 14, 49, 1987; *Met. Mater.,* 9, 491, 1992.
73. **Richards, K. J., George, D. B., and Bailey, L. K.,** *Advances in Sulphide Smelting,* Sohn, H. Y., George, D. B., and Zunkel, A. D., Eds., TMS, Warrendale, PA, 1983, 489.
74. **Themelis, N. J.,** *Met. Trans.,* 3, 2021, 1972.
75. **Shibasaki, T. and Hayashi, M.,** *JOM,* 43, September 20, 1991.
76. **Jiang, R. and Fruehan, R. J.,** *Met. Trans.,* 22B, 481, 1991.
77. **Roset, G. K., Matousek, J. W., and Marcantonio, P.,** *JOM,* 44 (4), 39, 1992.
78. **Matthew, S. P., McKean, G. R., Player, R. L., and Ramus, K. E.,** in *Lead-Zinc 90,* Minerals, Metals and Materials Soc., Warrendale, PA, 1990, 889.
79. **Cameron, A. M., Canham, D. L., and Aurich, V. G.,** *JOM,* April 46, 1990.
80. **Cameron, A. M.,** *User Aspects of Phase Diagrams,* Joint Research Center, The Netherlands, June 1990, Institute of Metals, 1991, 241.
81. **Korshunov, B. G.,** *Metall. Rev. of MMIJ,* 9, 1, 1992.
82. **Doddins, M. S. and Burnet, G.,** *AICHE J.,* 34, 1087, 1988.
83. **Sale, F. R. and Kosumco, Z.,** *User Aspects of Phase Diagrams,* Joint Resarch Center, The Netherlands, June 1990, Institute of Metals, 1991, 225.
84. **Abraham, K. P., Davies, M. W., and Richardson, F. D.,** *J. Iron Steel Inst.,* 196, 82, 1960.
85. **Schuhmann, R. and Ensio, P. J.,** *Trans. Am. Inst. Min. (Metall.) Engrs.,* 194, 718, 1952.
86. **Chang, L. C. and Derge, G.,** *Trans. Am. Inst. Min. (Metall.) Engrs.,* 172, 90, 1947.
87. **Elliott, J. F.,** *Trans. Am. Inst. Min. (Metall.) Engrs.,* 203, 485, 1955.
88. **Fuwa, T., Mizoguchi, S., and Karashima, K.,** in *Pretreatment of Blast Furnace Molten Iron,* Iron and Steel Society, Warrendale, PA, 1987, 117.
89. **Bodsworth, C. and Bell, H. B.,** *Physical Chemistry of Iron and Steel Manufacture,* 2nd ed., Langman, London, 1972, 401.
90. **Turpin, M. L. and Elliott, J.,** *J. Iron Steel Inst.,* 204, 217, 1966.
91. **Turkdogan, E. T.,** in *Perspectives in Metallurgical Development,* Metals Society, London, 1984, 49.
92. **Bishop, H. L., Grant, N. J., and Chipman, J.,** *Trans. Am. Inst. Min. (Metall.) Engrs.,* 200, 534, 1954.
93. Discussion meeting on "The removal of tramp elements in steelmaking," *Ironmaking Steelmaking,* 12, 284, 1985.
94. **Neuschutz, D.,** in *User Aspects of Phase Diagrams,* Hayes, F. H., Ed., Institute of Metals, Petten, The Netherlands, 1991, 199.
95. **Cramb, A. W. and Jimbo, I.,** *Steel Res.,* 60, 157, 1989.
96. **Plockinger, E.,** in *Clean Steel,* Iron Steel Institute, Special Report No. 77, 51, 1963.
97. **Ohno, R.,** *Met. Trans.,* 22B, 405, 1991.
98. **Knight, R., Smith, A. W., and Apelian, D.,** Application of plasma arc melting technology to processing of reactive materials, *Int. Metall. Rev.,* 36, 221, 1991.
99. **Wightman, P. and Hengsberger, E.,** *Met. Mater.,* 7, 676, 1991.

100. **Duckworth, W. E. and Hoyle, G.,** *Electro-slag Refining,* Chapman and Hall, London, 1969.
101. **Natarajan, K. A.,** *Met. Trans.,* 23B, 5, 1992.
102. **Smith, R. and Olson, G. J.,** *Metallurgical Processes for the Year 2000 and Beyond,* TMS, Warrendale, PA, 1989.
103. **Berezowsky, R. M. G. S., Collins, M. J., Kerfoot, D. G., and Torres, N.,** *JOM,* February 9, 1991.
104. **Fuerstenau, M. C.,** *Met. Trans.,* 17B, 415, 1986.
105. **Hansen, D. A., and Traut, D. E.,** *JOM,* May, 34, 1989.
106. **Rickelton, W. A., Flett, D. S., and West, D. A.,** *Solv. Extr. Ion Exch.,* 2, (6) 24, 1984.
107. **Stubina, M. M. and Distin, P. A.,** *Trans. Inst. Min. Metall.,* 96, C123, 1987.
108. **Kotze, M. H.,** *JOM,* May, 46, 1992.
109. **Winand, R. and Fontena, A.,** *Trans. Inst. Min. Metall.,* 92, C27, 1983.
110. **Bockris, J. O. M. and Reddy, A. K. N.,** *Modern Electrochemistry,* Plenum Press, New York, 1976.
111. **Pace, G. F. and Stauter, J. C.,** *Can. Min. Metall. Bull.,* 67, 85, 1974.
112. **Wiesner, K.,** *J. Electrochem. Soc.,* 136, 3370, 1989.
113. **Parker, P. D., Ed.,** *Chloride Electrometallurgy,* TMS, Warrendale, PA, 1982.
114. **Grjotheim, K. and Welch, B.,** *JOM,* November 12, 1989.
115. **Winand, R.,** *JOM,* April, 24, 1990.
116. **Schloen, J. H.,** *The Electrorefining and Winning of Copper,* TMS, Warrendale, PA, 1987.
117. **Chen, T. T. and Dutrizac, T. T.,** *JOM,* August, 39, 1990.
118. **Stern, A. C.,** *Air Pollution,* vol. 2, Academic Press, New York, 1962, 81.
119. **Forrest, D. and Szekely, J.,** *JOM,* December, 23, 1991.
120. **Post, W. M.,** *Am. Sci.,* 78, 310, 1990.
121. **Frost, M. T., Jones, M. H., Hart, R. L., Strode, P. R., Urban, A. J., and Tassios, S.,** *Trans. Inst. Min. Metall.,* 99, C117, 1990.
122. **Chapman, P. F. and Roberts, F.,** *Metal Resources and Energy,* Butterworths, London, 1983, 139.
123. **Pourbaix, M.,** *An Atlas of Electrochemical Equilibria in Aqueous Solutions,* Pergamon, Oxford, 1966, 478.
124. **Kawakami, M., Goto, K. S., and Matsuoska, M.,** *Met. Trans.,* 11B, 463, 1986.
125. **Balajee, S. R., Robertson, K. J., and Bradley, J. E.,** *Steel Tech. Int.,* 4, 71, 1992.
126. **Dammert, K.,** *Steel Tech. Int.,* 4, 37, 1992.
127. **Szekely, J.,** *Met. Trans.,* 19B, 525, 1988.
128. **The Iron and Steel Institute of Japan,** *Blast Furnace Phenomena and Modelling,* Elsevier, London, 1987.
129. **Mills, K. C. and Keene, B. J.,** *Int. Mater. Rev.,* 32 (No. 1–2), 1987.
130. **Cumming, M. J. and Thurby, J. A.,** *Ironmaking Steelmaking,* 17, 245, 1990.
131. **Hussain, M. M. and Morris, D. R.,** *Met. Trans.,* 20B, 97, 1989.
132. **Hahn, Y. B. and Sohn, H. Y.,** *Met. Trans.,* 21B, 945, 1990.
133. **Seo, K. W. and Sohn, H. Y.,** *Met. Trans.,* 22B, 791, 1991.
134. **Kyllo, A. K. and Richards, G. G.,** *Met. Trans.,* 22B, 153, 1991.
135. **Krishna Murthy, G. G. and Mehrotra, S. P.,** *Ironmaking Steelmaking,* 19, 377, 1992.
136. **Koria, S. C. and Shamsi, M. R. R. I.,** *Ironmaking Steelmaking,* 17, 401, 1990.
137. **Harris, R.,** *Can. Metall. Q.,* 27, 169, 1988.

INDEX

A

Acid practice, 29
Acid slag, 195
Activated complex, 98–99, 267
Activation, energy, 98, 104, 111, 116, 119, 149, 267
  polarization, 267–268, 272
Activity, evaluation from phase diagrams, 77–78
  in aqueous solution, 89, 93, 240, 242, 251
  in ideal solution, 59
  in multi-component solution, 70–72
  in non-ideal solution, 64, 77
  standard states for, 60, 67
Activity coefficient, 59, 64
  in dilute solution, 68
  in multi-component system, 70–72, 334–336
Agglomeration, 21–25, 135–140
AISI iron smelting process, 171
Aluminum, electrolytic extraction, 36, 273–277
  halide extraction, 187–188
  leaching of ore, 35, 239, 246, 248
  ore (bauxite), 6, 20, 35, 43
  -oxide-chloride predominance area diagram, 186
Amphoteric oxide, 195
Anode, 32
Antimomy, arsenic, as impurities, 23, 40, 43, 140, 150, 175, 221, 239, 245, 252, 279
AOD process, 211
Aqueous effluents, 293–294
Argon gas, 112, 193, 214, 215, 233
Arrhenius equation, 99
Atom fraction, conversion of wt% to, 69
Autogenous reaction, 133, 166, 76, 248

B

Bacterial leaching, 131, 247
Basic oxygen converter, see BOP
Basicity, 138, 153, 154, 201
Basic, practice, 29
  slag, 195
Bayer process, 248
Bauxite, 6, 20, 35, 246, 248
Beneficiation, 10–21, 31, 238, 248
Bessemer converter, 6, 38, 172, 211, 213
Bioleaching, 247
Bismuth, as an impurity, 41, 150, 190, 222, 229, 245, 279
Blast furnace process, 37, 38, 40, 42, 135, 137–163
  carbon requirements, 161, 162
  chemical reserve zone, 160
  heat balance, 141–144
  iron smelting, 38, 127, 144, 151–155
  lead smelting, 40, 145
  lead-zinc smelting, 42, 148, 150, 155–157, 220
  operating line, 159–161
  partitioning of elements in, 151–154
  temperature distribution in, 157–159
  thermal reserve zone, 158–160
  zinc smelting, 42, 147, 155–157
Blister copper, 39, 175
Blowholes, 232
Boltzmann function, 100
Bond energy and heat of formation, 64
BOP (BOS) converter process, 38, 211–214, 317
Boudouard reaction, 145, 149, 169
Boundary layer theory, 113–115
Bubbles, kinetic effects, 193
  flushing action, 121, 230
  nucleation, 120, 205, 207
Buffer stock, 47
Bulmer-Volmer equation, 267

C

Cadmium, as an impurity, 23, 42, 150, 230, 245, 252
Calcium, carbonate dissociation, 139, 157
  oxide activity, 197, 198
  sulfide, 134
Carbon reduction, 26, 38, 40, 42, 81–83, 126–128, 141–152, 170, 181, 186
Carbon dioxide "greenhouse" gas, 29, 287–291
Cathode, 32
Cementation, 32, 250–252, 262, 279
Chemical, analysis, 237, 306–310
  potential, 57
Chloridizing roast, 130, 184
Chlorination of ores, 184–185
Chromium, extraction, 43
  -iron-carbon equilibrium, 209–211

Classification, 12–14
Coke, 30, 42, 127, 135, 139, 140, 161–163, 165, 171, 184, 185, 289–290
Collectors, 16
Collision theory, 98–99
Comminution, 11, 19, 44
Computer control, 48, 305, 310–317
Concentrate, 10, 17, 20, 21, 27, 41, 130, 172
Contact angle, 14, 120
Copper, cementation, 250–252
  electrowinning, 263–266, 269, 270, 272, 337–338
  leaching, 32, 238, 244, 256
  -lead phase diagram, 116, 219
  -oxygen-sulfur equilibrium, 85–87
  precipitation from solution, 250–252
  refining, 39, 233, 279
  smelting, 39, 163, 172–179, 323–325, 335–336
  specifications, 9
  -water Pourbaix diagram, 239–242
COREX process, 170
Cost, factors affecting, 17, 19–20, 25, 29–31, 33–35, 43–46, 163, 176, 188, 190, 212, 237–238, 246, 248, 272, 281, 284, 294, 297, 302
Cowper stove, 144
Cupellation, 221
Current efficiency, 271, 276, 337–338
Cyanides, 17, 31, 294

D

Dead roasting, 130–134
Degassing, 232–234
Density of compounds, 8, 13–14, 313
Deoxidation, 38, 120, 222, 227
Diffusion, 100–104
  activation energy for, 104, 116
  and boundary layer theory, 113, 115
  and surface renewal theory, 114
  coefficient, 101
  Fick's laws for, 100–102, 108, 113
  molecular and Knudsen, 110, 112, 168
Diffusivity of gases through solids, 109–110
Dilute solution, activity with respect to, 67, 202
  multi-component, 70–72
Direct current electric arc furnace, 216
Directreduced iron(DRI), 26, 38, 165–170, 209, 322–323
Direct Iron Ore Smelting (DIOS) process, 171
Distillation, 229
Dross, 28, 40, 41, 201, 220, 222, 238, 245, 301, 329–331
Dust losses, 13, 176, 185, 213, 286

E

Electric arc furnace, 38, 215–216
  smelting, 39, 164–165
Electrochemical series, 33
Electrode potential, 90–92, 240, 251, 254, 261
Electro-slag refining, 235–236
Electrostatic separation, 17–18, 297
Electrolytes, aqueous, 263–273, 279
  molten, 273–278
Electrorefining, 262, 278–281
Electrowinning, 261–278, 337–338
Ellingham diagrams, 78–84
Embryo, 118
EMF measurement of oxygen, 309
EMF-temperature diagrams, 89
Emulsions, 178, 212
Endothermic reaction, 54, 62, 64, 112, 139, 170
Energy consumption, 11, 30, 34, 44–45, 127, 143, 161–163, 170, 171, 212, 275–278, 283–284, 289–291, 294–297, 300, 337
Enthalpy, 53
Entropy, 53
Environmental problems and constraints, 30, 34, 134–135, 140, 164, 232, 237, 273, 281, 283–294, 298–299
Excess thermodynamic quantities, 65
Exothermic reaction, 54, 64, 112, 168

F

Faraday unit, 89
Faraday's laws of electrolysis, 270–272, 337
Ferro alloys, 43, 141, 165
Ferrous oxide, activity in slags, 164, 197, 198, 200
Fick's laws of diffusion, 100–102, 108, 113
Fitness for purpose, 189
Fixed asset costs, 46
Flash roasting and smelting, 39, 40, 51, 109, 133, 176, 284, 323–325
Flocculents, 21, 249, 293
Flotation, 14–17, 43, 178, 180, 312
Fluidized bed, 109, 132–134, 165, 169, 171, 185
Fluorspar, 199, 217

Fluxes, 24, 27, 194, 199, 201
Foaming slag, in copper refining, 178–179
in steel refining, 212–213, 215
Fractional distillation, 185, 229–230
Free energy, 52
change and equilibrium, 60–62
of formation of oxides, 79–80
of formation of sulfides, 124–125
of mixing, 63
standard, 55, 68
-temperature diagrams, 78–84, 89, 94, 124, 136, 182
variation with temperature, 64
Frothing agent, 15
Fugacity, 58
Fume, 180, 286, 287, 302

G

Galvanic cell, 88, 261
Gangue, 7, 9, 23, 26, 123, 149, 161, 166, 169, 193
Gas, cleaning, 287
reforming, 166, 170
solubility in metals, 146, 174, 222–224
Gaseous reduction, 109–112, 159–161
Gibbs, absorption isotherm, 108
chemical potential, 57
-Duhem equation, 57, 66, 78
Gold, 33, 41, 221, 279, 298
leaching, 31, 245, 247, 248, 252
Gram equivalent, 270
Gross process, 188

H

Hafnium, separation from zirconium, 186, 257
Halides, 183–188, 230, 273–278
Half-cell reaction, 90, 251, 261
Hall-Heroult process, 274–277, 281
Hardhead, 41, 164
Harris process, 221
Heat, balance, 141–144, 161, 319–322
capacity, 53, 56, 143, 312
of reaction, 54
transfer, 104, 112, 140
Heavy media separation, 13–14
Henry's law, 66–68, 203, 233, 330
Hess law, 55, 142
Heterogeneous reaction, 95, 100
High Intensity Smelting (HISMELT) process, 171
Hydrogen, as a deoxidizer, 223–224
electrode, 89, 253
-ion concentration, 89, 255, 264
reduction, 26, 81–83, 128–129, 162, 167–168
reduction from leach solutions, 253–255, 259
overvoltage, 269, 270
HyL process, 165

I

Ideal solution, 59, 62, 64, 72, 74, 78, 197
Igneous intrusion, 3–4
Imperial Smelting process, 155–156
INCO process, 176
Integral molar free energy, 57
Interaction coefficient, 71–72, 205, 313, 334–336
Internal energy, 53
Ion exchange, 259–260, 294
Ionic structure of slags, 154, 195, 199
Ionization potential, 241
Iron, carbide production, 169
deoxidation of, 38, 224–227
-iron oxides equilibria, 146–148, 168
ores, 20, 23, 38, 43
-oxide-chloride predominance area diagram, 186
-oxygen phase diagram, 146–147
oxygen solubility in, 147
refining, 38, 201–218, 233, 299, 307–308, 333–335
smelting, 38, 151–155
-water Pourbaix diagram, 239–243
Iron and steel recycling, 38, 189, 201, 202, 208, 211, 213–215, 297–298, 301
ISASMELT process, 180

K

Kinetic energy, 54
Kinetics of reactions, gas-liquid, 106, 193
gas-solid, 108, 132, 149
liquid-liquid, 112, 193
liquid-solid, 249–250, 259
Kirchoff equation, 53
Knudsen diffusion, 110, 112, 168
Kroll process, 43, 187
Kroll-Betterton process, 222

L

Ladle refining, 38, 201, 208, 216–218

Latent heat, 53, 70, 77, 228
Leaching, 20, 30–31, 237–250
Lead, anodes, 263–264
 -copper phase diagram, 116, 219
 -iron and lead-zinc phase diagrams, 219
 leaching, 244
 recycling, 284, 299
 refining, 40–41, 220–222, 279, 329–333
 smelting, 40, 155–157, 180–181
LD converter process, 38, 211–214
Le Chatelier principle, 191, 232
Lime, activity in slags, 197, 198
 reactivity, 194
Liquation, 28, 41, 42, 220–222
Liquid immiscibility, 196–197
Lixiviant, 237

M

Magnesium, extraction, 42, 181–183, 277
 -oxide-chloride predominance area diagram, 186
 reduction of halides, 187
Magnetic separation, 18–19, 169, 297
Manganese, extraction, 43
 oxide activity, 198
 partitioning in iron smelting, 151–154
Mass transport, 105–107, 314
Matte, 27, 39, 43, 172–180, 245, 280, 307, 323–324, 335
Metallothermic reduction, 26, 43, 129, 181–183, 187
Midrex process, 165
Mineral, deposits, 3
 dressing, 10–21
Mitsubishi process, 177, 178
Models, dynamic, 311, 315–317
 empirical, 314–315
 fundamental, 311–314
 solution, 59–64
 static, 311, 317
Molality, molarity, mole fraction, 52

N

Nernst equation, 91, 251, 253, 266, 269, 337
Neural network, 314
Nickel, carbonyl, 231–232
 electrorefining, 280
 electrowinning, 269, 272, 273
 extraction, 43, 179–180
 leaching, 248
 -water Pourbaix diagram, 243
Nitrogen removal from steel, 215
Nitrous oxides, 292
Noranda process, 176–177
Nucleation, homogeneous, 119–121
 of gas bubbles, 120, 205, 268
 of precipitates, 116, 183, 224, 225, 231, 250, 255, 325–326
Nucleus, 118

O

OBM process, 213
Open hearth furnace, 35, 211, 290
Order of a reaction, 96–98, 107
Ore, formation, 3–6
 types, 8
Outokumpu process, 176
Overpotential, 266–267, 270
Overvoltage, anode, 268
 hydrogen, 269, 270
Oxygen, enrichment, 162, 175–176, 178
 -potential diagrams, 78–84, 94, 124, 127, 128, 141, 181
 solubility in metals, 146, 174, 222, 223

P

Parkes process, 41, 221, 329–331
Partial molar quantities, 56–58, 60, 77
Partition ratio, 151, 154, 191–192, 194
Passivation, 263
Pelletizing, 21–22, 109, 140, 167
Phase diagrams, complete solubility, 72
 eutectic and peritectic, 75–77
Phase rule, 84, 92
Phosphorus, deoxidation, 40, 224
 partitioninginiron refining, 206–207, 212–213, 215, 217
 partitioning in iron smelting, 153
Pidgeon process, 183
Pierce Smith converter, 39, 172, 175
Placer deposits, 5, 9, 12, 41
Plasma heating, 183, 235, 301
Polarization, 266–269, 272, 276
Pollution, 134–135, 283–294
Potential energy, 54
Pourbaix diagram, 88–94, 239–244, 265, 293, 298
Precious metals, 33, 34, 40, 81, 150, 179, 272, 279
Predominance area diagram, 84–87, 93, 94, 131, 146, 172, 185
Pressure leaching, 31, 247–249
Process control, 48, 305, 310–317

Pyrophoric powder, 167, 169, 182

Q

QBOP process, 213

R

Raffinate, 237
Raoult's law, 59, 66, 74
Rate, constant, 96
control, 107, 109, 111, 114, 130
Recycling, 46, 164, 194, 202, 215, 293–302
Reducibility, 138–140, 161
Reference state, 53, 55, 89
Refractories, 28–29, 40, 144, 224, 235, 273, 333
Regular solution model, 63–64, 78, 315, 331–332
Relative partial molar free energy, 60
Residual deposits, 6
Reverberatory smelting, 39, 163–164, 172, 320–322
Reversible potential (emf), 89
Reversible process, 52, 88
Reversion from slag to metal, 192, 218, 226
RH vacuum stream degassing, 234
Rist diagram, 159–161
Roasting, 39, 84, 88, 112, 130–135

S

Scrap metal recycling, 38, 46, 175, 189, 201, 202, 211, 213–215, 284, 294–302, 310
Secondary metal producers, 295
Sedimentary ores, 4–6
Self-fluxing burden, 24–25
Selling price of metals, 2–3, 7, 8, 32, 46–49, 190
Sensors, 48, 305–310, 315–317
Sievert's law, 66, 107, 122, 233
Silica activity in slags, 197–198
Silicon, monoxide, 151–152
-oxide-chloride predominance area diagram, 186
partition in iron smelting, 151–154
Silver, 33, 41, 221, 244, 252, 269, 279, 298, 329–331
Sintering, 22–25, 109, 111, 137–140, 167, 287, 312
Slag, acid, basic and neutral, 29, 195
activities, 164, 197–198, 200, 313
basicity ratio, 152, 154
disposal, 28, 286,
foaming, 178–179, 212–213, 215
ionic structure, 154, 195, 199
viscosity, 27–28, 105, 174, 194–195, 307, 313
volume, 27, 161, 169, 174, 194
Smelting Reduction process, 170–171, 290
Soderberg electrode, 274
Sodium electrowinning, 277
Softening of lead, 40
Solutions, dilute, 65–66
ideal, 59, 62, 72
multi-component, 70–72
regular, 63–64, 78, 315, 331–332
Solvent extraction, 255–159
Speiss, 40, 221
Specifications, 9, 190–191
Spontaneous process, 52, 54, 55
Stainless steel, 210, 301
Standard hydrogen electrode, 89
Standard electrode potentials, 89, 90, 240, 251–253, 261, 264, 266, 275
Standard state, 55, 60, 67–70, 89
State property, 52, 55, 56, 72
Steelmaking, 201–218
Stoichiometric composition, 80
Stoke's law, 13–14, 21, 225–226
Sulfating roast, 88, 130, 131, 248
Sulfide capacity of slags, 216–217
Sulfide ore reduction, 171–180
Sulfides; free energy-temperature plot, 124–125
Sulfur, in fuel, 30
gases, 84, 131, 134, 288
partitioning in iron smelting, 154
precipitation, 134–135, 244
removal from gas, 134, 291–292
removal from steel, 206–208, 215, 216
Sulfuric acid, 134, 178, 239, 263, 279, 291
Surface, adsorption, 108, 230, 252, 270
renewal theory, 114
tension, 14, 22, 108, 119, 120, 178, 179, 212, 223, 226, 274, 313
Surfactants, 11
System, 52

T

Tafel equation, 267
Thermal reserve zone, 158–159
Thermodynamic laws, 53–54
Thiobacillus ferrooxidans, 131, 247
Tin-arsenic and tin-iron phase diagrams, 219

Tin, extraction, 41, 164
  refining, 41, 221, 222
Tinplate, 47, 220, 298
Titanium extraction, 43, 187
Topochemical reaction, 112, 132, 249, 250
Tough pitch copper, 40, 224
Tramp elements, 9, 209
Transfer ladle refining, 38, 201, 208
Trouton's rule, 80, 227

U

Uranium leaching, 245–247, 259, 260

V

Vacancy diffusion, 22, 109
Vacuum melting and refining, 39, 122, 187, 193, 230, 233, 333
van't Hoff isochore, 61
Vapor pressure, 59, 64, 65, 148, 227–229
Variable costs, 46
Vat leaching, 247–249
Viscosity, 13, 26, 28, 105, 178, 185, 194, 195, 274, 307, 313
Volatilization, 26, 155, 184, 227–232, 301
  of impurities, 140, 150, 175, 212, 235

W

Waste materials, 285–294
Water, gas reaction, 129, 166–168, 327–329
  ionization, 264
Weight per cent standard state, 68, 202, 333–336
Wood charcoal, 139
Wustite, 80, 146, 147, 159

Z

Zinc, electrolytic extraction, 42, 269, 270, 272, 273
  leaching, 42, 238, 245, 248, 252
  refining, 230
  smelting, 42, 148, 155–157
  -water system, 91–94
  -zinc oxide equilibrium, 147, 148
Zirconium extraction, 43, 185, 257